# Two-Dimensional Quantum Materials

# Two-Dimensional Quantum Materials

By

Pramoda Kumar Nayak
and Kushal Mazumder

Cambridge
Scholars
Publishing

Two-Dimensional Quantum Materials

By Pramoda Kumar Nayak and Kushal Mazumder

This book first published 2024

Cambridge Scholars Publishing

Lady Stephenson Library, Newcastle upon Tyne, NE6 2PA, UK

British Library Cataloguing in Publication Data
A catalogue record for this book is available from the British Library



ISBN (10): 1-0364-0600-8
ISBN (13): 978-1-0364-0600-4

# CONTENTS

# PREFACE

The realm of material science has witnessed only a handful of discoveries and technological advancements with the potential to revolutionize our world, and the emergence of two-dimensional (2D) materials stands prominently among them. Originating from the isolation of graphene from graphite in 2004, this class of materials, characterized by their atomically thin nature and dominated by surface effects, has opened up a new frontier in material science. Research on 2D materials, including graphene and its counterparts like silicene, germanene, phosphorene, as well as transition metal dichalcogenides (TMDs), MXenes, and other layered semiconductors, has burgeoned into a global effort involving thousands of researchers across diverse fields such as physics, chemistry, engineering, and biology. The distinctive feature of 2D materials lies in their layered structure, comprising strong in-plane chemical bonds juxtaposed with weak out-of-plane coupling between layers. This structural arrangement allows for the cleavage of individual atomic layers, giving rise to extraordinary variations in electronic properties when the material's thickness is reduced to a single or few layers. This phenomenon, known as quantum confinement, imparts unique and often unexpected properties to 2D materials, fuelling exploration into novel applications and innovative avenues in various domains. As researchers delve deeper into the intricacies of these layered materials, it becomes increasingly evident that they hold the promise of unlocking unprecedented possibilities and paving the way for ground-breaking advancements in science and technology.

Welcome to "Two-Dimensional Quantum Materials," a comprehensive exploration of the fascinating world of 2D materials and their implications in the realm of quantum phenomena and applications. This book delves into the multifaceted aspects of 2D materials, from their fundamental properties to cutting-edge applications, promising to provide valuable insights into the forefront of nanoscience and technology.

In the introductory chapters, readers will embark on a journey through the classification of materials based on confinement, tracing the historical milestones leading to the discovery of graphene, and uncovering the vast landscape of other 2D materials such as transition metal dichalcogenides (TMDs), black phosphorus, and boron nitride. The significance and diverse applications of these materials across various fields will also be elucidated, setting the stage for a deeper exploration.

Subsequent sections delve into the crystal structure and properties of 2D materials, unravelling the atomic intricacies of their stacking configurations and polymerization phenomena. Readers will gain insights into the electronic band structure, as well as optical, electrical, and mechanical properties that underpin the unique characteristics of 2D materials.

Synthesis and fabrication techniques, ranging from top-down to bottom-up approaches, are meticulously elucidated, providing readers with a comprehensive understanding of the methods employed in the production of 2D materials. Characterization methods such as Raman spectroscopy, photoluminescence (PL) spectroscopy, and transmission electron microscopy (TEM) are also explored, shedding light on the techniques used to unravel the structural and optical properties of these materials at the nanoscale.

The book further delves into quantum phenomena exhibited by 2D materials, including quantum conductance, the quantum Hall Effect, and the emergence of Dirac and Weyl fermions. Exciton and multibody physics in 2D materials are also investigated, offering valuable insights into the quantum behavior of these materials.

The applications of 2D materials in nanoelectronics, optoelectronic devices, and energy storage and conversion are thoroughly examined, highlighting their potential to revolutionize various technological domains. Furthermore, the book explores emerging topics such as spintronics, valleytronics, twistronics, and the interplay between charge density waves (CDW) and superconductivity in 2D materials.

Finally, the challenges, future directions, and conclusions drawn from the collective insights presented in this book offer valuable reflections on the scalability, integration challenges, environmental considerations, and potential breakthroughs in the field of 2D quantum materials.

We hope this book serves as a valuable resource for researchers, students, and enthusiasts alike, fostering a deeper understanding of the intricate world of two-dimensional quantum materials and inspiring further exploration and innovation in this burgeoning field.

Pramoda Kumar Nayak
CNMS, Jain University, Bengaluru, India
& Dept. of Physics, IIT Madras, Chennai, India

Dr. Kushal Mazumder
CNMS, Jain University, Bengaluru, India

# CHAPTER 1

# INTRODUCTION TO QUANTUM MATERIALS

Quantum materials are substances that exhibit novel and exotic physical properties below the quantum confinement limit, where the behavior of particles is governed by the principles of quantum mechanics. These materials are characterized by unique electronic structures, such as the arrangement of electrons in energy bands or levels that lead to intriguing phenomena. On the microscopic level, four fundamental degrees of freedom *i.e.,* charge, spin, orbit and lattice become intertwined, resulting in complex electronic states.[1] Quantum materials often display emergent properties that arise from the collective behavior of electrons or other quantum particles. These properties can be very different from what one would expect based on the individual components of the material. In quantum materials, electron-electron interactions play a major role and these interactions can lead to phenomena like superconductivity, magnetic ordering, and fractional quantum Hall effect and so on.[2] These materials can undergo quantum phase transitions at very low temperatures, where small changes in external conditions lead to abrupt changes in the material's properties. They also exhibit unusual physical anomalies, such as non-Fermi liquid behavior, that challenge conventional theories of condensed matter physics. These distinctive physical properties are dictated by quantum mechanical effects, particularly when studied at length scales comparable to or smaller than the de Broglie wavelength of the constituent particles and the exciton Bohr radius.

## 1.1 De Broglie Wavelength

The de Broglie wavelength is a fundamental concept in quantum mechanics, and it relates the momentum of a particle to its wavelength, $\lambda = \frac{h}{mv}$ where $\lambda$ is the wavelength of the particle, m and v are the mass and velocity of the particle respectively, and h is the Planck constant.[3]

In quantum materials, when the characteristic length scale of the material approaches or falls below the de Broglie wavelength of the particles involved, quantum effects become increasingly prominent. This can lead to phenomena such as wave-particle duality, quantum confinement, and quantization of energy levels.

## 1.2 Exciton Bohr Radius

Excitons are bound electron-hole pairs that exist in certain materials, including quantum materials. The exciton Bohr radius is a crucial parameter that defines the typical spatial extent of these excitonic states within the material. In quantum materials, when the size of the material becomes comparable to or smaller than the exciton Bohr radius, the behavior of excitons can significantly impact the material's optical and electronic properties.

### Bohr radius

The equation for the Bohr radius of an electron is given by,[4]

$$a_0 = \frac{4\pi\varepsilon_0 \hbar^2}{mq^2} \tag{1}$$

Where $\varepsilon_0$= 8.85×10$^{-12}$ F/m (permittivity), $\hbar$ =1.054× 10$^{-34}$ J.s (Planck's constant over $2\pi$), $m_e$ = 9.11× 10$^{-31}$ kg (mass of a free electron) and $q$ = 1.602× 10$^{-19}$ C (charge). If you plug all the numbers in and do the math you come up with result

$$a_0 = 5.28\times 10^{-11} \text{ meters} = 0.528 \text{ Å}$$

This is the standard Bohr radius one sees all the time.

### Derivation

Basically we need to balance the centrifugal (outward) force of a carrier with the Coulomb attractive (inward) force.

$$\frac{mv^2}{r} = \frac{q^2}{4\pi\varepsilon_0 r^2} \tag{2}$$

Here we make use of the relation

$$2\pi r = n\lambda \tag{3}$$

Where, n is an integer. The de Broglie relation comes in by the wavelength $\lambda = \frac{h}{p}$, where h is Planck's constant and $p$ is the momentum of the particle ($p = mv$). Starting with the above equation we rearrange it to get

$$\lambda = \frac{2\pi r}{n} = \frac{h}{p} = \frac{h}{mv} \tag{4}$$

Solve $v$ for to get

$$v = \frac{nh}{2\pi mr} = \frac{n\hbar}{mr} \tag{5}$$

Replace this into the main equation

$$\frac{n^2\hbar^2}{mr} = \frac{q^2}{4\pi\varepsilon_0} \tag{6}$$

Rearrange this to get

$$r = \frac{4\pi\varepsilon_0\, n^2\hbar^2}{mq^2} \tag{7}$$

If $n = 1$ (the lowest orbit) this gives us the Bohr radius

$$a_0 = \frac{4\pi\varepsilon_0\, \hbar^2}{mq^2} \tag{8}$$

Which is the standard textbook equation we showed earlier.

At this point we note that if the electron or carrier is not in vacuum the equation should be modified to take into account the dielectric constant of the medium. (Instead of $\varepsilon$, replacing it with $\varepsilon\,\varepsilon_0$)

$$a_0 = \frac{4\pi\varepsilon_0\, \varepsilon\, \hbar^2}{mq^2} \tag{9}$$

Furthermore, for the case of an exciton (electron hole pair) in a semiconductor just replace the mass of the electron with the effective mass of the exciton.

$$\frac{1}{m_{eff}} = \frac{1}{m_e} + \frac{1}{m_h} \tag{10}$$

Where $m_e$ and $m_h$ are the effective masses of electron and hole in the material.

This equation gives the relation between the de Broglie wavelength and the exciton Bohr radius. So, in effect, our initial statements about confinement dealing with either the exciton Bohr radius or de Broglie wavelength are essentially one and the same. The de Broglie wavelength or exciton Bohr radius are therefore natural length scales by which to compare the physical size of a nanomaterial. In general objects with dimensions smaller than these natural length scales will exhibit quantum confinement effects.

## 1.3 Quantum Confinement

Quantum confinement is a fundamental concept in condensed matter physics that refers to the confinement of electrons or charge carriers within a space so small that their quantum mechanical properties dominate their behavior. This phenomenon becomes significant when the physical dimensions of a material, such as its thickness or particle size, become comparable to the de Broglie wavelength associated with quantum mechanics.[5] The concept of quantum confinement arises from size reduction in materials, leading to tighter confinement of electronic wave functions and resulting in changes in electronic and optical properties in nanomaterials. A smaller (or bigger) particle size results in a stronger (or weaker) confinement which gives rise to enhancement (or decrease) of the band gap and modifies the band structure of the material. These changes affect electron mobility, effective mass, relative dielectric constant, and optical properties, among others. Quantum confinement effects give rise to intriguing properties in metallic nanoparticles, such as color variations in colloidal suspensions with changing particle size, UV photoemission, and enhanced photoluminescence.

## 1.4 Density of States

Density of States (DOS) is a concept in condensed matter physics that describes the distribution of energy states available to particles, typically electrons, within a material. Quantum confinement, especially in nanoscale materials like quantum dots or nanowires, has a significant impact on the DOS, leading to discrete energy levels rather than a continuous band structure, as seen in bulk materials. Here's an explanation of DOS in the context of quantum confinement:

**Bulk vs. Nanostructured Materials**

Bulk Materials: In a bulk material (e.g., a large piece of semiconductor), electrons occupy energy states forming continuous energy bands. There is a continuous range of energy levels available for electrons to occupy, and this band structure is a fundamental concept in solid-state physics.

Nanostructured Materials: When the dimensions of a material become comparable to or smaller than the de Broglie wavelength of its electrons (quantum confinement regime), the DOS changes dramatically. In nanomaterials like quantum dots, quantum wires, or thin films, electrons are confined in three, two and one dimensions. In quantum dots, the electrons are confined in all three dimensions. This confinement results in the quantization of energy levels. Electrons can only occupy discrete energy states or energy levels. These discrete energy levels are often referred to as "energy shells" or "quantum levels." In the case of quantum wells (confinement in one dimension) or quantum wires (confinement in two dimensions), there is still quantization of energy levels, but the quantization occurs in fewer dimensions.

**Effect on DOS:**

Bulk DOS: In bulk materials, the DOS is continuous, forming energy bands. The number of energy states per unit energy interval is relatively constant within a band.

Nanostructure DOS: In nanostructured materials with quantum confinement, the DOS exhibits distinct peaks or steps corresponding to the quantized energy levels. These energy levels are determined by the size and shape of the nanostructure and are often referred to as "quantum energy levels" or "quantum states."

DOS also helps us understand the electronic structure of materials based on their dimensionality. Here's an explanation of 0D, 1D, 2D, and 3D materials in terms of DOS.

## 1.5 Classification of the Materials [0D, 1D, 2D and 3D] based on DOS

### 0D Materials (Zero-Dimensional)

0D materials are quantum dots or nanoparticles that have zero dimensions in space, meaning they are extremely small in all three dimensions. In 0D materials, DOS exhibits discrete energy levels due to quantum confinement. These discrete energy levels are often referred to as "energy states" or "quantum levels." The energy spacing between these levels is determined by the size of the quantum dot, with larger dots having wider energy spacings.

### 1D Materials (One-Dimensional)

1D materials are nanowires, nanotubes, or nanorods that have one dimension significantly larger than the other two. In 1D materials, DOS shows continuous energy bands along one direction (the dimension along which the material is extended). These energy bands are quantized along the other two dimensions, resulting in a 1D electronic structure with energy states that vary with wave vector along the extended direction.

### 2D Materials (Two-Dimensional)

2D materials, like graphene, transition metal dichalcogenides (TMDs), and black phosphorus, are composed of atomic layers with strong in-plane covalent bonds but weak van der Waals bonds between layers. In 2D materials, DOS displays two-dimensional behavior. There is a continuous band of energy states within the atomic layers (in-plane) and a quantization of energy states perpendicular to the layers. This quantization leads to the formation of discrete Landau levels in the presence of a magnetic field.

### 3D Materials (Three-Dimensional)

3D materials include conventional bulk materials like metals, semiconductors, and insulators, where all three dimensions are extended and of comparable size. In 3D materials, DOS exhibits the typical behavior seen in bulk materials. There are continuous energy bands in all three dimensions, resulting in a three-dimensional electronic structure with a high density of energy states. The DOS curve is smoother and denser compared to lower-dimensional materials.

In summary, the dimensionality of a material strongly influences its DOS behavior. 0D materials show discrete energy levels, 1D materials have continuous energy bands along one direction, 2D materials exhibit two-dimensional electronic behavior, and 3D materials have a bulk-like DOS with continuous energy bands in all three dimensions as depicted in Figure 1.

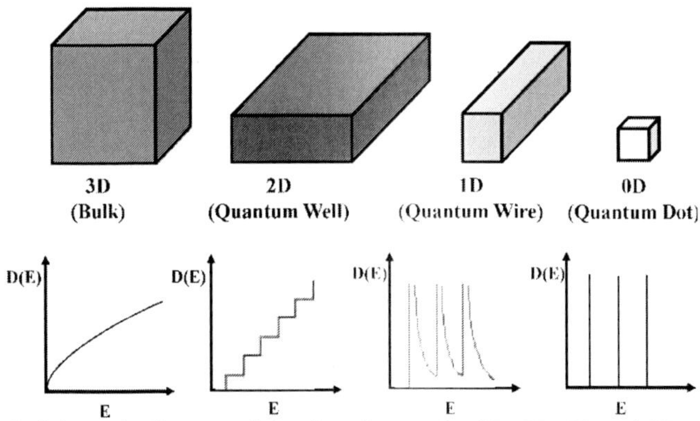


Figure 1. Schematic diagram of density of states for 3D, 2D, 1D and 0D systems. Reproduced with permission.[6]

Since this book is concentrating on 2D quantum materials, the following chapters and their constituent sections will mostly talk about 2D materials and related quantum phenomena responsible for those. Although there exists a plethora of 2D materials now a days, we must start with Graphene, the earliest discovered 2D quantum material which revolutionized the scientific community for its indigenous design and multifaceted applications.

## 1.6 Historical Overview and Discovery of Graphene

Graphene is a two-dimensional (2D) structure consisting of $sp^2$ hybridized carbon atoms covalently bonded in a hexagonal honeycomb lattice. It has the capability to form zero-dimensional (0D) fullerenes when wrapped, one-dimensional (1D) nanotubes when rolled, or three-dimensional (3D) graphite when stacked. Figure 2 displays the schematic of graphene and its unit cell. P.R. Wallace, a Canadian physicist, made significant theoretical contributions by conducting calculations on graphene.[7] Prior attempts to isolate graphene predominantly centered on chemical exfoliation. In this methodology, bulk graphite underwent an initial process of intercalation, whereby graphene layers were separated by intermediate layers of atoms or molecules. Brodie altered graphite through chemical treatment with strong acids and oxidants, resulting in the oxidation of graphite.[8] Brodie's approach was subsequently refined for the production of graphite oxide (GO) by Hummer and Staudenmaier.[9,10] In 1962, Boehm *et al.*[11] produced graphene by chemical reduction of GO.

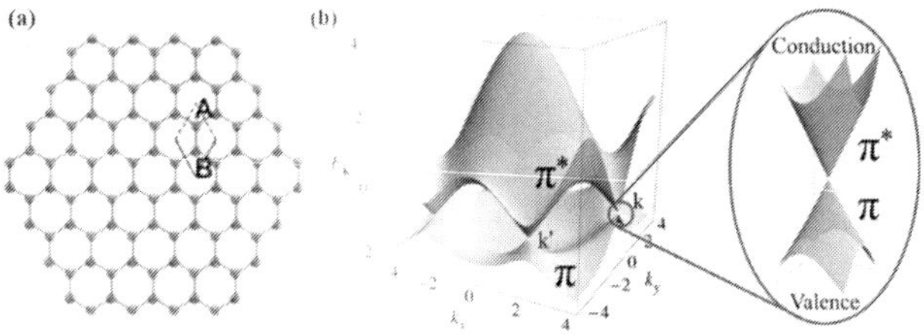


Figure 2. (a) Hexagonal honeycomb lattice of graphene including two atoms (A and B) in a unit cell;[12] (b) Electronic dispersion in the honeycomb lattice, in which the conduction band minima and valence band maxima meet at six Dirac points (left), and close view of the energy bands at one of the Dirac points (right). Reproduced with permission.[13]

Epitaxial growth of graphene has been achieved through chemical vapor deposition of hydrocarbons on metal substrates. In 1966,[14] hot single crystals of nickel was exposed to methane at a temperature of 900°C, resulting in the formation of thin crystalline films of graphite. Moving forward, in 1975, Lang *et al.*[15] successfully demonstrated the deposition of a monolayer graphitic structure on platinum using ethylene. Eizenberg *et al.*[16] in 1979 reported the formation of graphitic layer on Ni (111). A significant milestone occurred in 1992 when Land *et al.*[17] synthesized single layer graphite structures on platinum by decomposition of hydrocarbons. Additionally, thermal decomposition of silicon carbide has been another method yielding single layer and few layer graphene.[18,19].

In 1999, Ruoff *et al.*[20] achieved the production of graphite islands with a thickness of approximately 200 nm using oxygen plasma etching. However, the year 2004 marked a significant breakthrough when physicists from the University of Manchester and the Institute for Microelectronics Technology in Chernogolovka, Russia, led by Novoselov *et al.*, successfully isolated graphene films from highly oriented pyrolytic graphite HOPG using a technique known as mechanical exfoliation.[21] This ground-breaking work represented a pivotal moment in the extraction and identification of both few-layer and single-layer graphene from graphite. The process involved the repetitive cleaving of a graphite crystal using adhesive tape until it reached its maximum limit. Subsequently, the thinned-down graphite was transferred onto an oxidized silicon wafer (figure 3.). This achievement brought to fruition the long-standing goal of creating a two-dimensional carbon material. In recognition of their remarkable contribution to this advancement, A. Geim and K. Novoselov were jointly awarded the Nobel Prize in Physics in 2010. The extraordinary electronic, mechanical, and thermal properties of graphene, including its high electrical conductivity [20-23] and exceptional mechanical strength,[22,23] immediately, captivated the scientific community. These remarkable attributes can be attributed to the unique arrangement of carbon atoms in a honeycomb lattice. This pivotal discovery ignited widespread interest in exploring other two-dimensional carbon allotropes and materials with similarly exceptional properties.

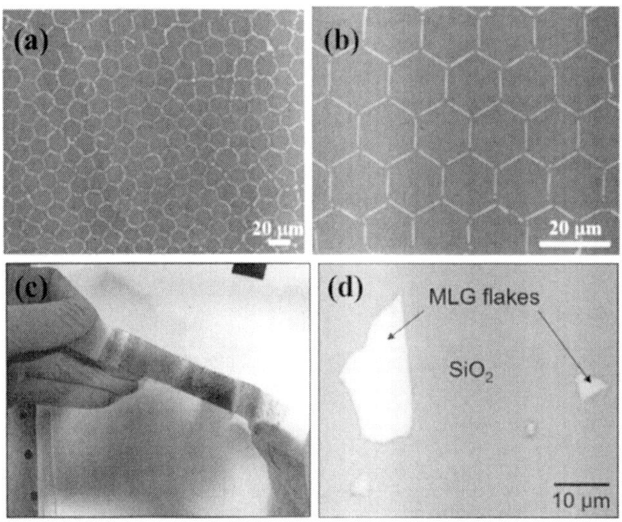


Figure 3. (a) SEM image of CVD grown hexagonal graphene flakes (b) the magnified image showing honeycomb lattice structure. Reproduced with permission.[24] (c) Mechanically exfoliated Graphene flakes and (d) optical microscope image for the exfoliated Multi-Layer Graphene flakes. Reproduced with permission.[25]

## 1.7 Other 2D Materials

There are a plethora of 2D materials apart from graphene that have garnered significant attention due to their unique properties and potential applications. Some of the notable examples are mentioned below.

### 1.7.1 Transition Metal Dichalcogenides (TMD)

TMDs with chemical formula $MX_2$ (M = transition metal and X = chalcogen) are a class of 2D quantum materials that have garnered significant attention in recent years due to their unique properties and versatile applications. These compounds consist of transition metal atoms, like molybdenum (Mo), tungsten (W), or tantalum (Ta), bonded to chalcogen atoms, typically sulfur (S), selenium (Se), or tellurium (Te).[26] They have a layered crystal structure where one metallic layer is sandwiched between two chalcogen layers. The resulting structures exhibit remarkable electronic, mechanical, and optical features.

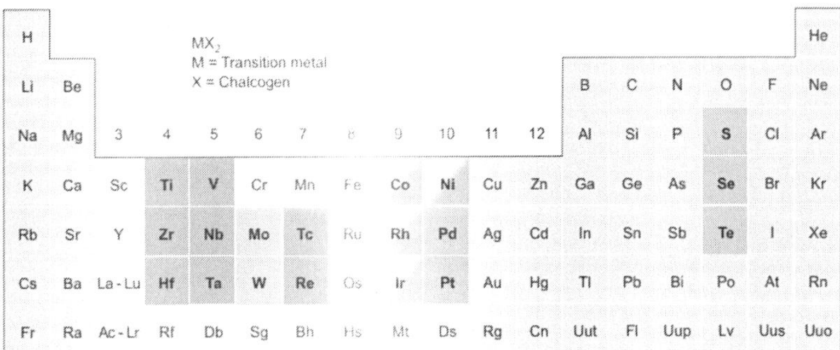


Figure 4. The transitional metals and three chalcogen elements compounds which crystallize in layered structure. The half-highlighted elements means only some of the dichalcogenides with these elements are in layered structures. Reproduced with permission.[26]

One of the key features of TMDs is their tunable bandgap. In the bulk form, these materials are indirect bandgap semiconductors, but when exfoliated down to monolayers, they become direct bandgap semiconductors. This property makes them suitable for optoelectronic applications. For instance, monolayer $MoS_2$, a widely studied TMD, has a direct bandgap around 1.9 eV, making it ideal for light-emitting diodes (LEDs), solar cells, and photodetectors.[27] The family of TMD materials also exhibits exceptional mechanical strength and flexibility. $MoS_2$, for example, is known for its robust intralayer covalent bonds and weak interlayer van der Waals interactions, resulting in extraordinary mechanical properties. This mechanical flexibility makes TMDs suitable for applications in flexible electronics and strain-sensitive devices.[28] Furthermore, TMDs are promising candidates for field-effect transistors (FETs). Due to their 2D nature, they can be easily integrated into semiconductor devices, offering excellent electrical performance. The 2D structure of TMDs enables gate-tunable carrier transport, allowing for efficient control of current flow.[29] In addition to their electrical and mechanical properties, TMDs exhibit unique optical characteristics. The direct bandgap transition leads to strong photoluminescence, which is crucial for LED applications. Their quantum confinement effects also enable efficient light-matter interactions.[30] They also display strong spin-orbit coupling, which is useful in spintronics.[31] The field of TMD research is rapidly evolving, with an increasing number of applications emerging, such as quantum dots, catalysts, and sensors. Their exceptional properties and tunable bandgaps position TMDs as a versatile class of materials with wide-ranging practical implications.

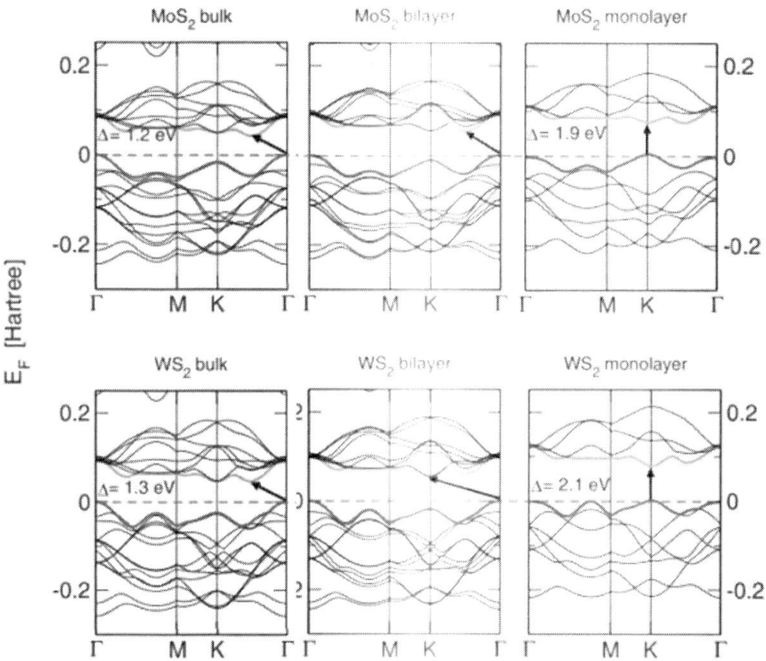


Figure 5. Band structures for bulk, bilayer, and monolayer $MoS_2$ and $WS_2$ suggesting transformation from indirect (bulk) to direct band (monolayer) gap. Reproduced with permission.[32]

## 1.7.2 Black Phosphorus (BP)

Black phosphorus or Phosphorene, is a unique 2D quantum material that has gained significant attention in the field of nanoscience and materials research. This material is distinct from its more common allotropic forms of white phosphorus and red phosphorus. BP was first isolated and characterized by the Scottish chemist Hugh Marshall in 1914. It naturally occurs in certain minerals like lazurite and Hapsburgite, but its isolation in pure form was a significant development. BP has a layered structure consisting of covalently bonded phosphorus atoms. These layers are held together by van der Waals forces. Each layer is made up of a puckered honeycomb lattice similar to graphene, but with each phosphorus atom covalently bonded to three neighbouring atoms. This puckered structure gives black phosphorus its anisotropic properties, making it a promising material for electronic and optoelectronic applications. BP exhibits anisotropic properties, with varying electronic and optical characteristics along different crystallographic directions. This anisotropy is advantageous for designing electronic devices and sensors. In contrast to graphene, black phosphorus has a direct band gap, making it suitable for optoelectronic applications such as photodetectors and LEDs. The band gap of BP is tunable through the number of layers, strain, and electrical gating, allowing for versatile device design. It has high carrier mobility, which is crucial for fast and efficient charge transport in electronic devices. It has strong light-matter interaction, making it promising for applications in photonics and optoelectronics. BP based FETs have shown promising results due to their high carrier mobility, making them suitable for next-generation electronic devices.[33] Its direct band gap and high carrier mobility make black phosphorus an ideal material for photodetectors, enabling efficient light absorption and charge separation.[34] BP based LEDs have the potential for use in displays and optical communications. Its anisotropic properties and strong light-matter interaction make black phosphorus sensors highly sensitive and suitable for various sensing applications.[35]

### 1.7.3 Hexagonal Boron Nitride (*h*-BN)

*h*-BN also known as white graphene, is a 2D material that has gained substantial attention in recent years due to its remarkable properties and versatile applications. Here, we provide an overview of *h*-BN, covering its origin, structure, properties, and applications. The existence of hexagonal boron nitride was first theorized by Hermann in 1910, but it was not until the 1940s that the compound was synthesized. The high-temperature, high-pressure synthesis of *h*-BN was independently reported by H. R. Verma and Robert N. Hall in 1945. Hexagonal boron nitride has a hexagonal lattice structure, similar to graphene. In this structure, boron and nitrogen atoms alternate in a two-dimensional honeycomb lattice. The arrangement of B and N atoms leads to the formation of strong covalent bonds within the layers. These layers are held together by weak van der Waals forces, allowing them to be easily exfoliated into single- or few-layer sheets.

*h*-BN is an excellent insulator with a wide band gap, making it ideal for use in electronic devices as an insulating layer. It exhibits exceptional thermal conductivity, even at the nanoscale, making it a valuable material for applications like thermal management in electronics.[36] It is chemically inert and thermally stable, even at high temperatures, making it a suitable material for extreme environments. In its bulk form, *h*-BN has been used as a dry lubricant due to its low friction properties. *h*-BN is used as a dielectric in electronic devices.

It is employed as an insulating material in various electronic devices, such as transistors, to prevent electrical leakage.[37] Its high thermal conductivity makes *h*-BN a promising material for thermal management in electronics, including heat sinks and thermal interface materials. In bulk form, *h*-BN serves as a solid lubricant in various mechanical applications to reduce friction and wear.[38] *h*-BN hosts color centres, such as the nitrogen-vacancy center, that are used in quantum optics and quantum information applications. It is utilized in photonics and nanophotonics for applications like the generation of single photons and as a platform for integrated photonic devices. The unique combination of electrical insulating properties, exceptional thermal conductivity, and chemical stability makes *h*-BN an attractive material for a wide range of applications in electronics, photonics, and quantum technologies. Researchers continue to explore and develop new applications for this versatile 2D material.

2D materials exhibit a diverse range of electronic properties, spanning from semimetals to insulators. Graphene is a semimetal, characterized by a remarkable electronic conductivity. It features a single Dirac cone-like electronic band structure where electrons behave as massless relativistic particles.[39] In contrast, materials like *h*-BN are insulators where the electronic structure is defined by a wide bandgap, which means it doesn't conduct electricity effectively.[40] This property makes *h*-BN a valuable insulating layer or substrate in 2D material-based electronic devices. Between these extremes, there are 2D materials with varying properties. TMDs like $MoS_2$ and $WSe_2$, are semiconductors with tunable bandgaps, making them ideal for electronic applications.[41] BP is a semiconductor with a moderate bandgap of 0.3 to 2 eV, allowing for versatile use in electronics and optoelectronics.[42] A schematic illustration of these materials' properties and band structures is depicted in figure 6. These diverse electronic properties enable engineers and scientists to design a wide array of 2D material-based devices, from high-conductivity graphene transistors to insulating layers of *h*-BN in heterostructures. The ability to fine-tune the electronic characteristics of 2D materials offers tremendous litheness for applications in nanoelectronics, photonics, and beyond.

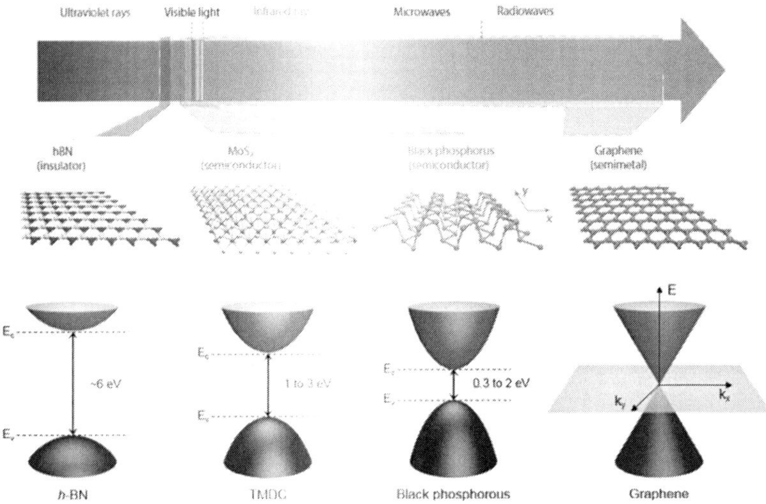


Figure 6. Energy spectrum of various 2D materials and their atomic crystal structures; Electronic band structures of hexagonal boron nitride (*h*-BN), transition metal dichalcogenides (TMDCs), black phosphorous, and graphene. Reproduced with permission.[43]

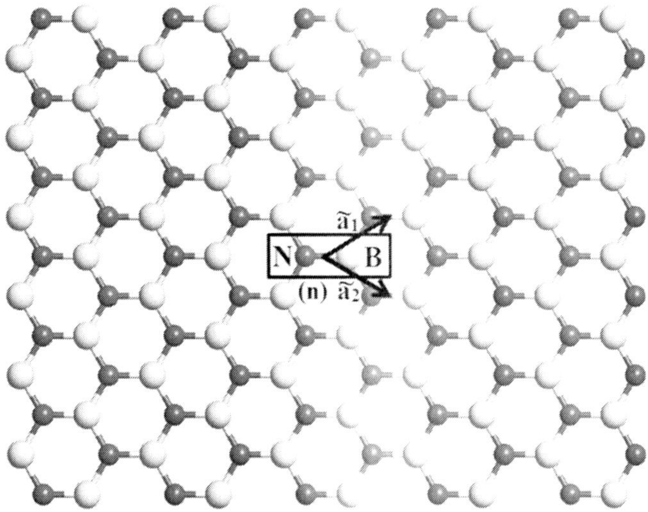


Figure 7. Schematic of a hexagonal boron-nitride nanosheet; $\tilde{a}_1$ and $\tilde{a}_2$ are primitive lattice vectors. Reproduced with permission.[44]

### 1.7.4 Stanene

Stanene is a fascinating two-dimensional material that has attracted significant attention in the field of condensed matter physics. Also known as 2D tin, it belongs to the family of 2D materials alongside graphene, phosphorene, and others. Stanene's remarkable properties and potential applications make it an exciting material for scientific probe and technological advancements. The concept of stanene was first proposed theoretically in 2013 by Shou-Cheng Zhang and his research team.[45] Stanene consists of a single layer of tin (Sn) atoms arranged in a hexagonal lattice, similar to graphene. The unique characteristic of stanene is its buckled structure, where the tin atoms form a honeycomb pattern, with every other atom slightly elevated above the plane. This buckling imparts distinctive electronic properties to stanene.

Stanene is a topological insulator, meaning it acts as an insulator in its interior but conducts electricity along its edges or surfaces.[46] This property has implications for the development of low-power electronics. It exhibits exceptionally high carrier mobility, which makes it promising for ultrafast electronic devices. Stanene is known for its potential to exhibit the quantum spin Hall Effect, a phenomenon crucial in the development of spintronic devices.[47] Stanene holds enormous promise for various applications, like low-power electronics, quantum computing, and high-temperature superconductors.[48] It could play a pivotal role in advancing next-generation electronic devices and quantum technologies. Researchers are continually investigating stanene's properties and potential applications, and while it remains challenging to synthesize in its purest form due to its extreme reactivity, the future of this 2D material looks promising. It may lead to ground-breaking developments in the world of materials science and technology.

### 1.7.5 Layered Double Hydroxides (LDHs)

LDHs, also known as hydrotalcite-like compounds or anionic clays, are a class of synthetic materials that belong to the larger family of layered materials. They are structurally similar to naturally occurring minerals and have a wide range of applications due to their distinctive properties. LDHs have a layered structure composed of positively charged brucite-like layers and interlayer anions, water molecules, or organic molecules. LDHs possess several noteworthy characteristics, including high anion exchange capacity, good thermal stability, and excellent reactivity. The tunability of their composition allows for control over various properties, making them versatile materials. LDHs find applications in diverse fields, such as catalysis,[49] drug delivery,[50] flame retardants,[51] and environmental remediation. They are excellent anion exchangers,[52] which is particularly useful in water purification[53] and wastewater treatment. In the pharmaceutical industry, LDHs are utilized for controlled drug release due to their ability to intercalate drug molecules in the interlayer space. Additionally, LDHs serve as efficient catalysts in various chemical reactions.

### 1.7.6 MXenes

MXenes are a novel class of 2D materials that have garnered significant attention in the field of materials science and engineering due to their exceptional properties and a wide range of potential applications. MXenes belong to the larger family of transition metal carbides and nitrides and have a general formula of $M_xM'_{2-x}T_x$, where M represents transition metals, M' stands for carbon and/or nitrogen, and $T_x$ denotes surface terminations, such as hydroxyl, oxygen, or fluorine groups. The discovery of MXenes can be traced back to 2011 when Yury Gogotsi and his team established a selective etching method to generate 2D nano-crystals through exfoliation of $Ti_3AlC_2$.[54] The resulting material consists of 2D layers with excellent electrical conductivity and mechanical properties. MXenes have a unique structure with strong in-plane metallic conductivity. They possess several remarkable properties like excellent electrical conductivity, high mechanical strength, and good thermal stability. They also exhibit good environmental stability and can withstand exposure to air and moisture. Furthermore, MXenes demonstrate tunable surface chemistry and can be modified to have different functional groups, making them versatile for various applications.

MXenes have a rapidly growing list of applications. They are considered promising candidates for energy storage and conversion devices, such as supercapacitors, lithium-ion batteries, and hydrogen evolution catalysts.[55,56] Additionally, their high conductivity and surface functionalization make them suitable for sensors,[57] electromagnetic interference shielding,[58] and water purification.[59] In the field of catalysis, MXenes exhibit potential for various reactions due to their active surface chemistry. Overall, the unique properties and versatile applications of MXenes have made them a subject of intense research, and their impact on various industries, including electronics, energy, and environmental technologies, is continually expanding. Researchers are exploring novel synthesis methods and functionalization techniques to unlock the full potential of MXenes for future applications.

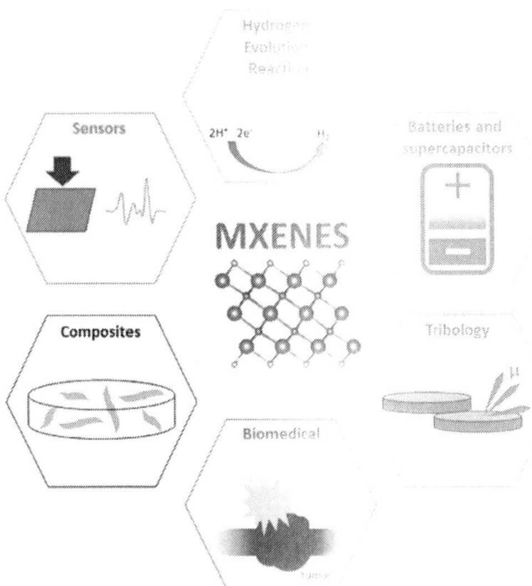


Figure 8. Schematics of crystal structure and multifunctional applications of Mxene. Reproduced with permission.[60]

### 1.7.7 Silicene and Germanene

Silicene is a 2D allotrope of silicon, akin to graphene, but with silicon atoms arranged in a honeycomb lattice. Theoretical predictions of silicene date back to the 1990s, but it wasn't until 2012 that it was first synthesized on a silver substrate through a process known as molecular beam epitaxy.[61] Silicene has unique electronic properties that arise from its buckled structure, where silicon atoms form a hexagonal lattice with one sub-layer slightly above the other. This buckling leads to a bandgap in the electronic structure of silicene, making it a potential candidate for semiconductor applications.[62]

Germanene is a 2D material composed of germanium atoms arranged in a hexagonal lattice, similar to graphene. Just like silicene, germanene also exhibits a buckled structure. While theoretical predictions for germanene existed for some time, it was successfully synthesized on gold substrates through molecular beam epitaxy in 2014.[63] Germanene's unique electronic structure and inherent semiconducting properties have led to its exploration for use in electronic devices, particularly FETs.[64]

Both silicene and germanene have characteristics that make them attractive for electronic applications. Their structural and electronic properties are tunable, and they exhibit good thermal stability. The buckled structure provides an intrinsic bandgap, allowing for the development of 2D transistors. Additionally, their compatibility with existing silicon-based electronics makes them viable candidates for future semiconductor technology. Further-more, these materials have potential applications in optoelectronics, where their 2D nature could be harnessed for photodetectors and light-emitting devices.[65] Their versatile properties and ongoing research into synthesis and functionalization make silicene and germanene exciting prospects in the field of 2D materials and nanoelectronics.

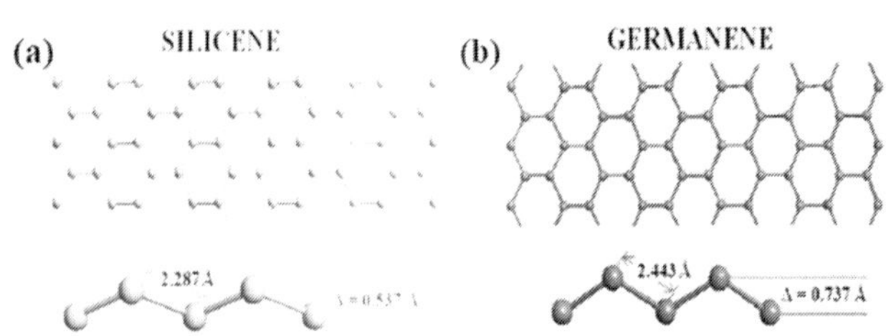


Figure 9. Crystal Structure of 2D materials (a) Silicene and (b) Germanene depicting their planar structure. Reproduced with permission.[66]

### 1.7.8 Phosphorene-like Materials

The group V elements Arsenic (As), Antimony (Sb), Bismuth (Bi) can form 2D materials with properties similar to phosphorene. For example, bismuthene (2D bismuth) has recently gained attention for its unique electronic properties and potential applications in topological insulators,[67] quantum computing,[68] and spintronics.[69]

These 2D materials have expanded the landscape of materials science and nanotechnology, offering a wide range of properties and applications. Researchers continue to investigate their synthesis methods, fundamental properties, and practical uses in various technological fields, shaping the future of materials science and electronics.

## 1.8 Importance and Applications in Various Fields

2D materials have sparked significant interest in the field of nano-electronics due to their exceptional electronic properties and the potential to revolutionize existing technologies. Here's an overview of their applications in key areas like Nanoelectronics, Optoelectronics, Energy storage and conversion as well as Quantum Computing.

### 1.8.1 Nanoelectronics

*High-Performance Transistors*: 2D materials, such as graphene and TMDs like $MoS_2$ and $WSe_2$, are being explored to create high-performance FETs.[70] Monolayer TMDs have a direct bandgap, making them suitable for FETs with excellent on-off current ratios.[71]

*Flexible Electronics:* The mechanical flexibility of 2D materials allows for the development of flexible and wearable electronics. Graphene-based FETs can be integrated into flexible substrates, enabling bendable and stretchable electronic devices.[72]

*Low-Power Electronics*: 2D materials enable the fabrication of low-power transistors. Their ultra-thin nature and excellent charge carrier mobility make them promising for energy-efficient electronics, which is crucial for portable devices and Internet of things (IoT) applications.[73]

### 1.8.2 Optoelectronics

2D materials have made significant strides in the field of optoelectronics, offering unique advantages for various devices. Here are their applications in key areas like photodetectors, light-emitting diodes (LEDs), and solar cells.

*High-Performance Photodetectors:* 2D materials such as graphene, $MoS_2$, and black phosphorus have been utilized to create high-performance photodetectors. These materials exhibit excellent electrical conductivity and light absorption properties, making them ideal for detecting light across a broad spectrum. [74,75,76]

*Broadband Sensing:* The tunable bandgap of 2D materials allows for the development of broadband photodetectors. By stacking different 2D materials or combining them with traditional semiconductors, engineers can design detectors capable of sensing light from ultraviolet (UV) to infrared (IR) wavelengths.[77]

*Ultrafast Response Times:* 2D material-based photodetectors offer ultrafast response times in the picosecond or even femtosecond range. This rapid response makes them suitable for high-speed communication systems and imaging applications.[78]

### 1.8.3 Light-Emitting Diodes (LEDs)

*Efficient Light Emission:* 2D materials like TMDs exhibit efficient light emission due to their direct bandgap nature. Monolayer TMDs, such as $MoS_2$ and $WS_2$, can emit light in the visible spectrum, making them promising for LEDs.[79]

*Flexible and Transparent LEDs:* The mechanical flexibility and transparency of 2D materials allow for the fabrication of flexible and transparent LEDs. These characteristics are vital for emerging technologies like flexible displays, wearable electronics, and smart windows.[80]

*Color Tunability:* By using different 2D materials or heterostructures, LED emission colours can be tuned precisely. This tunability is crucial for display technologies and RGB (red, green, blue) lighting applications.[81]

### 1.8.4 Solar Cells

*Enhanced Light Absorption:* 2D materials can enhance light absorption in solar cells. Graphene, for example, can be incorporated into organic photovoltaics to improve charge transport and increase the efficiency of converting sunlight into electricity.[82]

*Flexible and Lightweight Solar Panels:* The lightweight and flexible nature of 2D materials allows for the development of flexible and portable solar panels. These panels can be integrated into a variety of surfaces, from clothing to building materials, expanding the possibilities for solar energy harvesting.[83]

*Tandem Solar Cells:* Researchers are exploring the use of 2D materials in tandem solar cells, where multiple layers of different materials are stacked to capture a broader spectrum of sunlight. This approach has the potential to significantly increase the efficiency of solar energy conversion.[84]

### 1.8.5 Energy Storage and Conversion

2D materials have emerged as a revolutionary class of materials with immense potential for energy storage and conversion applications. In the realm of energy storage, 2D materials serve as excellent candidates for supercapacitors and batteries, offering high surface areas, good flexibility, good chemical and thermal stability, wide potential windows, rich surface chemistry, extraordinary electrical, thermal and mechanical properties, and rapid charge-discharge capabilities.[85,86,87,88] Moreover, their remarkable electrocatalytic properties make them ideal catalysts for fuel cells [89,90] and water splitting,[91] enabling efficient conversion of chemical energy into electricity and vice versa. As research in this field continues to advance, 2D materials hold the potential to revolutionize the energy landscape by offering more efficient and sustainable solutions for energy storage and conversion.

### 1.8.6 Quantum Computing

2D materials are at the forefront of quantum computing research, offering unique capabilities and opportunities in various aspects of this cutting-edge field:

### 1.8.6.1 Single Quantum Emission

*Quantum Light Sources:* 2D materials, such as single-photon emitters in hexagonal boron nitride (*h*-BN) or quantum dots in transition metal dichalcogenides (TMDs), serve as excellent sources of single photons. These sources are essential for quantum key distribution, quantum cryptography, and quantum communication.[92]

*Deterministic Emission*: Single-photon emitters in 2D materials can exhibit deterministic emission, ensuring that a single photon is produced on demand. This property is crucial for building quantum networks and quantum repeaters.[93]

### 1.8.6.2 Quantum Bits (Qubits) in 2D Materials

*Coherent Qubits:* Certain 2D materials, like superconducting qubits based on Josephson junctions, demonstrate long coherence times, a critical factor for performing complex quantum computations without losing quantum information.[94]

*Gate Operations:* Quantum gates can be implemented using the electronic states of 2D materials. Gate operations in 2D materials are highly precise and can be controlled through external electric and magnetic fields, enabling the manipulation of qubits for quantum algorithms.[95]

*Quantum Dot Qubits:* Quantum dots formed in 2D materials are potential qubits. Their discrete energy levels and strong Coulomb interactions make them suitable for quantum information processing. Researchers are exploring the use of TMDs and other 2D materials to create stable and scalable quantum dot qubits.[96]

### 1.8.6.3 Quantum Algorithms and Applications

*Quantum Simulation*: 2D materials can be employed to simulate complex quantum systems, aiding in the development and validation of quantum algorithms. This capability is instrumental in studying quantum phase transitions, quantum chemistry, and materials science.[97]

*Quantum Machine Learning:* Quantum algorithms running on quantum computers can enhance machine learning tasks. 2D materials contribute to the implementation of quantum machine learning algorithms, opening up new possibilities for solving optimization problems and data analysis.[98]

*Cryptography and Security:* Quantum computers built with 2D materials have the potential to break classical encryption schemes. Simultaneously, 2D materials-based quantum cryptography can offer secure communication methods, such as quantum key distribution, to protect sensitive information.[99, 100]

2D materials play a vital role in the development and advancement of quantum computing. They offer versatile platforms for generating single photons, serve as qubits with extended coherence times, and enable the execution of quantum algorithms and applications. As quantum computing continues to evolve, the integration of 2D materials will contribute to breakthroughs in quantum technologies with profound implications for various fields, including cryptography, simulation, and machine learning.

# CHAPTER 2

# CRYSTAL STRUCTURE AND PROPERTIES

Within the realm of the microscopic world there exists a fascinating category of materials that challenge conventional three-dimensional norms, known as Two-Dimensional Quantum Materials. These materials, often referred to as 2D quantum materials, possess unique properties that emerge due to their reduced dimensionality. Unlike bulk materials, which extend in three dimensions, 2D quantum materials are confined to just two dimensions, resulting in intriguing quantum phenomena. This chapter delves into the fundamental aspects of their existence, focusing specifically on the intricacies of crystal structure and the remarkable properties that emerge in two dimensions.

At the heart of this exploration is the archetype of 2D materials, graphene, a single layer of carbon atoms arranged in a hexagonal lattice. Beyond graphene, we navigate through the crystal structures of diverse two-dimensional wonders, such as transition metal dichalcogenides, black phosphorus, and hexagonal boron nitride. The hexagonal lattice, a recurring motif, becomes a crucial architectural element influencing the unique behaviors exhibited by these materials.

As we journey through the electronic, optical, mechanical, and thermal landscapes, we uncover the quantum effects that govern these materials. From Dirac cones in graphene to the quantum Hall effect, each property is intricately linked to the crystal structure, unveiling a tapestry of quantum phenomena. Moreover, we explore the implications of defects, doping, and growth methods in tailoring the properties of 2D materials. This chapter not only unveils the structural beauty of 2D materials but also uncovers the immense potential they hold for revolutionizing electronics, optics, and beyond. It serves as a gateway into the mesmerizing world, where the atomic arrangement dictates the rules of quantum engagement.

## 2.1 Atomic Structure and Bonding in 2D Materials

The atomic structure of 2D materials, governed by hexagonal lattices and unique bonding patterns, fortifies their extraordinary properties, offering a microscopic realm where quantum interactions shape material behaviors.

### 2.1.1 Graphene

As seen in Figure 1(a), the nucleus of a carbon atom is surrounded by six electrons, four of which are valence electrons. Carbon is the sixth element in the periodic table, with a ground-state electronic configuration of $1s^2 2s^2 2P_X^1 2P_y^1 2P_z^0$, as shown in Figure 1(b). These electrons in the valence shell of a carbon can form three types of hybridization, namely sp, $sp^2$ and $sp^3$. Figure 1(c) illustrates the formation of $sp^2$ hybrids. When carbon atoms share $sp^2$ electrons with their three neighbouring carbon atoms, they form a layer of honeycomb network of planar structure, which is also called monolayer graphene. The unit cell of a graphene crystal contains two carbon atoms as seen in Figure 1(d), and the unit-cell vectors $a_1$ and $a_2$ have the same lattice constant of 2.46 Å. The resonance and delocalization of the electrons are responsible for the stability of the planar ring. In a typical $sp^2$ hybridization of two neighbouring carbon atoms on the graphene layer (see Figure 1 (e)), an out-of-plane π bond is made up by $2p_z$ orbitals which are perpendicular to the planar structure, while an in-plane σ bond is formed by the $sp^2$ (2s, $2p_x$ and $2p_y$) hybridized orbitals. The resulting covalent σ bond has a short interatomic length of ~ 1.42 Å, making it even stronger than the $sp^3$ hybridized carbon–carbon bonds in diamonds, thus giving the monolayer graphene remarkable mechanical properties (e.g. a Young's modulus of 1 TPa and an intrinsic tensile strength of 130.5 GPa.[101] In monolayer graphene, the conduction band and valence band with zero band gap are formed due to the half-filled π band that permits free-moving electrons. Further, the π-bonds provide a weak van der Waals (vdW) interaction between adjacent graphene layers in bilayer and multi-layer graphenes.

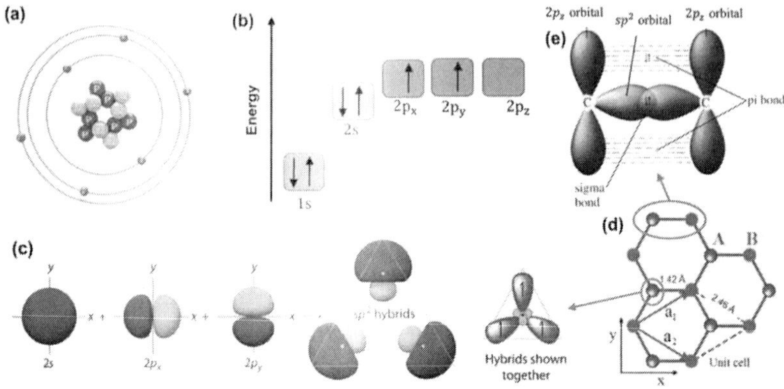


Figure 1. (a) Atomic structure of a carbon atom. (b) Energy levels of outer electrons in carbon atoms. (c) The formation of $sp^2$ hybrids. (d) The crystal lattice of graphene, where A and B are carbon atoms belonging to different sub-lattices, a1 and a2 are unit-cell vectors. (e) Sigma bond and pi bond formed by $sp^2$ hybridization. Reproduced with permission.[102]

Generally, graphene can be stacked in various ways to form different multilayered graphenes (MLG), which have different properties. Typical graphene stacking includes Ordered (AA) and Bernal (AB) stacking.[103] [104] [105] In Order stacking, all the carbon atoms of each layer are well-aligned in a straight line but for AB stacking, a cycle period is constituted by two layers of non-aligned graphene as shown in Figure 2. It has been suggested that graphene's properties like adsorption, band gap, and doping would be changed by the ways of graphene stacking.[106] [107] [108] Sylvain *et al.* found that charge carriers in few-layered graphites are highly sensitive to the layer numbers as well as the stacking of graphene layers, with ambipolar transport only existing in AB-stacked graphites.[107]

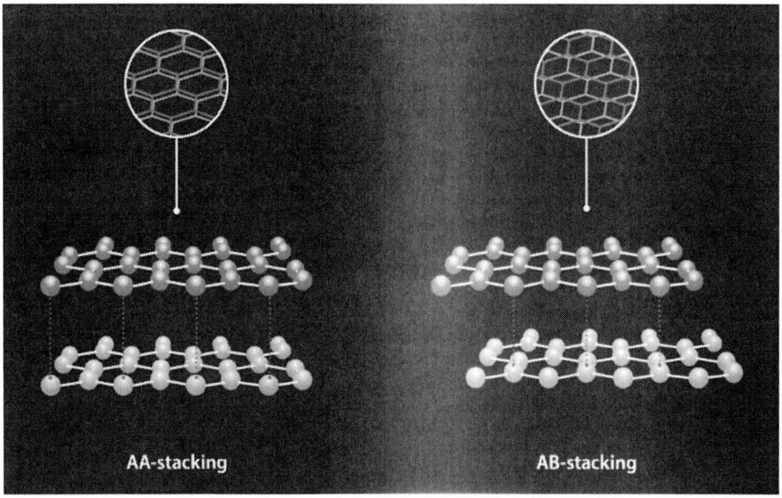


Figure 2. The crystal structures of AA-stacking and AB-stacking of graphene. Reproduced with permission.[109]

## 2.1.2 Hexagonal Boron Nitride (*h*-BN)

Hexagonal boron nitride (*h*-BN) exhibits a crystal structure of profound scientific interest, rooted in the arrangement of its constituent boron and nitrogen atoms. In Figure 3 (a), the atomic lattice of the *h*-BN monolayer is demonstrated, which shows the honeycomb structure as that of the graphene. It has a hexagonal crystalline lattice with point group $D_{6h}$ and space group $P6_3/mmc$.[110,111,112,113] In *h*-BN, each unit cell comprises two hexagonal planes stacked in an ABAB sequence along the vertical c-axis, indicative of a rhombohedral crystal lattice. The hexagonal symmetry is manifested in the coplanar alignment of the boron and nitrogen atoms within each plane, forming a $sp^2$ hybridized network. This planar, two-dimensional architecture extends to the entire crystal lattice, influencing the material's electronic and mechanical characteristics. The alternating boron and nitrogen atoms within the hexagon result in strong, covalent σ-bonds, while the in-plane π-bonds contribute to the delocalization of electrons, resulting in the unique electronic properties of *h*-BN. The hexagonal rings form basal planes with alternating B and N atoms bonded strongly by σ bonds and at a bond length of 1.45 Å. Furthermore, the rhombohedral stacking arrangement allows for vdW interactions between adjacent layers, affording *h*-BN its characteristic lubricating properties as shown in Figure 3 (b). The lattice constants are very similar to graphene, with $a = b = 2.5$ Å and $c = 6.6$ Å, also $\alpha = \beta = 90°$, and $\gamma = 120°$ and the interlayer spacing is of 3.3 Å.[114]

This intricate crystal structure, governed by hybridization, symmetry, and stacking, underpins the diverse and technologically relevant properties exhibited by *h*-BN.

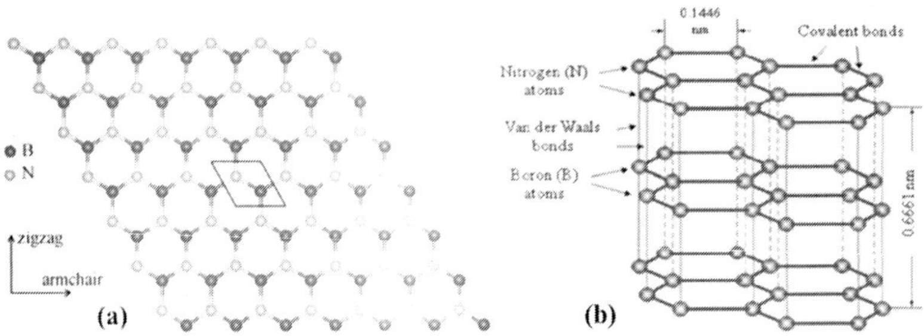


Figure 3. (a) Lattice structure of monolayer *h*-BN. The unit cell is shown as parallelogram contains one nitrogen and one boron atoms. Reproduced with permission.[115] (b) Schematic of multilayer *h*-BN. Reproduced with permission.[116]

### 2.1.3 Transition Metal Dichalcogenides (TMD)

Few TMDs exhibit a layered crystalline structure similar to graphite, which imparts significant anisotropy to their electrical, chemical, mechanical, and thermal properties.[117] Unlike Graphene which is chemically inert, single-layered 2D TMDs - whose generalized formula is $MX_2$, where M is a transition metal of groups 4-10 and X is a chalcogen as shown in Figure 4 (a), exhibit versatile chemistry. TMDs have layered structure where each layer typically has a thickness of 6~7 Å, which consists of a hexagonally packed layer of metal atoms sandwiched between two layers of chalcogen atoms. The intralayer M-X bonds are predominantly covalent in nature, whereas the sandwich layers are coupled by weak vdW forces thus allowing the crystal to readily cleave along the layer surface. The metal atoms provide four electrons to fill the bonding states of TMDs such that the oxidation states of the metal (M) and chalcogen (X) atoms are +4 and –2, respectively. The lone-pair electrons of the chalcogen atoms terminate the surfaces of the layers, and the absence of dangling bonds renders those layers stable against reactions with environmental species. The M-M bond length varies between 3.15 Å and 4.03 Å, depending on the size of the metal and chalcogen ions. As per the metal coordination layered TMDs can be either trigonal prismatic (2H) or octahedral (1T) as depicted in Figure 4 (b) and (c) respectively. Depending on the combination of the metal and chalcogen elements, one of the two coordination modes is thermodynamically preferred. For example in 2H-$MoS_2$, each Mo atom is prismatically coordinated to six surrounding

S atoms, forming a stable phase thermodynamically, whereas for the 1T-$MoS_2$ phase, six S atoms form a distorted octahedron around one Mo atom thus establishing a metastable phase. Interestingly, each phase can be easily transformed to the other via intralayer atomic gliding. In this context, 2H-$MoS_2$ can be converted into 1T-$MoS_2$ by intercalating Li or K.[118,119] Because 1T-$MoS_2$ is thermodynamically unstable it gradually converts back to the trigonal prismatic configuration (2H-$MoS_2$) at room temperature.[120]

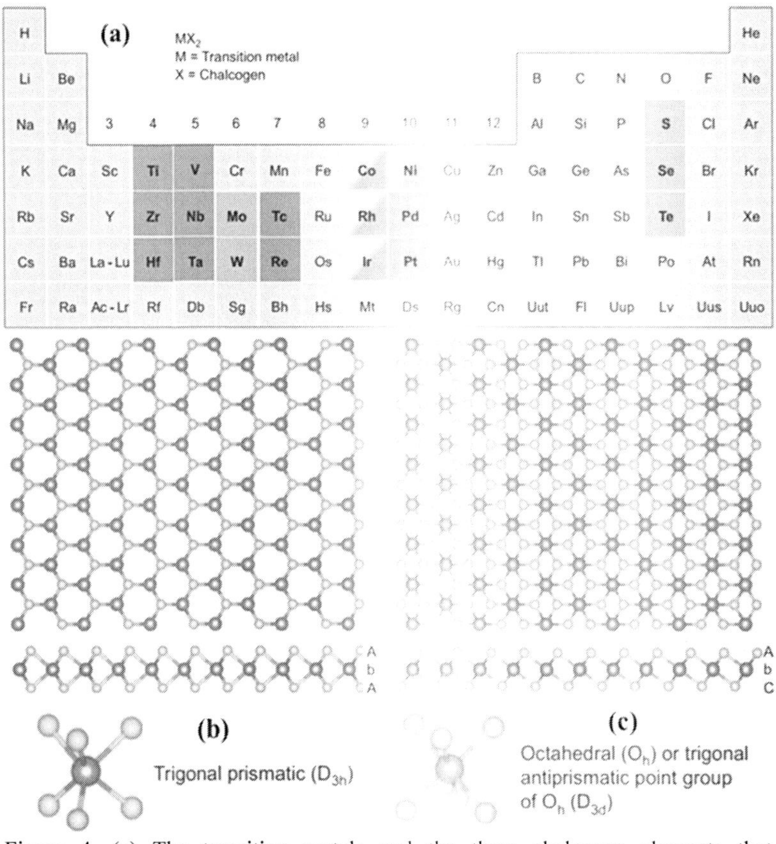


Figure 4. (a) The transition metals and the three chalcogen elements that predominantly crystallize in those layered structure are highlighted in the periodic table. (b, c) *c* - Axis view of single-layer TMD with (b) trigonal prismatic and (c) octahedral coordinations. The labels AbA and AbC represent the stacking sequence where the upper- and lower-case letters represent chalcogen and metal elements, respectively. Reproduced with permission.[121]

In contrast to graphite, TMDs exhibit a wide variety of polymorphs and stacking polytypes because an individual $MX_2$ monolayer, which itself contains three layers of atoms (X–M–X), can be in either one of the two phases.[121] Most commonly encountered polymorphs are 1T, 2H and 3R where the letters stand for trigonal, hexagonal and rhombohedral, respectively, and the digit indicates the number of X–M–X units in the unit cell (that is, the number of layers in the stacking sequence) as described in Figure 5. There are three different polytypes (that is, three different stacking sequences) for 2H polymorphs. A single TMD can be found in multiple polymorphs or polytypes, depending on the history of its formation. The 2H and 1T phases of $MoS_2$ is already discussed. However synthetic $MoS_2$, often contains the 3R phase where the stacking sequence is AbA CaC BcB.[122] In both cases of 2H and 3R, the metal coordination is trigonal prismatic as seen in Figure 5 (b).

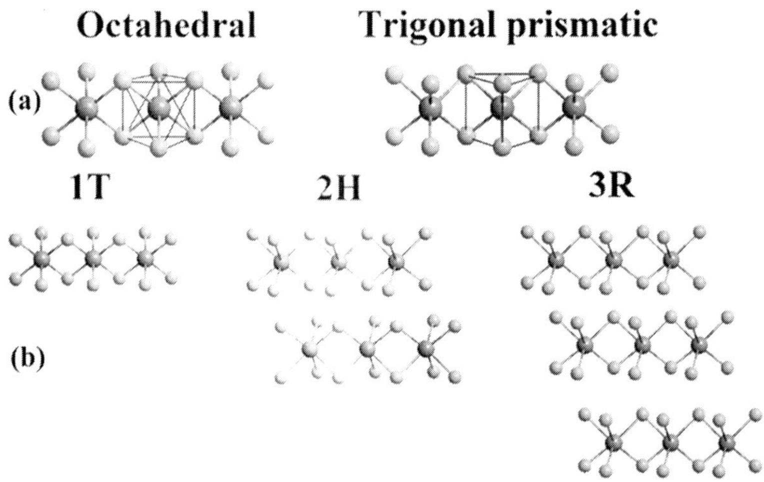


Figure 5. (a) Trigonal prismatic and octahedral transition metal coordination in TMDs. (b) The layer stacking in the most common polytypes of TMDs – 1T, 2H and 3R. The grey and yellow spheres represent transition metal and chalcogen atoms, respectively. Reproduced with permission.[123]

The stacking configurations in TMDs ($MX_2$) as shown in Figure 6, profoundly impact their structural and electronic characteristics. The AA stacking entails a direct alignment of metal atoms in consecutive layers, fostering strong interlayer interactions. Conversely, AA' stacking introduces a slight vertical displacement, inducing electronic decoupling between layers. AB stacking, marked by an alternating arrangement of metal and chalcogen atoms, results in weaker interlayer interactions than AA stacking.

A'B stacking involves a vertical shift with partial misalignment, contributing to an electronic asymmetry between layers. Similarly, AB' stacking, akin to AB stacking but with vertical displacement, introduces an electronic asymmetry. These stacking arrangements play a pivotal role in determining the material's electronic band structure, interlayer coupling strength, and overall physical properties, offering avenues for tailored applications in nanoelectronics and optoelectronics. Advanced characterization techniques, including scanning tunnelling microscopy and X-ray diffraction, are indispensable for elucidating and manipulating these stacking structures in TMDs.

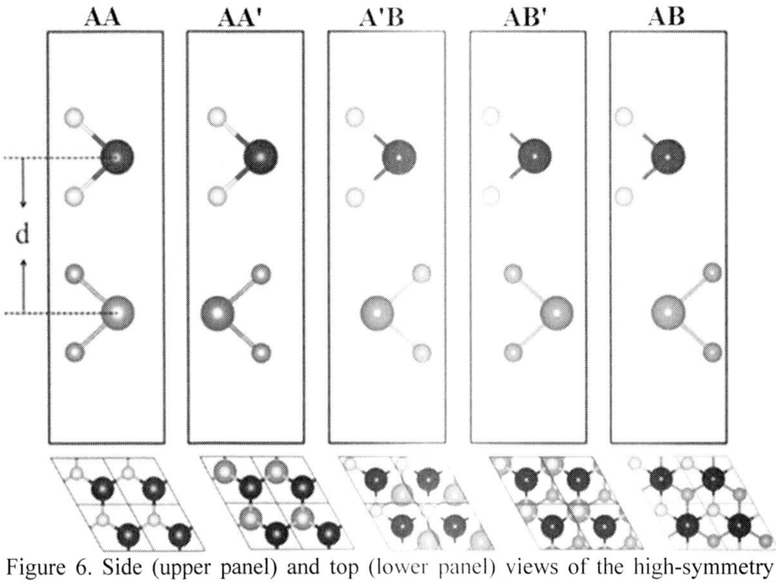


Figure 6. Side (upper panel) and top (lower panel) views of the high-symmetry stacking orders of bilayer $MX_2$. Large and small spheres represent the M and X atomic species, respectively. A grayscale (color coding) is used to distinguish the position of the atoms in the two $MX_2$ layers. The full lines demarcate the unit cell. Reproduced with permission.[124]

## 2.2 Electronic Band Structure

The electronic band structure of 2D materials serves as a foundational framework, delineating the intricate energy-momentum relationships of electrons within these atomically thin layers, profoundly influencing their electronic properties and potential applications. This intricate mapping of energy bands, facilitated by techniques like angle-resolved photoemission spectroscopy, unveils critical features such as Dirac cones in graphene, enabling a complete understanding of the quantum behaviors that define the electronic landscape of 2D materials.

### 2.2.1 Graphene

The electronic band structure of graphene, a prototypical two-dimensional material, derives its distinct features from the hexagonal arrangement of carbon atoms in a planar honeycomb lattice. The unit cell of graphene contains 2 equivalent atoms (A and B as shown in Figure 7 (a). This crystal structure results in the touching between bonding ($\pi$) and anti-bonding ($\pi^*$) orbitals at each energy valleys K and K′ when considering hopping only occurs between the nearest neighbour atomic sites.[125,126] Thus a linearly dispersive electronic band structure is occurred at low-energy regime around K and K′ creating a Dirac cone at the K-point of the Brillouin zone as seen in Figure 7 (b). The energy can be expressed by $E = \pm v_F k$, which indicates the low-energy carriers in graphene is zero-mass which is characterized by massless Dirac fermions and its speed is independent of energy. The carriers' speed, called as Fermi velocity, is as high as $10^6$ m/s (~1/300 $c$, where $c$ is velocity of light), and thus the electrical transport properties of graphene are remarkable ascribed to such photon-like relativistic Dirac fermions.

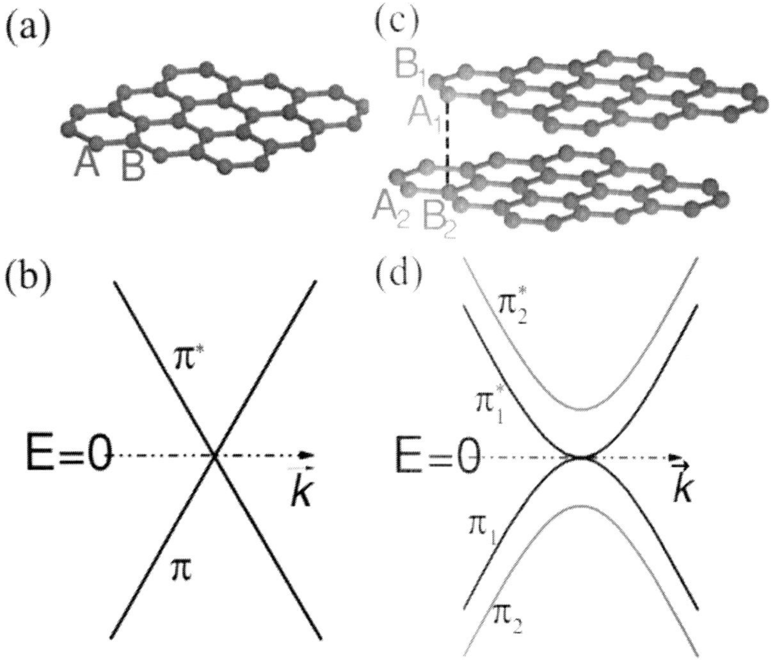


Figure 7. Crystal structures of graphene single layer (a) and AB-stacked bilayer (c). The unit cell of single layer graphene and bilayer graphene consist of 2 and 4 atoms in each unit cell, respectively. The hopping between nearest neighbour atom sites A and B resulting the linear dispersive electronic band structure (b). The interlayer coupling for bilayer graphene cannot be ignored as the hopping between A1 and B2 affected the electronic band structure and resulting two parabolic electronic bands (d). Reproduced with permission.[127]

Particularly, the gapless graphene displays ambipolar electric field effect with ballistic property presenting extremely high mobility at room temperature as observed experimentally.[128,129] The linear dispersion is a result of the π-electrons' delocalization over the hexagonal lattice, leading to a vanishing effective mass near the Dirac point.

The electronic behavior of graphene is also dependent on its stacking geometry. For bilayer graphene (BLG), the most energetically favourable one is stacked by the AB sequence. The AB stacked BLG consists of four atoms in a unit cell, labelled $A_1$, $B_1$, $A_2$ and $B_2$ (Figure 7 (c)), where each subscript numerical index represents the different layers. Different from that of single layer graphene (SLG) which only present one linear band at low-energy regime, two group of parabolic bands present (Figure 7 (d)) in BLG due to the interlayer coupling effect considering both the nearest neighbour hoping in-plane (from $A_1$ to $B_1$ and $A_2$ to $B_2$) as well as out-of-plane (from $A_2$ to $B_1$). Therefore, the electronic band structure of these parabolic bands can be modified by tuning the electronic coupling between layers. For example, breaking the unit-cell symmetry chemically or electrically, as well as rotating one layer respective to the other (Twistronics), can intentionally modify the corresponding electronic band structure.

The electronic band structure of graphene further reveals the presence of two distinct Dirac cones at the K and K' points of the Brillouin zone, corresponding to the two inequivalent valleys in the graphene hexagonal lattice. This valley degeneracy, a consequence of the hexagonal symmetry, plays a pivotal role in the material's electronic behavior and transport properties. Additionally, the chiral nature of the Dirac fermions in graphene results in unconventional quantum phenomena, such as the half-integer quantum Hall effect, providing a rich platform for exploring topological states of matter. The energy-momentum dispersion relations are sensitive to external perturbations, offering the potential for precise control of electronic properties through mechanisms like strain engineering or the application of external electric and magnetic fields. The electronic band structure of graphene, characterized by its unique Dirac cones and valley degeneracy, continues to captivate researchers, fostering advancements in both fundamental physics and technological applications.

### 2.2.2 Hexagonal Boron Nitride (*h*-BN)

The electronic band structure of *h*-BN, a quintessential two-dimensional insulator, is governed by the $sp^2$ hybridization of its boron and nitrogen atoms within a hexagonal lattice. The crystal structure, akin to graphene, is characterized by a rhombohedral stacking of boron and nitrogen planes, forming a repeating ABAB pattern along the c-axis. The *h*-BN band structure unveils a wide bandgap in the order of 5.97 eV, attributed to the significant energy separation between the valence and conduction bands. The distinct electronic properties arise from the alternating arrangement of boron and nitrogen, leading to strong σ-bonds within each hexagonal plane. The valence and conduction bands primarily consist of the localized π and σ orbitals of boron and nitrogen atoms. Importantly, the inherent insulating nature of *h*-BN renders it a crucial component in various electronic applications, including as a dielectric substrate and an insulating spacer in vdW heterostructures. Advanced spectroscopic techniques, such as angle-resolved photoemission spectroscopy (ARPES), facilitate a detailed understanding of the electronic band structure, paving the way for tailored applications in the burgeoning field of 2D materials.

### 2.2.3 Transition Metal Dichalcogenides (TMDs)

The electronic band structure of TMDs dictates their unique electrical properties, with a direct or indirect band gap depending on the number of layers. Mono-layer TMDs exhibit a direct band gap, making them promising for optoelectronic applications, while multilayer TMDs typically have an indirect band gap. Also the band structure of TMDs is characterized by a transition from a non-metallic state in bulk form to a semiconductor state in mono- and few-layer forms. This transition results from quantum confinement effects, leading to tunable band gaps and significant implications for their electronic and optical properties. Additionally, the band alignment at heterostructures formed between different TMD layers plays a crucial role in device applications such as photodetectors and transistors.

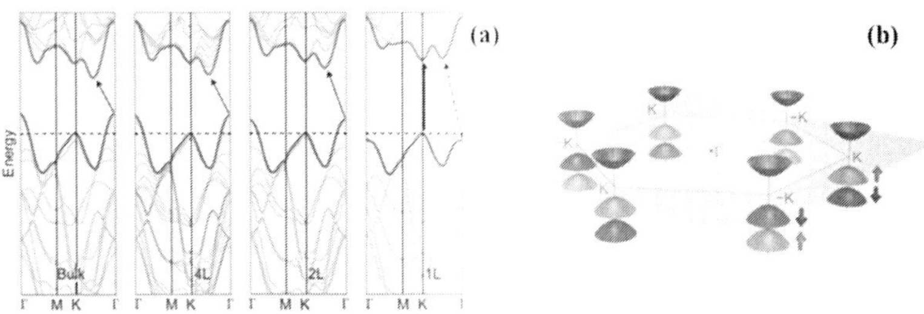


Figure 8. (a) Energy dispersion of bulk, quadrilayer (4L), bilayer (2L), and monolayer (1L) $MoS_2$, showcasing the transition from a non-metallic state in bulk to a semiconductor state in mono- and few-layer forms. The conduction and valence band edges are represented by red and blue lines, respectively, with the lowest energy transition being direct in the monolayer, as shown by solid arrows. (b) Band structure of $MoS_2$ reveals six valleys and opposite spin-orbit splitting of the valence band at the K and K' points, represented by red and blue surfaces, respectively, while green surfaces represent the conduction band minima or valleys. Reproduced with permission.[121]

Figure 8a illustrates the band structures of bulk and few- to mono-layer $MoS_2$, a group 6 TMD characterized by a trigonal prismatic structure. In bulk form, $MoS_2$ exhibits an indirect bandgap of approximately 1 eV, with a valence band maximum (VBM) at the Γ point and a conduction band minimum (CBM) at the midpoint along the Γ–K symmetry lines. Conversely, a monolayer of $MoS_2$ demonstrates direct bandgap behavior, with the VBM and CBM coinciding at the K-point. This transition from an indirect to direct bandgap in monolayer $MoS_2$ is a result of quantum confinement effects, leading to enhanced photoluminescence compared to multilayered forms of $MoS_2$, $MoSe_2$, $WS_2$, and $WSe_2$.[130,131,132,133] The monolayer bandgap of group 6 TMDs typically exhibits a ~50% increase compared to bulk materials.[134] Recent theoretical studies have uncovered intriguing phenomena related to spin and valley coupling in monolayer group 6 TMDs,[135] where valley quantum numbers become crucial for manipulating information akin to spintronics and electronics. The broken inversion symmetry and spin–orbit coupling characteristic of monolayer group 6 TMDs facilitate access to the valley degree of freedom via circularly polarized light. These materials feature two inequivalent momentum valleys (K and K′ or -K) among the six found at the corners of their first Brillouin zone, as depicted in Fig. 8b. Strong spin-orbit coupling induces a splitting of the valence band at these valleys. In monolayer $MoS_2$, the time-reversal symmetry dictates opposite spin splitting at different valleys, denoted by blue and red in Fig. 8b, resulting in a coupled spin and valley degrees of freedom. When excited by circularly polarized light, electrons with a specific spin populate only one of the two valleys. [136,137,138] Upon relaxation to the ground state, these electrons emit circularly polarized light, indicating the preservation of valley polarization during the process. Controlled valley polarization detection, not only with light but also via other methods like electric fields, could pave the way for novel "valleytronic" device designs.

## 2.3. Optical, Electrical, and Mechanical Properties

2D materials exhibit fascinating optical, electrical, and mechanical properties due to their unique atomic structures and confinement in a single layer. In terms of optical properties, 2D materials like graphene, TMDs, and phosphorene demonstrate remarkable light-matter interactions, including strong absorption and emission, as well as tunable band gaps through layer thickness or strain engineering. Electrically, these materials showcase exceptional conductivity, with graphene standing out for its high electron mobility and low resistance. TMDs offer a range of electrical behaviors, from semiconducting to metallic, depending on the number of layers and chemical composition. Additionally, 2D materials possess remarkable mechanical properties, such as exceptional flexibility, high tensile strength, and low mass density, making them promising candidates for flexible electronics and composite materials. Understanding and harnessing these properties are crucial for various applications, including optoelectronics, sensors, energy storage, and nanoelectromechanical systems. Continued research into the fundamental properties and fabrication techniques of 2D materials hold promise for further advancements in diverse technological fields.

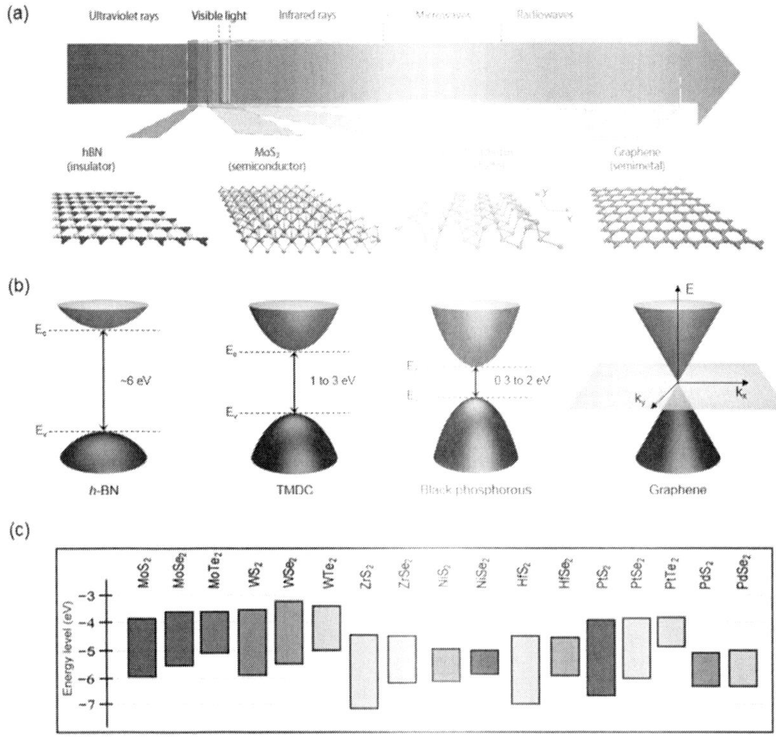


Figure 9. (a) Energy spectrum of various two-dimensional (2D) materials and their atomic crystal structures (b) electronic band structures of *h*-BN, TMDs, black phosphorous, and grapheme; and (c) energy level diagrams of the selected semiconducting TMDs. Reproduced with permission.[139]

Indeed, the natural world boasts a plethora of 2D materials, encompassing the graphene family, metal chalcogenides, and metal oxides, as illustrated in Figure 9 a, b.[140,141] TMDs, such as those from group IV (Mo and W), stand out among 2D semiconductors due to their sizable bandgap, making them suitable for electronic devices like field-effect transistors (FETs) and memories.[142,143] Besides, these semiconducting TMDs exhibit promising optical characteristics,[144] spanning from visible to near-infrared wavelengths,[145] featuring intriguing properties like the indirect-to-direct bandgap transition in monolayers,[146] intricate spin-valley interactions,[147] and robust light-matter interactions,[148] alongside significant exciton binding energies.[149,150] Recently,[151] black phosphorus has emerged as a notable 2D semiconductor with a tunable direct bandgap, varying from 0.3 eV to ~2.0 eV with thickness reduction.[152] Additionally, *h*-BN stands out among large bandgap insulators, boasting a bandgap of ~6 eV.[153] Its atomically flat and smooth surface, devoid of dangling bonds, makes *h*-BN an ideal substrate for other 2D conducting channels like graphene and TMDs.[154] Furthermore, 2D metal oxide nanosheets with diverse compositions have been explored as high-k dielectric layers. Lastly, numerous metallic TMDs, including group V compounds, and graphene are prominent among 2D metals.[121]

The presence of 2D materials with diverse compositions and electronic structures facilitates precise stacking of these layers, yielding artificial vdW heterostructures distinct from orthodox material heterostructures.[155] This capability presents a thrilling avenue for new materials design in both fundamental research and device applications.[156] Particularly, as depicted in Figure 9c, the disparate band gap energies and work functions of semiconducting TMDs allow for the creation of atomically thin semiconductor heterojunctions and/or superlattices with tailored band alignment.[157,158] This paves the way for exploring untapped optical and optoelectronic properties at the ultimate 2D limit, while also inspiring novel device concepts and technologies for next-generation optoelectronics. This article primarily reviews the distinctive features of 2D semiconductors, encompassing electronic band structures and optical properties, and underscores recent advancements in fabricating innovative optoelectronic devices using 2D semiconductor vdW heterostructures.

### 2.3.1 Optical Properties

2D materials exhibit remarkable optical properties owing to their unique electronic structures and quantum confinement effects. The absorption and emission spectra of 2D materials, such as graphene, TMDs, and phosphorene, are intricately linked to their band structures. Quantum confinement in the lateral dimension leads to discrete energy levels, resulting in tunable absorption edges. For instance, graphene, with its zero bandgap, absorbs light across a broad spectrum. In contrast, TMDs possess finite bandgaps, allowing for optical transitions in the visible range. Moreover, the strong light-matter interactions in 2D materials give rise to enhanced photoluminescence and Raman scattering, making them promising candidates for optoelectronic applications. The optical properties of 2D materials are highly influenced by their band structures, and their interactions with light play a pivotal role in numerous applications. Graphene, being a zero-bandgap material, exhibits unique optical conductivity over a broad wavelength range. On the other hand, TMDs, such as $MoS_2$ and $WS_2$, possess finite bandgaps, leading to distinct absorption peaks in the visible spectrum. The strong light-matter interaction in 2D materials is not only confined to absorption but extends to phenomena like surface-enhanced Raman scattering SERS and second-harmonic generation (SHG). These materials also find applications in nonlinear optics and quantum optics due to their size-dependent electronic structures and nonlinear response to intense light.

### 2.3.2 Electrical Properties

The electronic properties of 2D materials vary widely, contributing to their versatility in electronic devices. Graphene, characterized by its Dirac cones, displays high electron mobility and conductivity. TMDs, with their tunable bandgaps, find applications in transistors and optoelectronic devices. The thickness-dependent electronic structure, known as the quantum Hall effect, is a distinctive feature in 2D materials. Furthermore, the presence of vdW heterostructures, formed by stacking different 2D layers, introduces novel electronic phenomena and enables the engineering of band alignments for tailored electronic functionalities. The electrical properties of 2D materials are diverse, offering a rich platform for electronic applications. Graphene's high electron mobility, ballistic transport, and ambipolar field effect make it a key player in next-generation electronics. TMDs, with their tunable bandgaps, are essential for designing transistors with tailored electronic characteristics. The advent of topological insulators among 2D materials has introduced novel electronic states, promising for spintronics and quantum computing. The formation of heterostructures and superlattices with different 2D layers allows for the creation of designer electronic devices with engineered electronic band structures and tailored transport properties.

### 2.3.3 Mechanical Properties:

The mechanical properties of 2D materials are notable for their strength, flexibility, and lightweight nature. Graphene, recognized for its exceptional mechanical strength, is stronger than steel and extremely flexible. The planar structure of 2D materials contributes to their inherent mechanical anisotropy. TMDs exhibit layer-dependent mechanical properties, with single layers being more flexible than multilayers. Mechanical strain significantly influences the electronic band structure of 2D materials, offering avenues for strain engineering in device applications. The combination of remarkable strength and flexibility positions 2D materials as promising candidates for applications in flexible electronics, sensors, and nanocomposites. The mechanical properties of 2D materials are captivating, showcasing exceptional strength, flexibility, and strain sensitivity. Graphene, a single layer of carbon atoms in a hexagonal lattice, possesses an ultimate intrinsic tensile strength and Young's modulus, making it ideal for applications requiring high mechanical integrity. The flexibility of 2D materials, coupled with their lightweight nature, is exploited in flexible electronics and wearable devices. The mechanical behavior of these materials under strain is particularly intriguing, influencing their electronic properties through strain-induced bandgap modulation and revealing their potential in flexible and stretchable electronics. The synergy between outstanding mechanical and electronic properties makes 2D materials invaluable for advancements in materials science and technology.

In summary, the optical, electrical, and mechanical properties of 2D materials underscore their multifaceted nature and immense potential for a wide array of applications, from high-performance electronics to advanced optical devices and flexible materials. Ongoing research continues to unveil new facets of their behavior, paving the way for innovations in materials science and technology.

# CHAPTER 3

# SYNTHESIS AND FABRICATION TECHNIQUES

The synthesis and fabrication Techniques of 2D materials involves the creation and integration of 2D materials using various scientific methods. These materials are typically produced through various techniques such as chemical vapor deposition (CVD), mechanical exfoliation (ME), liquid phase exfoliation (LPE), and solution-based chemical synthesis. Each 2D material has unique synthesis methods, and the resulting samples can vary in form and feature size, ranging from single crystal flakes to dispersed nanosheets and large-area films. The synthesis of high-quality 2D materials is the first step for their on-chip integration, and the techniques used have the potential for scalable and industrial applications. [159,160]

In the top-down techniques, a 2D material is fabricated by carving larger or bulk solid material in a controlled process and material removal.[161] The bottom-up techniques, on the other hand, involve methods such as sol-gel, CVD, and biosynthesis for 2D material fabrication.[162] These techniques offer scalability, low cost, and versatility for the synthesis of 2D materials, making them suitable for various applications. The unique properties of 2D materials, such as their high electrical conductivity, mechanical strength, and flexibility, make them promising for a wide range of applications in electronics, energy, and healthcare. The understanding of the synthesis and fabrication techniques of 2D materials is essential for harnessing their potential in various scientific and industrial fields.

## 3.1 Top-Down Technique

The top-down synthesis techniques involve the fabrication of 2D materials by reducing larger or bulk solid materials to the nanoscale through controlled processes and material removal. These methods play a crucial role in producing high-quality 2D materials with specific properties for various applications. The top-down approach encompasses several subcategories, including micromechanical exfoliation, ultrasonic exfoliation, lithium-intercalated exfoliation, and ion-exchange exfoliation.

### 3.1.1 Micromechanical Exfoliation

Micromechanical exfoliation is a fundamental top-down technique used for the synthesis of 2D materials, such as graphene, by isolating single or few-layer nanosheets while retaining the crystal structure and properties of the original material. This method involves the physical separation of atomic layers from a bulk material, typically using adhesive tape, to produce high-quality 2D nanosheets. The following provides a detailed description of micromechanical exfoliation, including its process, advantages, and limitations.

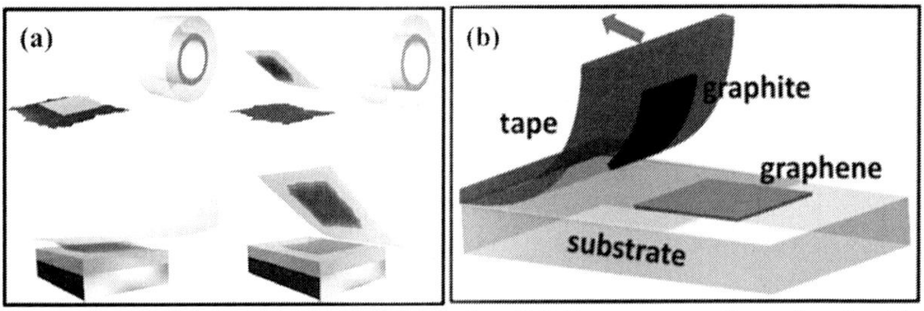


Figure 1. (a) Schematic of a mechanical exfoliation process. Reproduced with permission.[163] (b) Schematic of the synthesis of graphene from graphite by using the same process. Reproduced with permission.[164]

Process of Micromechanical Exfoliation: The process of micromechanical exfoliation, also known as the mechanical cleavage method, begins with the selection of a suitable precursor material, such as graphite. A small piece of the precursor material is attached to an adhesive tape, and the tape is repeatedly folded and peeled to separate atomic layers from the bulk material as shown in figure 1 (a). This iterative peeling process results in the production of single or few-layer 2D nanosheets e.g. graphene from graphite as depicted in figure 1 (b), with the number of layers controlled by the number of peeling steps. The resulting nanosheets retain the crystal structure and properties of the original material, making them highly desirable for various applications.

Advantages of Micromechanical Exfoliation:

- ➢ High-quality samples: Micromechanical exfoliation produces high-quality 2D nanosheets with specific properties, making them suitable for fundamental research and development purposes.

- Retention of crystal structure: The resulting nanosheets retain the crystal structure of the precursor material, ensuring the preservation of its unique properties.

- Simple and cost-effective: This method is relatively simple and cost-effective, as it does not require complex equipment or extensive chemical processing.

Limitations of Micromechanical Exfoliation:

- Low throughput: The process of micromechanical exfoliation has a low throughput, making it unsuitable for large-scale production of 2D materials.

- Lack of scalability: While effective for producing high-quality samples, this method is not easily scalable to meet industrial demands.

- Material waste: The iterative peeling process can lead to material waste and is not conducive to efficient material utilization.

### 3.1.2 Lithium-Intercalated Exfoliation

Lithium-intercalated exfoliation is a top-down technique used for the synthesis of 2D materials, such as graphene and TMDs, by introducing Li ions between the layers of a bulk material to weaken the interlayer forces, facilitating the exfoliation process as shown in Figure 2. The process involves the intercalation of lithium ions between the layers of the bulk material, followed by the exfoliation to obtain 2D nanosheets. This method is particularly useful for materials with strong interlayer bonding, as the intercalation of lithium weakens the interlayer forces, making it easier to separate the layers and obtain 2D nanosheets.

In the case of $MoS_2$, the formation of $Li_xMoS_2$ compound is a key step in the lithium intercalation process, and this reaction can be tuned to control the yield of single layers. The intercalation of lithium ions into the bulk material increases the interlayer spacing between individual layers, weakening the interlayer adhesion and reducing the energy barrier to exfoliation. This ultimately increases the accessible surface areas in 2D materials, making them more suitable for various applications.

Lithium-intercalated exfoliation is a versatile and promising method for the large-scale synthesis of 2D materials, offering advantages such as high yield, scalability, and the ability to produce single layers. This method has been proven to be a cost-effective and straightforward approach, with a larger and more uniform material yield compared to conventional methods. It also accommodates a large array of precursor materials, making it suitable for a wide range of applications.

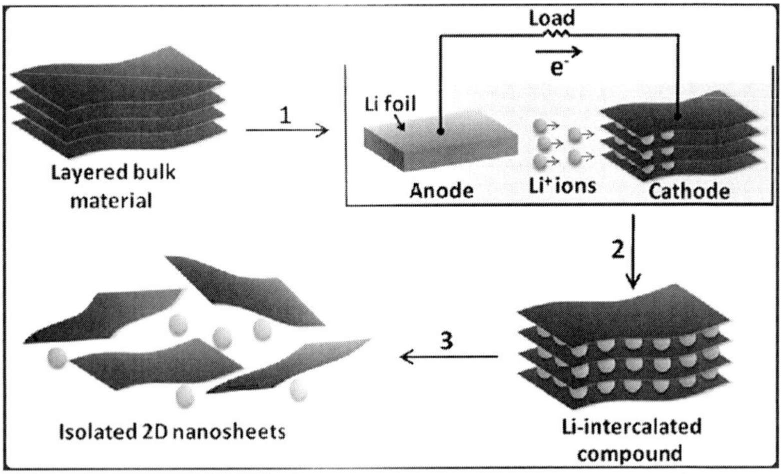


Figure 2. Schematic Representation of synthesis of 2D nanosheets from bulk material through Li-ion intercalation. Reproduced with permission.[165]

### 3.1.3 Ion-Exchange Exfoliation

Ion-exchange exfoliation is a top-down technique used for the synthesis of 2D materials, such as graphene, by replacing interlayer ions in a bulk material with larger ions, leading to the expansion and subsequent exfoliation of the material into mono-to-few layers. This method offers a degree of control over the exfoliation process by selecting appropriate intercalating ions, and it has been applied to various layered materials to produce 2D nanosheets with tailored properties. This process involves the replacement of interlayer ions in a bulk material with larger ions, leading to the expansion of the interlayer spacing and weakening of the interlayer forces. This expansion facilitates the exfoliation process, resulting in the production of 2D nanosheets. The choice of intercalating ions is critical, as it determines the degree of expansion and the properties of the resulting 2D material. The exfoliation process can be further optimized by controlling the concentration and reaction time of the intercalating ions. Ion-exchange exfoliation is a versatile and promising method for the large-scale synthesis of 2D materials, offering advantages such as high yield, scalability, and the ability to produce 2D nanosheets with tailored properties. This method has been proven to be a cost-effective and straightforward approach, with a larger and more uniform material yield compared to conventional methods. It also accommodates a large array of precursor materials, making it suitable for a wide range of applications.

In summary, top-down synthesis techniques play a vital role in the production of 2D materials with unique properties and functionalities. While these methods have contributed to significant advancements in the field of 2D materials, ongoing research and development are focused on enhancing the scalability, controllability, and cost-effectiveness of top-down approaches to meet the growing demand for 2D materials in diverse applications.

These methods have several advantages and disadvantages, which are discussed below.

**Advantages:**

- Ease of Manufacturing: Top-down approaches are relatively easier to implement for reducing bulk materials into nanostructures or nanoparticles, facilitating the development of desired structures efficiently.

- Minimization of Waste: Top-down methods minimize waste generation during the synthesis process, contributing to cost-effectiveness and environmental sustainability compared to some bottom-up approaches.

- High-quality samples: Top-down techniques can produce high-quality samples with specific properties, making them suitable for research and development purposes.

- Large Material Reduction: They are effective for large material reduction or separation, making them suitable for creating the desired structure with acceptable qualities in a straightforward manner.

**Disadvantages:**

- High energy consumption: Top-down techniques require high energy consumption, making them less energy-efficient compared to bottom-up techniques.
- Limited scalability: While top-down techniques are scalable, there are limitations to the size and quantity of 2D materials that can be produced.
- Material fracture: The material removal process in top-down techniques can lead to material fracture, which can affect the quality of the resulting 2D materials.
- Use of toxic chemicals: Some top-down techniques involve the use of toxic chemicals, which can pose safety and environmental concerns.

## 3.2 Bottom-Up Technique

The Bottom-up techniques for the synthesis of 2D materials involve the construction of nanoscale materials from atomic or molecular precursors, allowing them to react or self-assemble into more complex structures. These methods present facile approaches for the mass production of 2D materials by growing or assembling them from the fundamental atomic or molecular level. Unlike top-down approaches, which involve the cleaving of atomically thin layers from bulk crystals, bottom-up methods are capable of producing 2D materials in mass amounts, making them highly attractive for various applications.[166, 167]

### 3.2.1 Wet Chemical Strategy

Wet chemical strategy is a bottom-up synthesis method that includes hydro/solvothermal synthesis and template synthesis. This method is promising for preparing all types of 2D materials owing to its high-yield, low-cost, and mass production. Wet chemical synthesis strategies offer advantages such as controllable thickness, the ability to assemble architectures from nanosheets, and the creation of composite nanostructures, providing new opportunities for material science. Additionally, the wet chemical strategy is versatile and suitable for the synthesis of a wide range of 2D materials, making it a promising approach for large-scale production.

Hydro/solvo-thermal synthesis is a common strategy used for the synthesis of inorganic materials. This method involves the use of high-pressure and high-temperature conditions to promote the growth of 2D materials. Many factors play key roles in the synthesis of ultrathin nanostructures, including temperature, reaction time, reactant ratio, and so on. Many 2D materials have been prepared by this method, such as $MoS_2$, $TiO_2$, ZnO, $Co_3O_4$, and $MnO_2$. Template synthesis is another wet chemical strategy used for the synthesis of 2D materials. This method involves the use of a template, such as a surfactant or a polymer, to control the growth of 2D materials. The template acts as a scaffold for the growth of the 2D material, allowing for the precise control of the material's size, shape, and thickness. This method has been used to synthesize various 2D materials, such as graphene, transition metal dichalcogenides, and metal oxides. Wet chemical techniques also allow materials to be doped by introducing different chemicals. Two traditional wet chemical methods for producing chalcogenides are the solvothermal method and the hydrothermal method. These methods have been used to synthesize various 2D materials, such as $MoS_2$, $WS_2$, and $MoSe_2$.

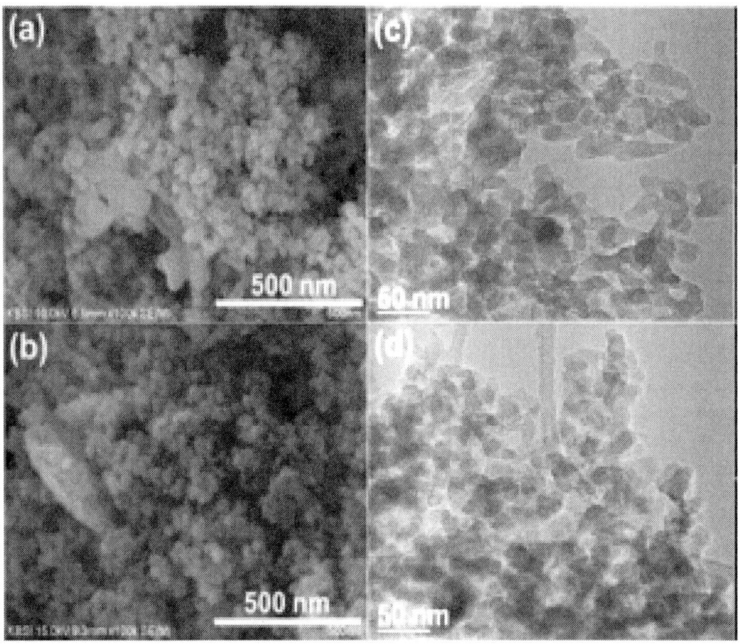


Figure 3. FESEM (a and b) and TEM (c and d) images of the as synthesized $MoS_2$ (a and c) and $WS_2$ (b and d) samples. Reproduced with permission.[168]

In summary, the wet chemical strategy is a versatile and promising bottom-up technique for the synthesis of 2D materials. This method offers advantages such as controllable thickness, the ability to assemble architectures from nanosheets, and the creation of composite nano-structures, providing new opportunities for material science. Additionally, the wet chemical strategy is versatile and suitable for the synthesis of a wide range of 2D materials, making it a promising approach for large-scale production.

Figure 4. (a) SEM image of a CVD grown near-perfect 2D lattice composed of hexagonal graphene flakes (HGF). (b) SEM image of large-sized HGF showing that the average sizes are ~ 120 μm.[170] (c, d) SEM images of CVD grown single-crystal twisted bilayer graphene (t-BLG).[171] Reproduced with permission.[170, 171]

### 3.2.2 Chemical Vapor Deposition (CVD)

CVD is a versatile and scalable method for the synthesis of various 2D materials.[169] This technique involves the deposition of a thin film of material onto a substrate through the reaction of gaseous precursor compounds. CVD offers precise control over the growth process, allowing for the production of high-quality and large-area 2D materials. It has been widely used for the synthesis of graphene, TMDs, and other 2D materials proving its versatility and scalability for industrial applications.

Geng *et al.* used a CVD approach that allows the direct synthesis of uniform single-layered, large-size (10000 $\mu m^2$), spatially self-aligned, and single-crystalline hexagonal graphene flakes (HGFs) and their continuous films on liquid Cu surfaces. Employing a liquid Cu surface enables self-assembly of HGFs into compact and ordered structures.[170] Figure 4 (a) shows the SEM image of a CVD grown near-perfect 2D lattice composed of similar-sized HGFs. Varying the growth parameters can result in very large sized flakes ~ 120 µm as seen in figure 4 (b). Figure 4 (c) and (d) displays SEM images of CVD grown single-crystal t-BLG by Yeh *et al.*[171] The large grains refer to the primary layers and the small grains, which grow from the center of large grains, refer to the secondary layers. The linear dimension of primary layer is ~ 10 µm, and that of secondary layer is ~ 5µm with both exhibit hexagonal shapes.

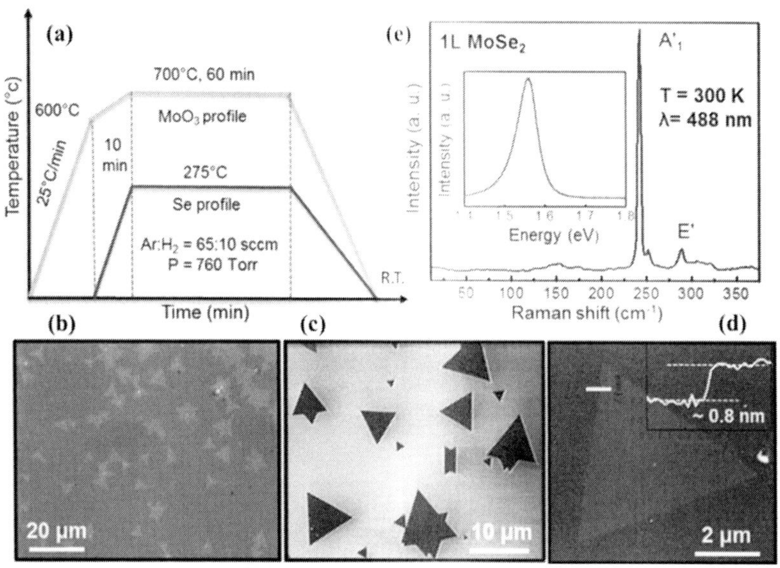


Figure 5. (a) Schematic of $MoSe_2$ growth process using CVD. (b) OM image of as prepared $MoSe_2$ grains on sapphire substrate. (c) SEM image of a triangular grains showing very sharp edge. (d) AFM image with height profile of a triangular grain, showing a step height of ~ 0.8 nm corresponds to monolayer MoSe2 thickness. (e) Raman and PL characterization of a triangular MoSe2 flake using 488 nm laser. Reproduced with permission.[172]

CVD provides a scalable route to produce large-area 2D materials, which is essential for many applications. The process involves the introduction of gaseous precursors into a reaction chamber, where they undergo chemical reactions to form the desired 2D material on a substrate. The control over the reaction conditions, such as temperature, precursor concentration, and pressure, allows for the precise tuning of the material properties and the growth of high-quality 2D materials such as TMDs. Figure 5 (a) exhibits the growth process of monolayer $MoSe_2$ by CVD where the temperature profiles of precursor powders and gas flow rates are displayed.[172] The OM image in figure 5 (b) suggests that the grains are mostly triangular but some truncated triangular grains are also present. SEM image in figure 5 (c) of a triangular grains showing very sharp edge, implying good crystallinity of prepared sample. Figure 5 (d) shows AFM image with height profile of a triangular grain, showing a step height of ~ 0.8 nm corresponds to monolayer $MoSe_2$ thickness. Raman spectra of a triangular $MoSe_2$ flake using 488 nm laser showing the characteristics Raman vibration modes $A_1'$ at 240 $cm^{-1}$ (out of plane mode) and $E'$ at 287 $cm^{-1}$ (in plane mode) as shown in figure 5 (e). The absence of the layer number sensitive mode $A_2''$ at 353 $cm^{-1}$ confirms that prepared $MoSe_2$ flakes are monolayer. The PL spectrum exhibits peak at 1.56 eV (shown as inset), which is the characteristics of monolayer $MoSe_2$.

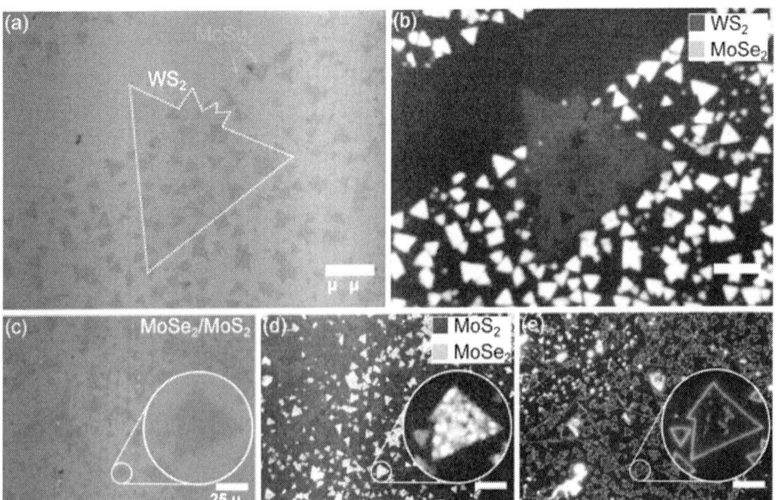


Figure 6. (a) Photoluminescence imaging of CVD-grown TMD heterostructures. (a) Bright-field image of a $MoSe_2/WS_2$ heterostructure on a $SiO_2/Si$ substrate. (b) PL image of the sample in panel a showing that the two materials have distinctly

different colours. (c, d, and e) Bright-field, PL, and dark-field images of a $MoSe_2/MoS_2$ heterobilayer on a $SiO_2/Si$ substrate. Reproduced with permission.[173]

Figure 6 (a) shows a bright-field image of a $MoSe_2/WS_2$ HSs sample on a $SiO_2/Si$ substrate composed of CVD grown monolayers. The large triangle in the center of the image corresponds to single layer $WS_2$, while smaller flakes around it are $MoSe_2$ monolayers.[173] Although both materials have similar appearance in the bright-field image, they can be easily distinguished by their distinctly different colours according to the wavelength of their PL bands: red for $WS_2$ and a false pink for $MoSe_2$ emitting in the near-infrared. The high crystalline quality of the $WS_2$ flake is evident from its bright and uniform PL. In contrast, the CVD grown $MoS_2$ flakes within the $MoSe_2/MoS_2$ sample exhibit weaker PL with significant intensity variations both within individual crystals and across the substrate (Figure 6d). Additionally, the $MoSe_2$ monolayers in the same sample display non-uniform PL emission. The observed granularity in the PL is likely attributed to organic residues from the transfer process, which are trapped between the flakes or between the flakes and the substrate. Despite the clear visibility of emission intensity variation within the grainy pattern in the PL image, there is limited correlation with features observed in the dark-field image (Figure 6e), and it remains entirely imperceptible in the bright-field image (Figure 6c).

One of the key advantages of CVD is its ability to produce large-area 2D materials with high uniformity and quality. This is essential for many applications, such as electronics and optoelectronics, where the performance of the devices is directly related to the quality and uniformity of the 2D materials. Additionally, CVD is a highly versatile technique that can be used to grow a wide range of 2D materials, including graphene, transition metal dichalcogenides, and other compound materials. The recent advances in the CVD growth of 2D materials, such as graphene, boron nitride, and transition metal dichalcogenides, have demonstrated the potential of this technique for large-scale production. The ability to achieve large area preparation and single crystal growth, as well as the mechanistic insights gained from these studies, have further established CVD as a reliable and important method for the synthesis of 2D materials.

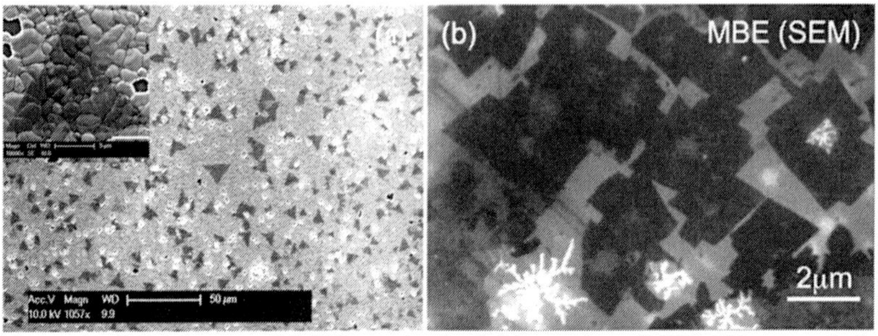


Figure 7. SEM image of MBE synthesised (a) graphene/*h*-BN HSs on cobalt substrate. (b) Graphene flakes on Ge (001) substrate. Reproduced with permission.[174, 175]

### 3.2.3 Molecular Beam Epitaxy (MBE)

MBE is an advanced thin-film deposition technique that has proven to be outstanding in the synthesis of 2D materials. This method involves the deposition of material atoms onto a substrate in a high vacuum environment, allowing for the precise control of the growth process at the atomic level. MBE has been widely used as a fundamental research tool in both academia and industry for the synthesis of high-quality 2D materials, including graphene, TMDs, and other compound materials.[176] Figure 7 depicts the MBE grown graphene and its HSs with *h*-BN where in figure 7 (b), the SEM images show the conductive graphene in dark contrast over Ge (001) wafer. Here the MBE prepared flakes have long straight edges compared to conventional CVD grown counterparts. On the other hand in figure 7 (a), one can see the SEM image of HSs of graphene with *h*-BN, where large triangular domains of ~20 μm are observed. Despite the underlying very rough Co substrate surface, which consists of small grains due to heat treatment, wafer-scale graphene has been grown, followed by the formation of the triangular shaped *h*-BN domains as seen in the inset.

One of the key advantages of MBE is its ability to produce high-purity, uniform, and atomically thin 2D materials. The precise control over the growth process allows for the tailoring of the material structure and properties, making MBE a versatile technique for the synthesis of a wide range of 2D materials. The high vacuum environment in which MBE operates minimizes the presence of impurities, resulting in the production of high-quality and uniform 2D materials. MBE growth of 2D materials has demonstrated the potential of this technique for large-scale production. The ability to achieve large area preparation and single crystal growth, as well as the mechanistic insights gained from these studies, have further established MBE as a reliable and important method for the synthesis of 2D materials. The true power of this technique comes from its ability to send a very slim amount of atoms, as low as entire growth takes place at pressures as low as $10^{-9}$ Torr to create high quality and pure materials. [177] MBE is a highly versatile and precise technique for the synthesis of 2D materials. Its ability to produce high-purity, uniform, and atomically thin materials makes it an essential method for many industrial applications, and the recent advances in this field have further established its potential for large-scale production.

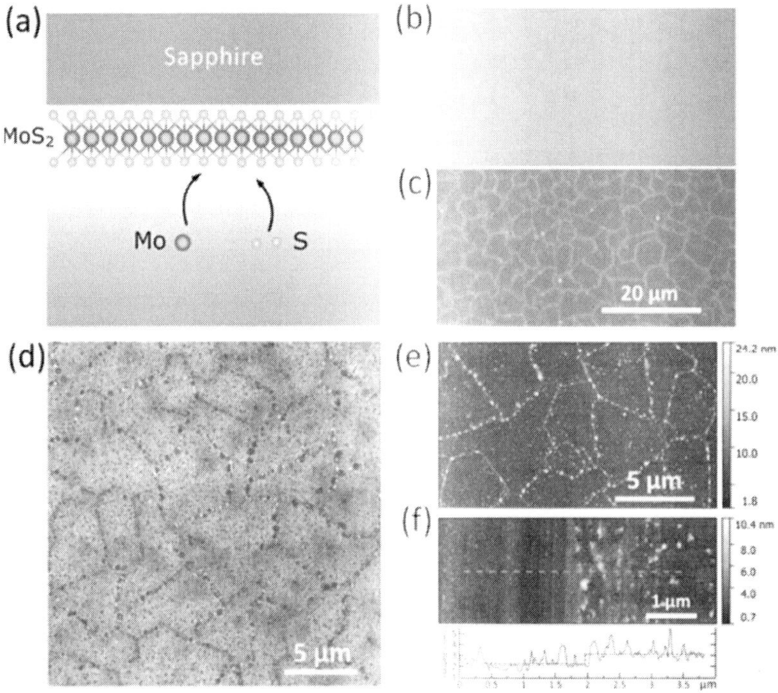


Figure 8. (a) A schematic diagram showing the concept of MBE growth monolayer $MoS_2$ on a sapphire substrate. (b,c) Optical image of the epitaxial $MoS_2$ taken in bright field (b) and dark field (c) regimes. (d) SEM image of the epitaxial $MoS_2$ revealing the high crystallinity of the samples (e) The AFM topography map of $MoS_2$ surface with a scan area of $17.5 \times 10\ \mu m^2$. (f) The AFM topography map and the cross-sectional profile of the edge of epitaxial $MoS_2$ along the green line, giving the $MoS_2$ layer thickness of ~ 0.9 nm. Reproduced with permission.[178]

Monolayers of $MoS_2$ were synthesized via MBE by Ermolaev *et al.*, as illustrated in Figure 8 (a).[178] Evaluation of the samples through optical, SEM, and AFM imaging, as depicted in Figure 8 (b-e), affirmed their uniformity and high crystallinity. The epitaxial $MoS_2$ monolayer exhibited consistent coverage over the double-polished sapphire substrate, featuring an average crystallite size of 6 μm, thus attesting to the exemplary quality of the samples.[179] Subsequently, the atomic thickness of the grown $MoS_2$ was ascertained via AFM. The topographical analysis in Figure 8 (f) revealed a film thickness of 0.9 nm, in alignment with monolayer MoS2. [180, 181]

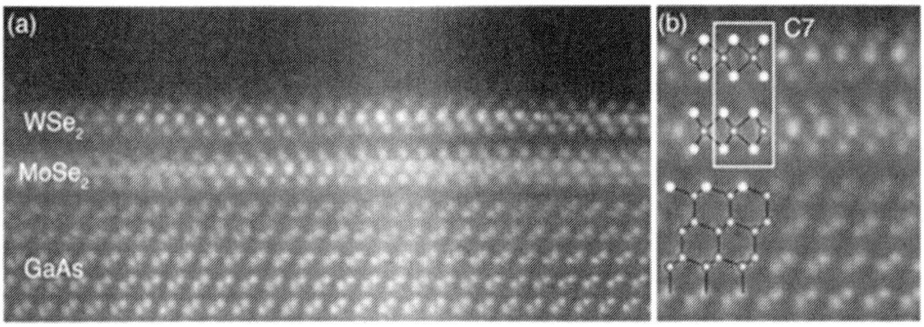


Figure 9. (a) HAADF-STEM image for the bilayer heterostructure grown on GaAs substrate by MBE (b) Magnified image of (a) where the atomic positions are indicated.[182]

The HAADF-STEM image in Figure 9 (a) illustrates the well-defined layered structure of the $WSe_2/MoSe_2$ HSs grown by MBE technique.[182] An atomically abrupt interface between the TMD layers and the GaAs substrate is evident, devoid of misfit dislocations, consistent with prior findings.[183] The interlayer distance between Mo and W layers measures 0.65 nm, closely aligning with bulk values for $MoSe_2$ and $WSe_2$. The magnified view in Figure 4b confirms the $WSe_2/MoSe_2$ heterostructure with a stable C7-stacking configuration, as determined in previous studies.[184]

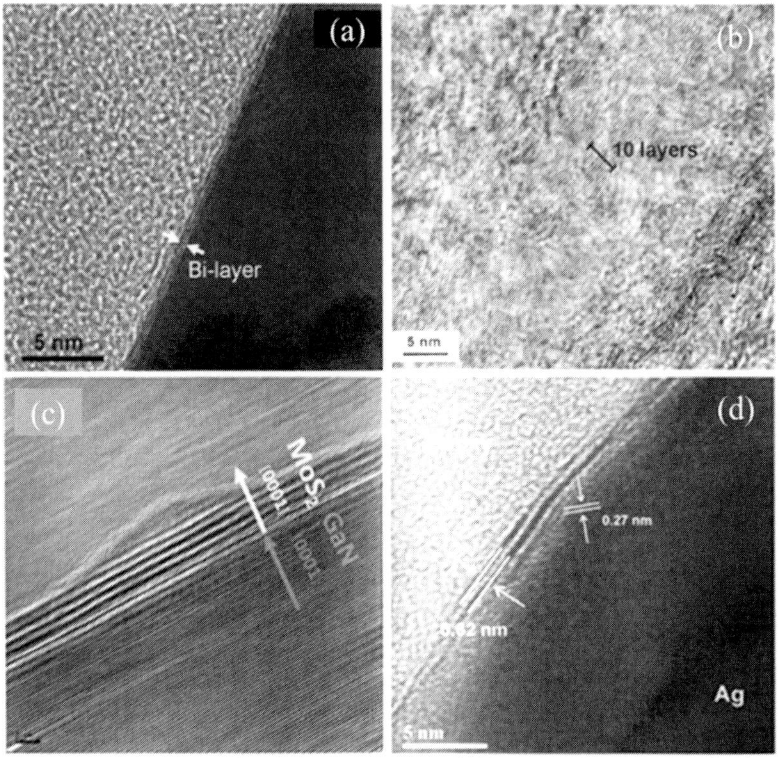


Figure 10. (a) Cross-sectional HRTEM image of bilayer graphene produced using PLD.[185] (b) HRTEM images of graphene produced using PLD. Reproduced with permission.[186] (c) Cross-sectional TEM image of the $MoS_2$ film grown on GaN substrate by PLD method. Reproduced with permission.[187] (d) Cross-sectional TEM image of PLD-grown $WS_2$ films on Ag substrate. Reproduced with permission.[188]

### 3.2.4 Pulsed Laser Deposition (PLD)

PLD is a physical vapor deposition technique that has gained significant attention for the synthesis of 2D materials. PLD involves the use of high-energy pulsed laser beams to ablate a target material, leading to the deposition of the ablated material onto a substrate. This method has been widely used for growing graphene, TMDs and complex oxide thin films and has been proven to be an alternative way to the synthesis of 2D materials due to its notable advantages.[189] Ultrathin graphene films (less than five layers) were synthesized by PLD method on Ni film and the result is shown in HRTEM image (Fig 10 a). The cross-sectional TEM image in figure 10

(b) confirms that the film obtained by direct PLD growth on Si substrate by Qian *et al.*,[184] corresponds to around 10 layers of graphene. Highly crystalline $MoS_2$ and $WS_2$ thin films can be obtained by PLD as confirmed by the cross-sectional TEM images (Fig. 10 c and d). Recent advances in the synthesis of 2D materials by PLD have demonstrated its potential for large-scale production.[190] The ability to achieve large-area preparation and single crystal growth, as well as the mechanistic insights gained from these studies, have further established PLD as a reliable and important method for the synthesis of 2D materials. In a recent study, autonomous synthesis of a contemporary 2D material was demonstrated by combining the highly versatile PLD with in situ and real-time spectroscopy, a high-yield methodology and cloud connectivity to enable autonomous synthesis workflows with PLD.[191] This approach has the potential to greatly accelerate the discovery, optimization, and understanding in materials synthesis, addressing the grand challenges in synthesis. The use of PLD for the synthesis of 2D materials has shown promising results, and its potential for autonomous synthesis and high-throughput methodologies makes it an attractive technique for the large-scale production and accelerated discovery of a vast number of materials.[192]

In summary, bottom-up techniques for the synthesis of 2D materials offer versatile and scalable approaches for the mass production of high-quality materials with tailored properties. These methods, such as the wet chemical strategy, CVD, MBE, and PLD, have been instrumental in advancing the field of 2D materials, offering promising opportunities for various applications in electronics, energy, and healthcare.

# CHAPTER 4

# CHARACTERIZATION METHODS

The characterization of 2D quantum materials necessitates the utilization of advanced and multifaceted techniques that can discern their intricate structural, electronic, and quantum properties at the atomic and subatomic scales. Central to these efforts are scanning probe microscopy methods, such as atomic force microscopy (AFM) and scanning tunnelling microscopy (STM), which enable high-resolution imaging and manipulation of individual atoms and molecules on 2D surfaces. Spectroscopic techniques, including angle-resolved photoemission spectroscopy (ARPES) and Raman spectroscopy, offer insights into the electronic band structures and vibrational modes, respectively, unravelling the materials' quantum behaviors. Quantum transport measurements, carried out using techniques like the quantum Hall effect, provide a means to probe the exotic electronic states that emerge in these quantum materials. Furthermore, advancements in synchrotron-based techniques, such as X-ray photoelectron spectroscopy (XPS) and X-ray diffraction (XRD), contribute invaluable information regarding the chemical composition, electronic structure, and crystallographic properties of 2D quantum materials. The amalgamation of these sophisticated characterization methodologies serves as a cornerstone in elucidating the fundamental quantum phenomena and harnessing the unique properties of 2D materials for innovative applications in quantum information science and electronics.

## 4.1 Raman Spectroscopy

The application of Raman spectroscopy to the investigation of 2D materials has emerged as an indispensable tool for probing their structural, electronic, and vibrational characteristics at the nanoscale. In the realm of 2D materials, such as graphene, TMDs, and phosphorene, Raman spectroscopy offers a non-destructive and highly sensitive means to discern the number of layers, detect lattice defects, and unveil the distinct phonon modes inherent to these materials. The characteristic Raman signatures, attributed to in-plane and out-of-plane vibrational modes, provide crucial insights into the materials' structural integrity and electronic properties. The unique layer-dependent Raman shifts and intensities, as well as the emergence of additional modes in few-layer and monolayer configurations, contribute to the depth of information that Raman spectroscopy imparts. This methodology not only facilitates the characterization of 2D materials but also serves as an essential diagnostic tool for monitoring their structural evolution, offering valuable contributions to the burgeoning field of nanomaterials.

The Raman spectroscopy of graphene provides a nuanced insight into its structural characteristics, with distinctive changes in Raman peaks attributed to the number of layers. In the monolayer graphene, the prominent G band, associated with the in-plane $E_{2g}$ vibrational mode, is observed at approximately 1580 $cm^{-1}$, reflecting the $sp^2$ carbon-carbon bonding. With an increase in the number of layers, a new feature, the 2D band, appears and gains intensity. The 2D band, centered on 2700 $cm^{-1}$, is indicative of the two-phonon double resonance process, representing a unique signature of monolayer graphene. The ratio of the intensity of the 2D band to that of the G band ($I_{2D}/I_G$) becomes a crucial parameter for discerning the layer thickness, where a higher $I_{2D}/I_G$ ratio corresponds to mono to fewer layers. The emergence of additional features, such as the D band associated with defects and disorder, further underscores the sensitivity of Raman spectroscopy to subtle changes in the layer structure of graphene. Figure 1 (a) compares the 514 nm Raman spectra of graphene and bulk graphite. The two most intense features are the G peak at 1580 $cm^{-1}$ and the 2D band at 2700 $cm^{-1}$. Figures 1 (b) and (c) plot the evolution of the 2D band as a function of layers for 514.5 and 633 nm excitations. These immediately indicate that a bilayer has a much broader and up-shifted 2D band with respect to graphene. This band is also quite different from bulk graphite. Figure 1 (b) and (c) show that a further increase in layers leads to a significant decrease of the relative intensity of the lower frequency peaks. For more than 5 layers the Raman spectrum becomes hardly distinguishable from that of bulk graphite. Thus Raman spectroscopy can clearly distinguish a single layer, from a bilayer from few (less than 5) layers. This layer-dependent evolution of Raman peaks provides a robust means for non-destructive characterization and quantification of graphene layers, thereby facilitating precise control and optimization of graphene-based applications.

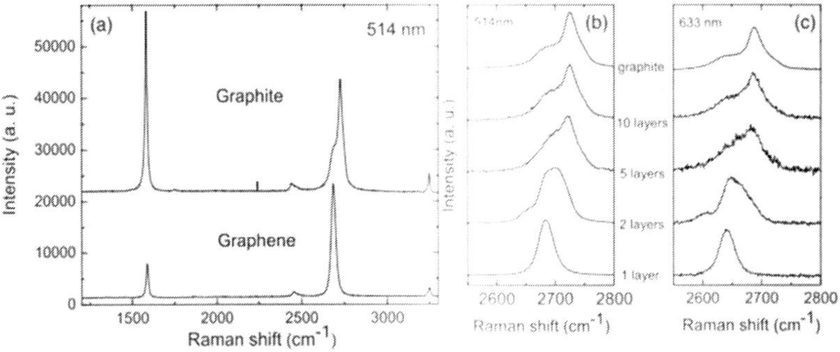


Figure 1. (a) Comparison of Raman spectra at 514 nm for bulk graphite and graphene. They are scaled to have similar height of the 2D peak at ~ 2700 cm$^{-1}$. (b) Evolution of the spectra at 514 nm with the number of layers. (c) Evolution of the Raman spectra at 633 nm with the number of layers. Reproduced with permission.[193]

The Raman spectroscopy of TMDs such as $MoS_2$, $MoSe_2$, $WS_2$, and $WSe_2$ serves as a powerful tool for elucidating their structural and vibrational characteristics, offering insights into layer thickness and material quality. In monolayer $MoS_2$, the prominent Raman peaks include the in-plane vibrational mode $E^1_{2g}$, at approximately 385 cm$^{-1}$, and the out-of-plane mode $A_{1g}$, at around 405 cm$^{-1}$ as shown in figure 2 (a). With an increase in the number of layers, there is a change in the peak positions of these modes, and additional features, such as the second-order Raman modes, may emerge. For $MoSe_2$, $WS_2$, and $WSe_2$, the Raman spectra exhibit analogous features, including the $E^1_{2g}$ mode and $A_{1g}$ mode in the low-frequency range. The $E^1_{2g}$ mode appears at approximately 240 cm$^{-1}$ for monolayer $MoSe_2$, 350 cm$^{-1}$ for monolayer $WS_2$, and 248 cm$^{-1}$ for monolayer $WSe_2$. The $A_{1g}$ mode, representing the out-of-plane vibration, occurs around 290 cm$^{-1}$ for monolayer $MoSe_2$, 419 cm$^{-1}$ for monolayer $WS_2$, and 253 cm$^{-1}$ for monolayer $WSe_2$. Similar to $MoS_2$, the Raman peak positions of these modes show a modest layer dependence, providing a means for layer identification. The evolution of Raman peaks with the number of layers in TMDs is characterized by the emergence of additional modes and shifts in peak positions. The layer-dependent changes are attributed to interlayer interactions, symmetry breaking, and modifications in the electronic structure. The combination of precise peak assignments and layer-dependent behaviors in TMD Raman spectra establishes Raman spectroscopy as an invaluable technique for non-destructive and quantitative characterization of these 2D materials.

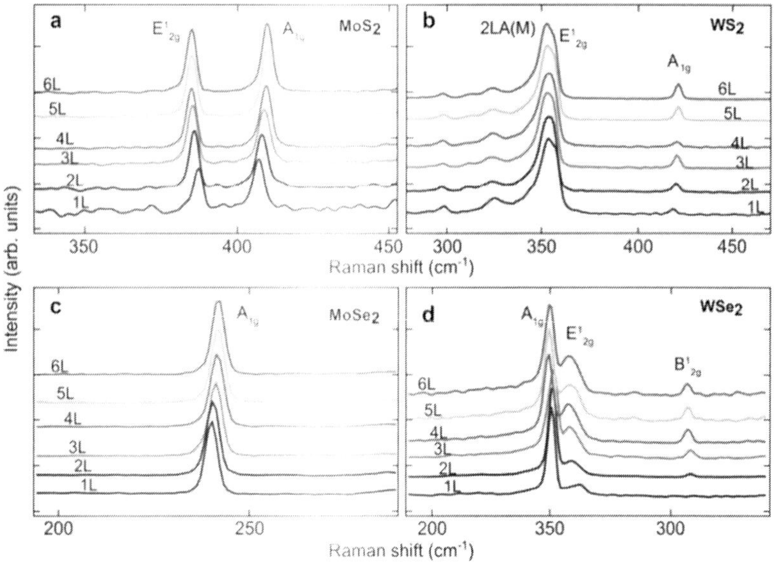


Figure 2. Raman Spectra of (a) $MoS_2$, (b) $WS_2$, (c) $MoSe_2$, and (d) $WSe_2$. Reproduced with permission.[194]

The Raman spectroscopy results of TMD heterostructures, specifically $MoS_2/WS_2$ and $MoSe_2/WSe_2$, offer valuable insights into their structural and vibrational properties. In these heterostructures, the constituent monolayers exhibit distinct Raman signatures, providing a sensitive means to characterize their crystalline quality and interlayer coupling. For $MoS_2/WS_2$, the Raman spectra reveal characteristic modes such as the in-plane $E^1_{2g}$ mode associated with $MoS_2$ and the $A_{1g}$ mode associated with $WS_2$. The observed frequency shifts and intensity variations of these modes reflect the interlayer interactions and potential modifications in the electronic structure arising from the generation of heterostructures. As shown in Fig. 3 (a) and (b), Raman spectra collected from the light purple area (points 1 and 2) show only the $E^1_{2g}$ (at 383.9 $cm^{-1}$) and $A_{1g}$ (at 405.3 $cm^{-1}$) peaks of the $MoS_2$ monolayer[195, 196] confirming the bottom layer is $MoS_2$.[197] In the bilayer region (points 3 and 4 in the dark purple area), however, two additional peaks located at 418.5 $cm^{-1}$ and 356.8 $cm^{-1}$ are observed, which can be assigned to the $A_{1g}$ mode and the overlapping 2LA (M) and $E^1_{2g}$ modes, respectively, of the top $WS_2$ monolayer. [198, 199]

In Figure 3 (c) to (e), the phonon vibrations evolution of the $MoSe_2/WSe_2$ vdW heterostructures (HSs) is probed at varying twist angles through

Raman spectroscopy. The Raman active modes A1′ and E′, alongside the higher-order out-of-plane mode A2″ (inactive in monolayers), are considered. The 2LA (M) peak arises from a two-phonon scattering process. The in-plane shear mode (SM) and out-of-plane layer breathing mode (LBM) reflect interlayer phonon vibrations of the entire 2D layers. Low-frequency (LF) Raman modes (observed at <40 $cm^{-1}$) are instrumental in probing vdW interlayer coupling. Notably, SM and LBM exhibit variations at twist angles of 0°, 56°, and 60°, possibly due to lattice mismatch. High-frequency (HF) Raman spectra in Figure 3 (d) depict high-frequency Raman spectra of the $MoSe_2/WSe_2$ vdW HSs for twist angles of 0-60° as well as characteristic peaks for monolayers of $MoSe_2$ at 240.5 $cm^{-1}$ (A1′) and 287.2 $cm^{-1}$ (E′) and $WSe_2$ at 250 $cm^{-1}$ (E′ and A1′) and 262 $cm^{-1}$ (2LA (M) mode). Interestingly, vibrational modes in the heterostructures show minimal dependence on twist angle, with frequencies remaining constant. Layer-number sensitive modes, such as A2″ for $MoSe_2$[200,201] and $WSe_2$,[202,203] exhibit activity in the vdW HSs, providing insights into interlayer stacking. HF modes provide less twist-angle-dependent information, underscoring the higher sensitivity of LF modes to interlayer stacking and coupling.[204,205]

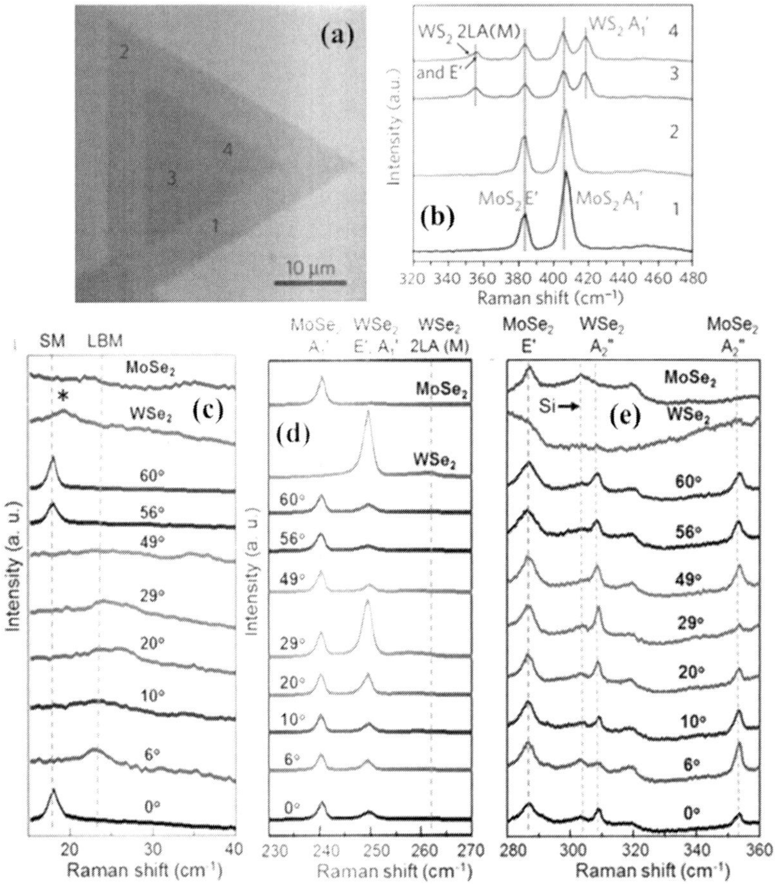


Figure 3. (a) Optical image of a $WS_2/MoS_2$ heterostructure used for Raman characterization. (b) Raman spectra taken from the four points marked in a, showing that the monolayer region is pure $MoS_2$, whereas the double layer area is the superposition of $MoS_2$ and $WS_2$ monolayers. The dashed lines are guides to the eye showing the position of the $MoS_2$ and $WS_2$ Raman peaks. Raman signatures of the $MoSe_2/WSe_2$ HSs. (c) LF Raman spectra of HSs with different twist angles ($0° \leq \theta \leq 60°$) and those of the individual TMD layers at room temperature, measured using a 532 nm laser. The positions of the shear mode (SM) and layer-breathing mode (LBM) are shown by the vertical dashed lines. HF Raman spectra of the HSs for different twist angles for frequencies of 230-270 $cm^{-1}$ (d) and 280-360 $cm^{-1}$ (e) along with those of the individual monolayers. The positions of the Raman active modes A1′, E′, 2LA (M) and the inactive modes A2″ along with that of the Si peak are marked with vertical dashed lines. Reproduced with permission. [206,207]

The evolution of Raman peaks with varying excitation energies provides additional information about the nature of the heterostructure, elucidating subtle changes in the phonon dispersion relations and electronic band structures. In summary, Raman spectroscopy serves as a powerful probe for characterizing TMD heterostructures, enabling researchers to unravel the intricacies of interlayer interactions, stacking order, and electronic coupling effects in these atomically thin materials. These findings contribute to the fundamental understanding of 2D materials and their applications in emerging nanoelectronic and optoelectronic devices.

## 4.2 Photoluminescence Spectroscopy (PL)

The photoluminescence (PL) spectra of 2D materials serve as a powerful tool for probing their optical properties. By analysing the emitted light upon exposure to external stimuli, such as photons or electrons, researchers gain valuable insights into the electronic structure, excitonic behavior, and other fundamental characteristics of these materials, contributing to advancements in optoelectronics and quantum technologies.

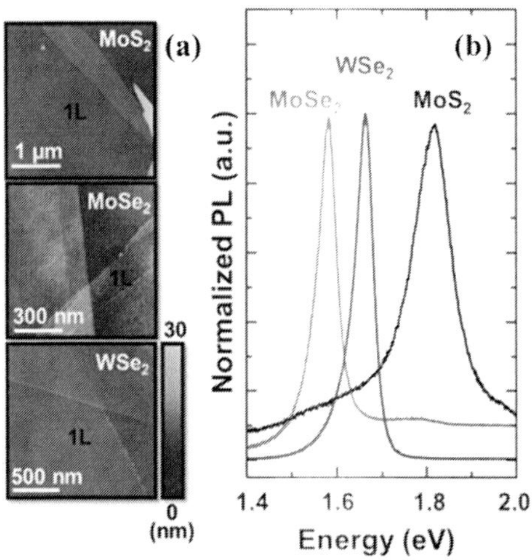


Figure 4. (a) AFM images taken on mechanically exfoliated monolayer of $MoS_2$, $MoSe_2$, and $WSe_2$. (b) Room-temperature normalized PL for monolayer $MoS_2$, $MoSe_2$, and $WSe_2$. Reproduced with permission.[208]

Figure 4 (a) shows the AFM images of monolayer $MoS_2$, $MoSe_2$, and $WSe_2$ flakes which were mechanically exfoliated from bulk crystals onto 90 nm $SiO_2$ substrates. AFM measurements on monolayers yield a height of ~ 0.7

nm corresponding to the thickness of a single unit cell. The PL signal of a 2D material is greatly enhanced by orders of magnitude from the bulk to monolayer due to the indirect to direct bandgap transition.[209, 210, 211] As depicted in figure 4 (b), at room temperature, monolayer $MoS_2$, $MoSe_2$, and $WSe_2$ show a strong PL peak at 1.84 eV, 1.56 eV and 1.65 eV, respectively.

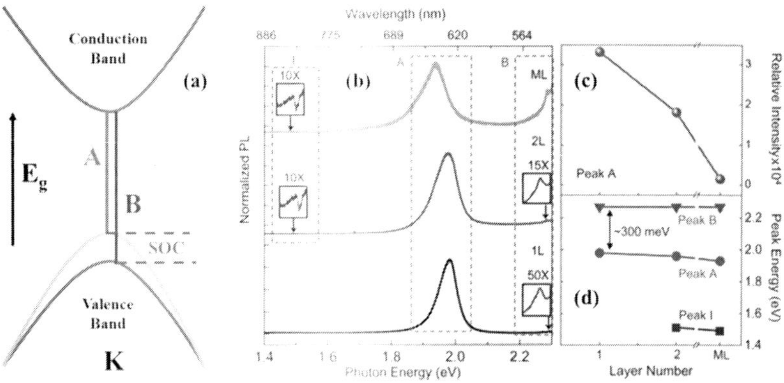


Figure 5. (a) Schematic of a monolayer TMD band structure with two distinct emission spectra due to spin orbit coupling (SOC). (b) Normalized PL spectra as a function of photon energy for $WS_2$ films including monolayer, bilayer, and multilayer. A and B peaks are the direct gap transitions from the split valence band edge to the conduction band edge at K points. Luminescence from indirect gap inter band transitions is labelled I. Spectra in the zoom windows have been multiplied by a factor as indicated for clarity. (c) The relative PL intensity of peak A as a function of film thickness. (d) The peak energy of A, B, and I transitions as a function of layer number. The difference in peaks A and B shows a nearly constant energy of ~ 300 meV, which corresponds to splitting of the valence band edge. Reproduced with permission.[212]

The PL spectra in monolayer TMDs are governed by the recombination of electrons in the conduction band and holes in the spin-orbit split valence bands. Two distinct emission features, denoted as the A-peak (representing the ground state exciton) and B-peak (indicating a higher spin-orbit split state), are evident as illustrated in Figure 5 (a). The transition of $WS_2$ from an indirect band gap (~1.2 – 1.4 eV in multilayer and bulk forms) to a direct band gap (~1.9 – 2.01 eV in monolayer form) is predicted and confirmed in Figure 5 (b).[213,214,215] Monolayer $WS_2$ exhibits a pronounced PL peak at 625 nm (~1.98 eV), corresponding to the direct excitonic transition at the K point.[216] The emission intensity undergoes a red-shift and substantial reduction for bilayer and multilayer due to an indirect band gap transition (Figure 5 (c)). Additionally, a weak PL peak labelled as B is observed at higher energy for all $WS_2$ film thicknesses. The energy difference (~300 meV) between A and B peaks for monolayer $WS_2$ indicates the spin-orbital interaction strength, slightly increasing with additional layers due to interlayer hopping and spin-orbital coupling.[217] In Figure 5 (d), a longer-wavelength peak labelled I is observed for bilayer $WS_2$, shifting to lower energy with an increase in the number of layers, originating from an indirect band gap transition.

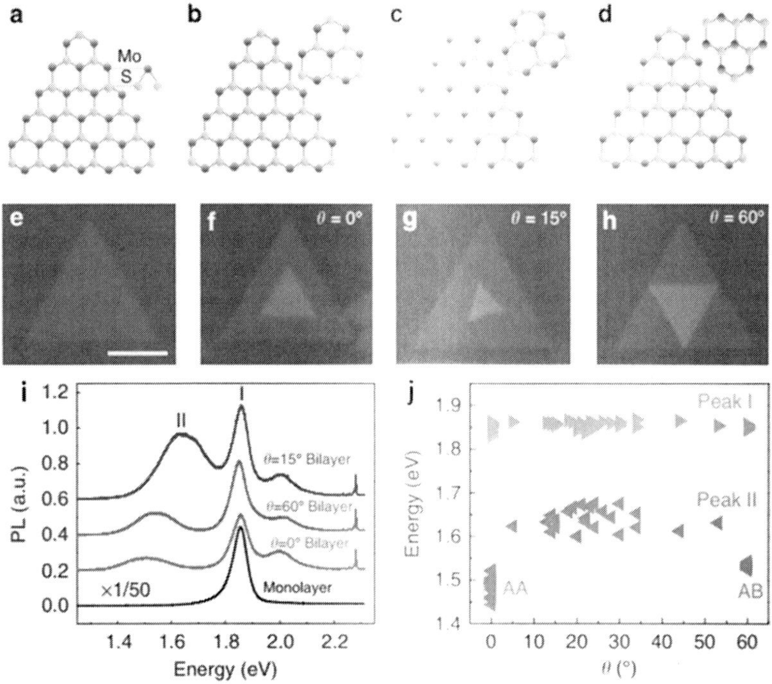


Figure 6. Schematics of the atomic structure of (a) a $MoS_2$ monolayer and bilayers with twist angles of (b) $\theta = 0°$, (c) $\theta = 15°$ and (d) $\theta = 60°$. (e – h) Optical reflection images of a $MoS_2$ monolayer and twisted bilayers corresponding to (a–d). Scale bar is 10 μm. (i) PL spectra of $MoS_2$ monolayer and bilayers with twist angles of $\theta = 0°$, 15°, and 60°. (j) Dependence of PL peak energies on the twist angle for 44 $MoS_2$ bilayers. The peak I energy is almost identical for all bilayers. The peak II energy is lowest for AA- and AB-stacked bilayers, and is higher but at a nearly constant value for other twist angles. Reproduced with permission.[218]

Each Mo atom in $MoS_2$ forms six covalent bonds with S atoms in a trigonal prismatic arrangement (Fig. 6 (a)). The orientation of the triangle-shaped layers, terminating with Mo at zigzag edges, correlates directly with the microscopic crystal orientation as seen in figure 6 (b-d). In Figure 6 (e) – (h), optical reflection images of $MoS_2$ monolayers and bilayers with twist angles (θ) of 0°, 15°, and 60° on fused silica substrates are depicted. Bilayer samples exhibit doubled optical contrast due to the smaller top layer positioned above the larger bottom layer. The twist angle of $MoS_2$ bilayers is determined by the relative orientation of the triangle-shaped layers. The twist angle of a $MoS_2$ bilayer is discerned from the relative rotation of the vertically stacked triangles. The interlayer interaction significantly influences the electronic band structure and phonon vibrations in $MoS_2$ bilayers. PL spectra in Figure 6i demonstrate the evolution of electronic coupling in bilayers with twist angles of 0°, 15°, and 60°. The monolayer PL spectrum features a prominent peak (~1.85 eV) corresponding to the exciton transition at the K-valley. In all bilayers, the direct exciton transition retains but with significantly reduced PL intensity. A new peak at lower energy (peak II) emerges, signifying an indirect bandgap resulting from interlayer electronic coupling. The energy of peak II reflects the interlayer coupling strength, indicating that the lower the indirect bandgap, the stronger the coupling. Detailed studies of 44 $MoS_2$ bilayers with various twist angles reveal interesting dependence on the twist angle. Figure 6j displays the transition energies of peaks I and II in all measured bilayers. While peak I energy remains virtually unchanged, peak II energy is lowest for perfectly registered (AA- and AB-stacked) bilayers, signifying the strongest interlayer electronic coupling in these configurations.

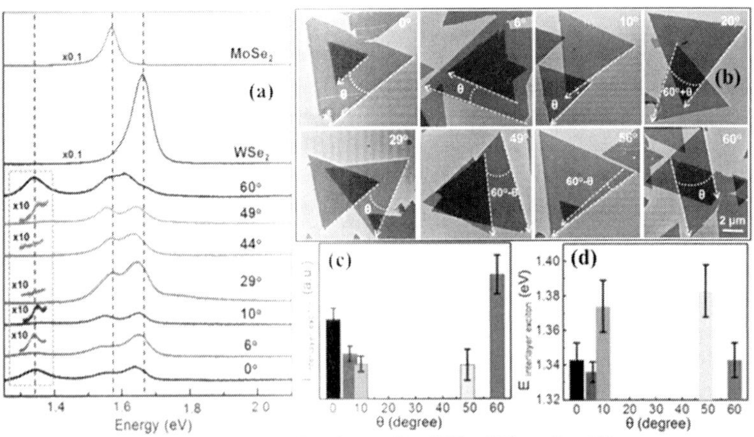


Figure 7. Probing interlayer coupling in $MoSe_2/WSe_2$ HSs using PL spectroscopy. (a) Room-temperature PL spectra of monolayered MoSe2 and monolayered WSe2, and their HSs for various twist angles (0° ≤ θ ≤ 60°) under 50 μW laser excitation at 2.54 eV (488 nm laser). The direct excitonic peak positions for $MoSe_2$ (1.57 eV) and

$WSe_2$ (1.66 eV) are marked with vertical black dashed lines. A low-energy peak was observed at ~1.35 eV for a few twist angles; it is marked with a blue dashed line. Inset show high-magnification images of these low-energy peaks for some angles. (b) SEM characterization of $MoSe_2$/WSe2 HSs. SEM images of MoSe2/WSe2 HSs with variety of twist angles in the range $0° \leq \theta \leq 60°$. The flakes with lower contrast are the WSe2 layer (bottom layer) and with higher contrast are MoSe2 layer (top layer). In some places, top layer is perfectly overlapped with bottom layer, where as in some places, it is partially overlapped. Twist angle (θ) has been determined by considering the clockwise orientation of the top MoSe2 triangle with respect to bottom WSe2 triangle with an error of about 2°. Due to three fold symmetry, the maximum angle is 60° between the two triangles. When the angle exceeds 60° or lags 0°, it is counted as (60°+ θ) or (60°-θ), where θ is considered as the angle between the two layers. The error has been determined by the differences of the three twist angles determined from each edge (total three edges) in two triangles. (c) Intensity of the interlayer exciton peak versus the twist angle. (d) Energy of the interlayer exciton peak versus the twist angle. The error bar was determined by considering three different samples at each angle. Reproduced with permission.[207]

Exciton evolution in $MoSe_2$/$WSe_2$ van der Waals heterostructures (vdW HSs) is probed through PL measurements. In Figure 7 (a), PL spectra of vdW HSs for various twist angles and individual monolayers of $MoSe_2$ and $WSe_2$ are presented. Corresponding SEM images indicating twist angles from 0° to 60° are shown in Figure 7 (b). The excitonic PL peaks of $MoSe_2$ and $WSe_2$ monolayers, observed at 1.57 and 1.66 eV, respectively, show 1 order of magnitude lower PL intensities in the vdW HSs compared to individual monolayers. The lowered PL intensities suggest strong coupling between $MoSe_2$ and $WSe_2$ layers in the vdW HSs, and PL quenching is attributed to PL quantum yield in bilayer systems. [219,220,221] The direct excitonic peaks exhibit a slight redshift in the vdW HSs, indicating changes in the band structure due to the influence of the adjacent layer. In addition to the direct excitonic peaks, an extra peak at lower energy (~1.35 eV) is observed in vdW HSs, representing an interlayer exciton peak. This peak is characteristic of heterobilayer systems and arises from interlayer radiative recombination of spatially separated excitons.[222,223] Interestingly, the interlayer exciton peak is observed selectively at certain twist angles (0°, 6°, 10°, 49°, and 60°) and disappears at other intermediate angles (10-50°). The angle-dependent evolution of the interlayer exciton peak is linked to interlayer coupling strength, correlated with interlayer distance.[224,225] For twist angles of 0° and 60°, where the vdW HS is in high-symmetry-stacking configuration, the strongest interlayer coupling occurs, resulting in high PL intensity. Conversely, at intermediate twist angles (10°-50°) with highly mismatched atomic alignment and larger interlayer distance, interlayer coupling strength is weak, leading to a decrease in interlayer exciton population. Consequently, the interlayer exciton peak appears at twist angles close to 0-60° and vanishes at other intermediate angles, influencing intralayer PL yield. Figure 7 (d) displays the energy of the interlayer exciton peak for different twist angles, showcasing variations from 1.33 to 1.38 eV.

## 4.3. Transmission Electron Microscopy (TEM)

TEM serves as a powerful tool for investigating the structural characteristics of 2D materials at the nanoscale. Through TEM studies, researchers gain valuable insights into the layer thickness, crystallographic orientation, and defects within these materials. The technique enables precise imaging of individual atomic layers, providing a detailed understanding of their morphology and structural arrangements. Moreover, TEM allows for the observation of features such as stacking order, grain boundaries, and the presence of heterostructures, contributing significantly to the comprehensive characterization of 2D materials for diverse applications in nanotechnology and materials science.

In Figure 8 (a), an exemplary pristine region of single-grain monolayer graphene is depicted, characterized by mostly single-layer composition (> 90% of the sample area) and high quality. Subsequently, Figure 8 (b) presents the diffraction pattern of bilayer graphene (BLG), revealing two sets of hexagonal diffraction spots, each originating from a distinct layer. Notably, these two sets exhibit a relative rotation of 21°, as discerned in Figure 8 (b). For rotated double-layer graphene samples, clean interfaces and discernible atomic-scale Moiré patterns are evident, as illustrated in Figure 8 (c). Further scrutiny, in Figure 8 (d–g), involves High-Resolution Transmission Electron Microscopy (HRTEM) images of twisted BLGs (tBLG) with varying twist angles of 2.9°, 9.4°, 18.3°, and 30°, respectively. These images highlight clear tBLG super-lattices, attesting to their high crystalline quality. The Moiré patterns, notably with a period of ~ 4.9 nm for the 2.9° tBLG, are distinctly observed. The 30° tBLG exhibits a unique 12-fold rotational symmetry, characterized by Stampfli tiles such as equilateral triangles, squares, and rhombuses, suggesting the formation of a high-quality quasi-crystalline system with rotational order but lacking translational symmetry.[226,227]

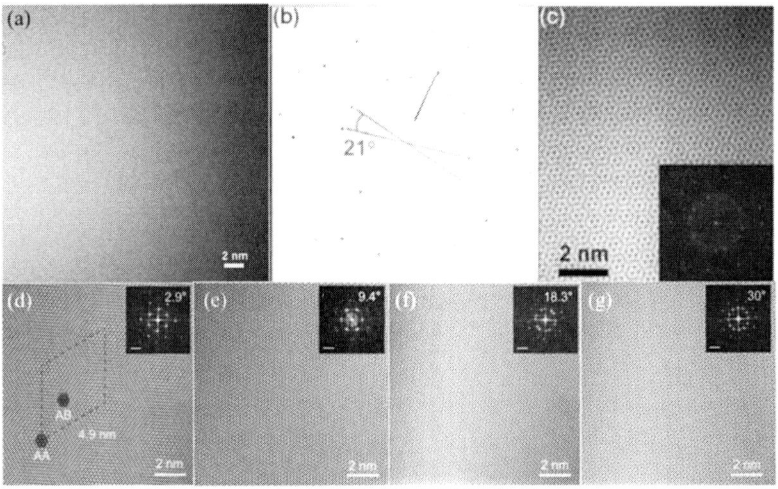


Figure 8. (a) Single-shot atomic resolution TEM image of large-area monolayer graphene grown by CVD synthesis on copper foil. (b) Diffraction pattern of a bilayer graphene (BLG) sample; the two sets of hexagonal patterns are relatively rotated by 21°. (c) Atomic resolution TEM image of BLG with rotational-angle of 21° showing a Moire′ pattern. Inset shows a fast Fourier transform (FFT) of the image. (d – g) HRTEM images of twisted BLG with clear Moiré patterns. Insets: FFTs of the corresponding HRTEM images; scale bars: 5 $nm^{-1}$. Reproduced with permission. [228,229,230]

In Figure 9(a), the depicted graphene-like $MoS_2$ layers are a product of molybdic acid treatment with an excess of thiourea in an $N_2$ atmosphere at 773 K. Notably, the separation between these $MoS_2$ layers falls within a narrow range of 0.65-0.7 nm, showcasing a controlled synthesis. The high-resolution image in Figure 9 (c) provides a detailed view of the hexagonal arrangement formed by Mo and S atoms, revealing a specific Mo–S distance of 2.30 Å. This structural insight is crucial for understanding the unique properties of these $MoS_2$ layers. Moving to Figure 9(b), it showcases $MoS_2$ layers obtained through a different method - the hydrothermal reaction between $MoO_3$ and KSCN. The resulting layers maintain a similar separation range of 0.65-0.7 nm, indicating the reproducibility and versatility of the synthesis methods.

Examining $WS_2$, Figures 9(d) and (e) present TEM images of $WS_2$ obtained through two distinct procedures. Firstly, bulk $WS_2$ undergoes lithium intercalation and exfoliation in water, a process resulting in the formation of lithium hydroxide and hydrogen gas. This reaction leads to the separation of sulfide layers, causing a loss of periodicity along the c-axis. The second method involves treating tungstic acid with an excess of thiourea in an $N_2$ atmosphere at 773 K. In both cases, the resulting $WS_2$ layers predominantly consist of bilayers and single layers, showcasing the potential for controlled layer thickness. The observed separation between $WS_2$ layers in the bilayer sample aligns with the range of 0.65-0.70 nm, reinforcing the precision of the synthesis methods. Moreover, the $WS_2$ layers obtained through the thiourea method exhibit a slightly larger interlayer spacing of 0.9 nm, suggesting nuanced variations in the synthesis conditions can influence the structural characteristics of the final material. These insights are crucial for tailoring the properties of 2D materials for specific applications.

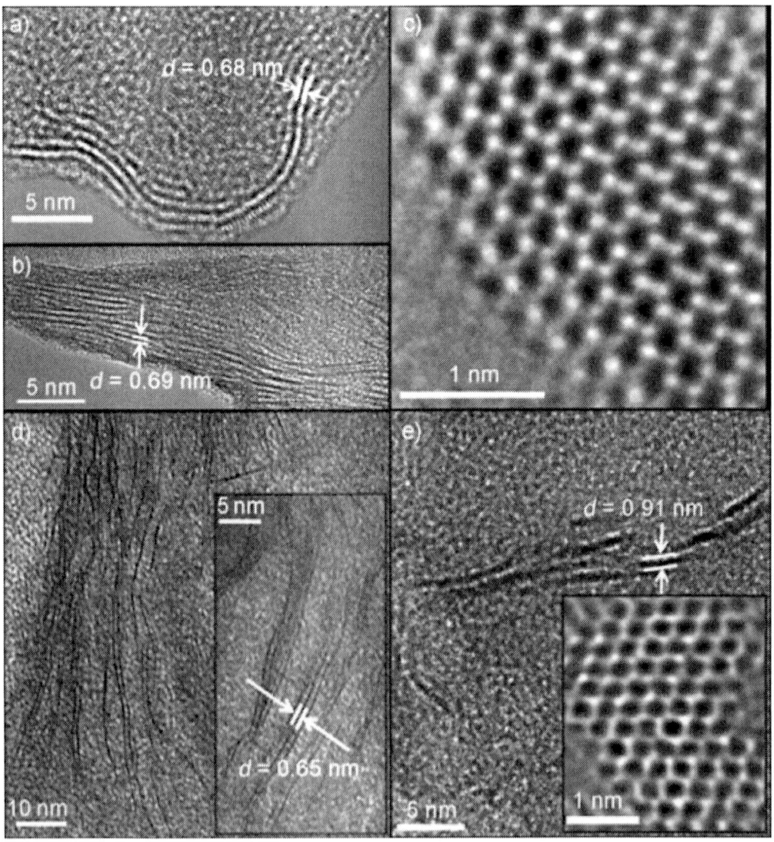


Figure 9. TEM images of $MoS_2$ layers obtained by (a) high temperature chemical reaction in $N_2$ atmosphere; and (b) the reaction between $MoO_3$ and KSCN under hydrothermal conditions. (c) HRTEM image of layered $MoS_2$ prepared by

hydrothermal method. TEM and HRTEM Images of $WS_2$ layers from (d) ion intercalation and then exfoliation from bulk and (e) high temperature chemical reaction. Reproduced with permission.[231]

In Figure 10(a), a cross-sectional bright-field HRTEM micrograph of a stacked $MoS_2/WS_2$ film provides insight into its structural composition. The film comprises horizontally-grown 2D $MoS_2$ and $WS_2$ layers, each consisting of approximately 4-5 layers. The inset offers a closer view of the periodically-stacked 2D $MoS_2$ in hexagonal (002) basal planes, emphasizing the ordered structure. Notably, atomically-sharp interfaces between $MoS_2$ and $WS_2$ are evident, confirming the structural integrity of the stacked film without alloying. The dark-bright contrast, where $WS_2$ appears darker than $MoS_2$, indicates the heavier nature of W compared to Mo. This contrast validates the ordered stacking with seamless heterointerfaces. Figure 10(b) presents an annular dark-field scanning TEM (STEM) image of a cross-sectioned $MoS_2/WS_2$ film. The brighter contrast for $WS_2$ over $MoS_2$, consistent with previous observations in other in-plane 2D heterostructures, further underscores the distinctive composition. To characterize the chemical compositions, energy dispersive x-ray spectroscopy (EDS) in STEM mode was employed. In Figure 10(c), the EDS-STEM elemental mapping image of the region marked in red in Figure 10(b) unveils a highly localized spatial distribution of Mo and W on either side of the $MoS_2/WS_2$ stack, providing valuable compositional insights.

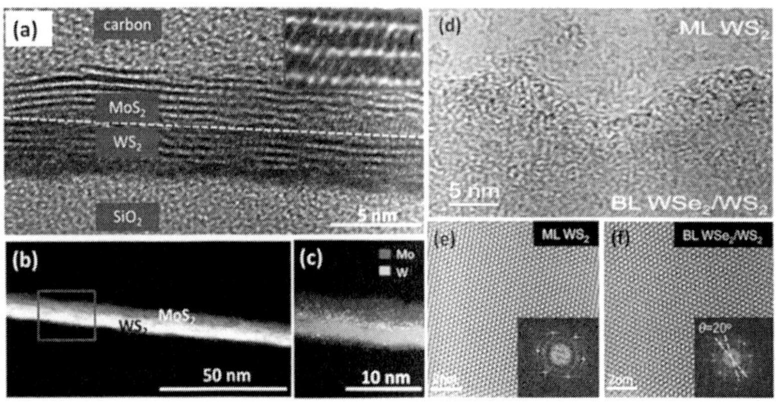


Figure 10. (a) Cross-sectional HRTEM of a vertically-stacked $MoS_2/WS_2$ HSs flm, revealing a nearly clean heterointerface of $MoS_2/WS_2$. (b) Cross-sectional ADF-STEM image of a vertically-stacked $MoS_2/WS_2$ HSs, and (c) its corresponding EDS elemental mapping image. (d) HRTEM image taken from $WSe_2/WS_2$ bilayer. (e and

f) FFT filtered images of monolayer $WS_2$ and $WSe_2/WS_2$ bilayer in (d). Insets show the corresponding FFT patterns. Reproduced with permission.[232,233]

Figure 10(d) presents a TEM image of a stacked $WSe_2/WS_2$ bilayer on the boundary. In comparison to the straightforward hexagonal lattice observed in the $WS_2$ monolayer (Figure 10(e)), a FFT filtered HRTEM image (Figure 10(f)) reveals Moiré fringes with a spatial envelope periodicity approximately 3 to 4 times the lattice constant of $WSe_2$, consistent with the observation in the $WS_2/WSe_2$ junction area. Measurement of the two hexagonal reciprocal lattices in Figure 10(f) indicates a twist angle of 20° for this specific bilayer. Theoretically, this should result in a periodicity of 3.5 in the Moiré pattern, aligning well with observations in the HRTEM images. Lastly, the presence of two sets of well-defined spots associated with $WS_2$ and $WSe_2$ monolayers in the FFT images dismisses the possibility of chemical reactions occurring at the interface during annealing.

# CHAPTER 5

# QUANTUM PHENOMENA

The exploration of a quantum phenomenum in 2D materials has emerged as a captivating frontier in modern condensed matter physics and materials science. The unique electronic, optical, and mechanical properties exhibited by atomically thin layers have opened up unprecedented avenues for investigating quantum effects in these materials. From graphene, the pioneering member of the 2D material family, to TMDs, black phosphorus, and beyond, a diverse array of 2D materials has been the subject of intense scrutiny owing to their intriguing quantum behaviors. In this chapter, we embark on a journey through the quantum landscape of 2D materials, delving into their remarkable phenomena and applications. We begin by elucidating the fundamental principles underlying quantum mechanics and their relevance to the behavior of electrons and other quasiparticles in 2D materials. Subsequently, we explore a myriad of quantum phenomena manifested in these materials, including quantum confinement, quantum Hall effect, quantum spin Hall Effect, and many-body effects such as excitons and plasmons.

Moreover, we delve into the role of quantum mechanics in shaping the electronic band structure of 2D materials, giving rise to phenomena such as Dirac and Weyl fermions, topological insulators, and valleytronics. We also discuss the pivotal role of quantum mechanics in governing the optical properties of 2D materials, enabling phenomena such as photoluminescence, Raman scattering, and nonlinear optics. Furthermore, we examine the quantum mechanical aspects of mechanical properties in 2D materials, elucidating phenomena such as phonons, flexural modes, and strain engineering. Throughout this exploration, we highlight the profound implications of quantum phenomena in the design and development of novel electronic, optoelectronic, and quantum devices based on 2D materials. From quantum computing and spintronics to quantum sensing and quantum communication, the rich tapestry of quantum phenomena in 2D materials holds promise for revolutionizing various fields of science and technology.

## 5.1 Quantum Conductance

Quantum conductance (QC) refers to the conductance of electric charge through a system at the quantum level, where the behavior of electrons is governed by quantum mechanics. In QC, electrons exhibit wave-like behavior, and their propagation through a material is subject to quantum phenomena such as wave interference, tunnelling, and quantization. QC is a fundamental property in quantum mechanics that describes the flow of electrical current through nanostructures at room temperature. It is a quantum mechanical phenomenon that becomes significant when the size of the electrical element approaches the nanoscale. In classical conductance, the current is determined by the voltage potential applied, but in QC, the conductance is dependent on the dimensions of the nanostructure. The Landauer Formula, a key concept in QC, relates the conductance of a quantum conductor to its quantum properties, [234] with the conductance quantum denoted as $G_0$. At the quantum level, the conductance of a system is quantized, meaning it can only take on certain discrete values determined by fundamental constants such as Planck's constant $h$ and the electron charge $e$. The QC is defined by the equation $G_0 = \frac{2e^2}{h}$, with a value of $\approx 7.75\times10^{-5}$ Siemens.[234] QC has been observed in various materials, including carbon nanotubes, and plays a crucial role in fields like nano-chemistry, electrochemistry, and quantum mechanics. This quantization of conductance is a consequence of the wave nature of electrons and the confinement of their motion in nanoscale systems.

Electrical conductance quantifies the ease with which electricity moves through a given path within an electrical component, representing the reciprocal of electrical resistance, which gauges the opposition to an electric current's flow. The initial observation of quantum conductance occurred during an experiment probing the electron characteristics of a two-dimensional electron gas.[235] Subsequent investigations have delved into quantum conductance across diverse materials and methodologies, remaining a pivotal subject within quantum mechanics and electrochemistry. Furthermore, it holds potential for applications such as contaminant detection in materials[236] and advancements in molecular electronics.[237] This phenomenon is attributed to two main conjectures: the discrete nature of atomic contacts in nano-sized junctions, where atomic rearrangements modify contact force, or the contact's role as an electron waveguide, where changes in contact size correspond to discrete shifts in the number of permitted electron modes.

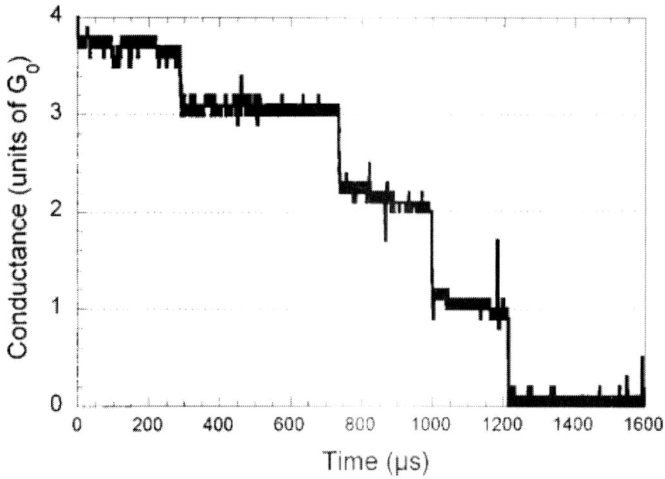


Figure 1. Quantized conduction during the breaking of a contact between two iron wires. Reproduced with permission.[238]

Fig. 1 depicts a standard graph illustrating the temporal evolution of conductance during the disconnection of two iron wires, each with a purity of 99.99% and a diameter of 0.5 mm.[238] The distinct quantized behavior of conductance is evident, characterized by a series of plateaus separated by $G_0$. Deviations are notable among different break records, both in the observed quantized values and the duration of individual plateaus. Occasionally, instances of conductance values deviating slightly from integral multiples of $G_0$ are also detected; for instance, the step near $G_0$ in Fig. 1.

Despite ongoing debate among physicists regarding the precise mechanism, this phenomenon underscores a fundamental property of nanostructures, detectable even at ambient temperatures. Understanding and controlling quantum conductance is essential for the development of advanced electronic devices with enhanced performance and novel functionalities, including quantum computers, single-electron transistors, and quantum sensors.

## 5.2 Quantum Hall Effect

Edwin Hall's discovery unveiled a fascinating phenomenon: when an electric field is applied to a metal concurrently with a perpendicular magnetic field B, the transverse resistance $R_H = \frac{\mathrm{B}}{qn_{el}}$ where $q$ is the charge carrier and $n_{el}$ is the 2D carrier density,[239] a profoundly intriguing outcome with manifold applications across various domains. The Quantum Hall Effect (QHE) emerges as a quantum mechanical extension of the classical Hall Effect, observable in 2D systems at extremely low temperatures, marking a relatively recent breakthrough in condensed matter physics.

In 1980, Klitzing, Dorda, and Pepper unearthed a quantized iteration of the Hall Effect, earning them the Nobel Prize.[239] Upon subjecting a 2D system to low temperatures and a robust magnetic field, the Hall resistance achieves quantization, $R_H = \frac{h}{ne^2}$, where n represents an integer. Unlike its linear dependence on the magnetic field, the Hall resistance exhibits plateaus (refer to Figure 2). Notably, the Hall resistance remains unaffected by material properties, solely contingent upon fundamental constants - $h$ and $e$. This revelation led to a novel resistance standardization, employed since 1990 as $R_{K\text{-}90} = h/e^2 = 25812.807\ \Omega$,[239] and offers a valuable means of measuring the fine structure constant.[240]

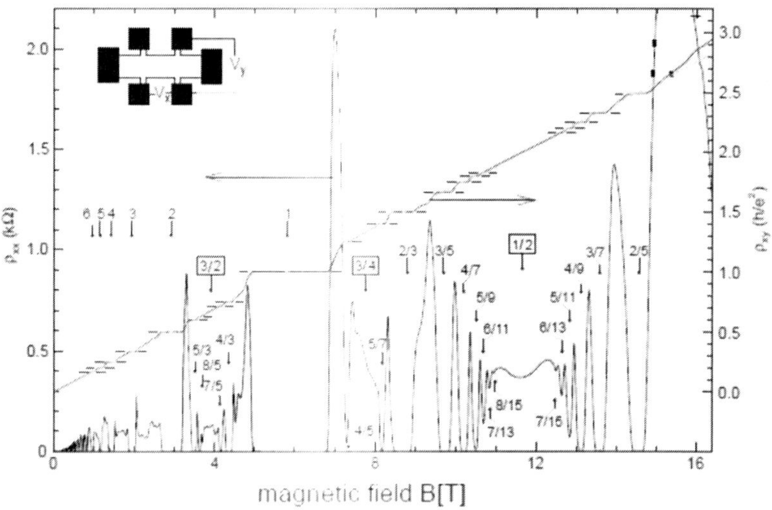


Figure 2. A measurement conducted by J. Smet of MPI-Stuttgart depicting both Integer and fractional Quantum Hall Effects. The plateau pattern is evident, where the longitudinal resistance approaches zero. Reproduced with permission.[239]

In classical theory, it was anticipated that the longitudinal resistivity would remain unaffected by the magnetic field. However, in 1930, the Shubnikov-de Hass effect was discovered, revealing that the longitudinal resistivity, or resistance, oscillated with changes in the magnetic field (refer to figure

3(a)).[239] This phenomenon, known as Landau quantization, occurs due to the quantization of energy experienced by a 2D electron under the influence of a strong magnetic field. When the magnetic field is sufficiently intense for electrons to complete an orbit without collision, their kinetic energy becomes quantized into energy levels termed Landau levels, given by En = $\hbar\omega_c(n + 1/2)$, where n is an integer.[239] The orbits' radii, known as cyclotron radii, also undergo quantization.

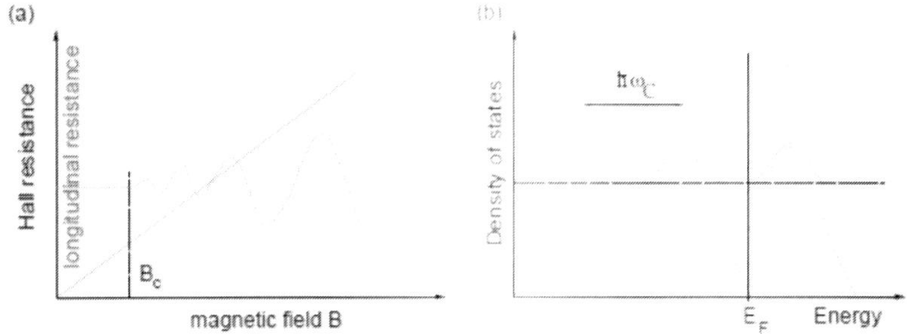


Figure 3: (a) The oscillation of the longitudinal resistance as a function of the magnetic field. Here, $B_c$ represents the critical magnetic field where the Hall resistance begins to oscillate while the longitudinal resistance remains linear. (b) The density of states, with EF representing the Fermi energy. The broadening of peaks is attributed to impurities. The horizontal line indicates the summation of overlapping peaks. Reproduced with permission.[239]

In addition to the distinctive electron energy band structure and electronic polarization induced by quantum effects, the phenomena of electron transport in 2DLMs, such as the QHE, are also of significant importance. Discovered in the 1980s, QHE stands as one of the most remarkable occurrences in condensed matter physics, showcasing the profound influence of quantum mechanics on a macroscopic scale.[241] Exclusive to 2D systems like 2D electron gases (2DEGs), QHE manifests when a 2DEG encounters a strong perpendicular magnetic field, leading to the quantization of electron energy levels into discrete Landau levels (LLs). Consequently, the Hall conductance $\sigma_{xy}$ exhibits plateaus when the Fermi level $E_F$ aligns between LLs and steps when $E_F$ intersects a LL. The QHE phenomenon is consistently accompanied by Shubnikov–de Haas (SdH) oscillations, with the Hall plateau aligning with the disappearance of longitudinal resistance $\sigma_{xx}$. The fundamental experimental characterization of the QHE involves the quantization of Hall conductance as $\sigma_{xy} = \frac{ve^2}{h}$ and the nearly negligible dissipation $\sigma_{xx} \rightarrow 0$ in a 2D electron system exposed to strong perpendicular magnetic fields and low temperatures. Here, v represents the filling factor, e denotes the elementary electron charge, and h signifies the Planck constant. The dimensionless factor ν can take either an integer value (Integer QHE, IQHE) or a "simple" fraction p/q (Fractional QHE, FQHE), where p and q are integers. Initially, QHE studies focused on artificially constructed 2DEGs in quantum well structures like GaAs/AlAs heterostructures grown via molecular beam epitaxy (MBE) with minimal disorder, observable only at low temperatures. However, with the advent of graphene, a bona fide 2D material with remarkably high mobilities, observing the QHE became much more feasible.[242,243]

### 5.2.1 QHE in Graphene

In 2005, Geim and Kim's research team first reported the experimental observation of Integer QHE (IQHE) in graphene.[242,244] Illustrated in Figure 4a is the establishment of Landau Levels (LLs) in graphene, comprising electron and hole series of levels. The energy of LLs is represented by

$$E_n = sgn\,(n)\sqrt{2e\hbar v_F^2|n|B} \tag{1}$$

Where the integer n denotes the LL index for either electron-like (n>0) or hole-like (n<0) states. Remarkably, graphene possesses an additional zero-energy Landau Level (zLL) $E_0 = 0$, which remains robust irrespective of the magnetic field, provided the particle-hole symmetry is maintained.[245246] Consequently, the Hall conductance in graphene exhibits an unconventional half-integer behavior,

$$\sigma_{xy} = \pm 4\left(n + \frac{1}{2}\right)\frac{e^2}{h} \tag{2}$$

Where the coefficient "4" originates from the fourfold degeneracy of LLs attributed to spin and valley degeneracy. Additionally, the factor of "1/2" stems from the persistent presence of the zLL at n = 0, distinct sharply with orthodox 2D systems featuring integer Hall plateaus.[245,246]

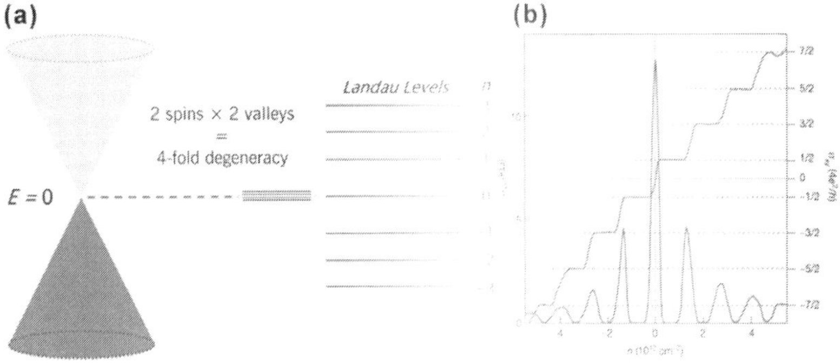


Figure 4. Representation of the QHE in graphene. (a) Schematic depiction illustrating the formation of LLs in graphene, denoted by the Landau index (n). LLs exhibit non-uniform spacing, with positive/negative n values corresponding to the electron/hole series of LLs. Notably, a zero-energy LL at the neutral point (E = 0) is associated with a LL. The LLs are fourfold degenerate due to the spin and valley pseudospin degrees of freedom. (b) Observation of integer Hall plateaus ($\sigma_{xy}$) and Shubnikov-de Haas (SdH) oscillations ($\rho_{xy}$) in graphene, with carrier density (n) adjusted accordingly. The absence of a plateau at n = 0 indicates the presence of a LL at the neutral point. Reproduced with permission.[266]

Figure 4b illustrates a typical instance of the IQHE alongside the associated SdH oscillations observed in a Single-Layer Graphene Hall-bar device. At high magnetic fields (above 20 T), the zLL may split into two subbands at $\nu = \pm 1$ by lifting the sublattice degeneracy, indicating a many-body effect associated with Landau Levels near the charge-neutral Dirac point.[247] Like many other quantum phenomena, observing the QHE necessitates low temperatures, typically below those of liquid helium.[248] Graphene is notable for extending the QHE temperature range up to unprecedented levels, including liquid-nitrogen and even room temperatures, owing to the unique behavior of charge carriers in graphene as massless Dirac fermions with ultrahigh mobilities.[249] Conversely, the experimental observation of the FQHE demands much more extreme conditions, such as ultraclean samples, extremely low temperatures, and very strong magnetic fields. To date, the observation of FQHE has been demonstrated in suspended single-layer graphene,[250] bilayer graphene,[251,252,253] as well as twisted bilayer graphene,[254] although the underlying physical mechanism is complex and beyond the scope of this book.

**QHE beyond Graphene:**

In extensively studied 2DLMs apart from graphene, including TMDs, InSe, and BP, exploring quantum transport phenomena poses additional challenges due to their relatively moderate carrier mobilities. Typically, these mobilities range from 0.1 to 100 $m^2 s^{-1}$ for monolayer TMDs and around 1000 $m^2 s^{-1}$ for multilayer BP at room temperature, diminishing further at lower temperatures.[255,256] High carrier mobility at low temperatures is crucial for observing quantum mechanics-enhanced transport phenomena; for instance, the onset magnetic field for observing SdH oscillations typically correlates inversely with carrier mobility.[257] Consequently, under current laboratory conditions, the SdH oscillations in these 2DLMs, let alone the QHE, remain unobservable. The limited mobilities stem from disorders, intrinsic defects, interfacial charge traps, etc. To address this, researchers have proposed various strategies to enhance carrier mobilities, including dielectric screening,[258] chemical doping,[259,260] and *h*-BN encapsulation.[261,262]

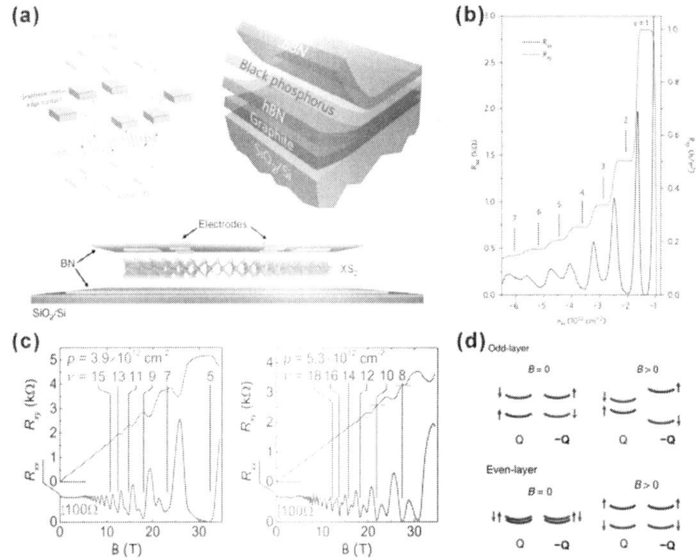


Figure 5. QHE observed in BP and TMDs. a) Illustrations depict typical device configurations aimed at achieving high carrier mobility and low contact resistance in BP and TMDs. [263,264,265] b) Quantum Hall plateaus ($R_{xy}$) and SdH oscillations ($R_{xx}$) plotted against carrier density ($n_H$) in a few-layer BP device. The negative sign of $n_H$ indicates the hole-doped regime in BP. The integers signify the Quantum Hall State with filling factor ν from 1 to 7, indicating the complete lifting of LL degeneracy in BP. [264] c) $R_{xy}$ and $R_{xx}$ plotted against magnetic field in a bilayer $WSe_2$ device at different p-doping levels: (left) $p = 3.9 \times 10^{12}$ $cm^{-2}$ demonstrating quantum Hall states primarily at odd filling factors and (right) $5.3 \times 10^{12}$ $cm^{-2}$, demonstrating quantum Hall states primarily at even filling factors. d) Schematic representations

illustrating the valley Zeeman splitting and spin Zeeman splitting at ±Q valleys in odd- and even-layer MoS2 devices, respectively. For odd-layer MoS2, the sub-band at the Fermi level is non-degenerate at B = 0; at B > 0, the degeneracy between Q and −Q valleys is resolved by the valley Zeeman Effect. For even-layer $MoS_2$, the sub-band at the Fermi level is spin-degenerate at B = 0; at B > 0, the degeneracy between spin-up and spin-down states at ±Q valleys is resolved by the spin Zeeman effect. Adapted with permission.[266]

Figure 5a illustrates various typical device configurations designed to achieve high carrier mobilities and minimize contact resistance in 2D materials using sandwiched heterostructures. By sandwiching 2DLMs between *h*-BN layers, interfacial scattering from substrate phonons and charge traps is minimized, while screening from surface adsorbates is increased, leading to significant enhancements in carrier mobility.[261,263] In $MoS_2$, combining *h*-BN encapsulation with graphene contacts effectively reduces contact resistance to as low as 100 Ω μm, resulting in the highest measured low-temperature mobilities to date: 34,000 $cm^2V^{-1}s^{-1}$ for 6-layer $MoS_2$ and 1020 $cm^2V^{-1}s^{-1}$ for monolayer $MoS_2$.[263] For few-layer BP, the *h*-BN/BP/*h*-BN sandwiched structure achieves record-breaking hole mobilities of 5200 $cm^2V^{-1}s^{-1}$ at room temperature and 45,000 $cm^2V^{-1}s^{-1}$ at low temperatures.[267] These significantly enhanced low-temperature mobilities enable experimental observations of both SdH oscillations and QHE under realistic magnetic field conditions in such materials.

In p-type BP, quantum transport investigations are typically conducted within the *h*-BN/BP heterostructure or *h*-BN/BP/*h*-BN encapsulation system.[264,268] Through local back gate tuning, Li *et al.* revealed the 2D nature of negatively gated BP thin flakes, where holes are confined within a 2-nm quantum well (equivalent to 2 atomic layers) along the out-of-plane direction.[264] This 2D BP system, coupled with its remarkably high carrier mobility and adaptable charge carrier types, facilitates the observation of QHE in both electron and hole-doped regimes (see Figure 5b).[264,269] Unlike graphene, LLs in BP are twofold degenerate and can be fully lifted at high magnetic fields, leading to the manifestation of QH states at both even and odd filling factors (ν). The effective g-factor is characterized through tilted-field and temperature-dependent measurements. Long *et al.* reported a significant g* value of 2.8, enhanced by exchange interactions between spins in opposite directions,[269] while Li *et al.* reported a g* value of 2 for ν = 5 to 15, suggesting a negligible enhancement.[264]

In 2DEGs, the sequence of observed QH states consistently reflects the energy ratio between the Zeeman energy ($E_Z = g^*\mu_B B$) and the cyclotron energy ($E_c = \hbar\omega_c$), where $\mu_B$ represents the Bohr magneton and $\omega c$ denotes the cyclotron frequency. As depicted in Figure 5c, the QH plateau sequence in bilayer $WSe_2$ is influenced by the Ez/Ec ratio, which can be density-dependent. When Ez/Ec approximates an even/odd integer, the filling factors predominantly exhibit odd/even values. The g value experiences significant enhancement with decreasing carrier density due to robust electron-electron interactions.[266] In multilayer WSe2, Xu *et al.*[270] observed an unconventional QH state sequence characterized by predominantly odd-integer filling factors under moderate magnetic fields. Unlike p-type WSe2, QH transport in n-type MoS2 and WS2 is governed by Q-valley electrons, as the CBM states in multilayer Mo/WS2 are situated at ±Q points rather than ±K points.[265,271,272] These six Q valleys, depicted in Figure 5d, are connected by threefold rotational symmetry, forming three pairs of ±Q valleys. In even-layer configurations, each pair of ±Q valleys is additionally linked by spatial inversion symmetry and time-reversal symmetry (TRS), necessitating Kramers degeneracy.

**Dirac and Weyl fermions:**

Dirac and Weyl fermions are fundamental particles in quantum field theory, each with distinct properties and significance in physics. Dirac fermions are particles described by the Dirac equation, a relativistic wave equation that accounts for the behavior of spin-1/2 particles.[273] They have mass and exhibit both left-handed and right-handed chiralities. The Dirac equation was formulated by Paul Dirac in 1928 to describe spin-1/2 particles within relativistic quantum mechanics. Mathematically, any Dirac fermion can be decomposed into two Weyl fermions of opposite chirality coupled by the mass term. Dirac fermions have implications in various fields, including condensed matter physics, where they play a crucial role in understanding electronic properties in materials.[273]

Weyl fermions are massless particles described by the Weyl equation, a relativistic wave equation for massless spin-1/2 particles. They exhibit only one-handed chirality, either left-handed or right-handed, without mass. Hermann Weyl introduced the Weyl equation in 1929 as a simplified version of the Dirac equation. In condensed matter physics, materials can exhibit quasiparticles behaving as Weyl fermions, leading to the concept of Weyl semimetals. These materials have unique electronic properties and are studied for their potential applications in novel electronic devices.[273] Both Dirac and Weyl fermions are essential in understanding the behavior of elementary particles and have significant implications in various branches of physics, from high-energy particle physics to condensed matter systems.

Their distinct characteristics contribute to the richness of quantum field theory and provide insights into the fundamental nature of matter and energy at the quantum level.

**Dirac and Weyl fermions in 2D Materials:**

So far, researchers have identified over hundreds of 2D materials, encompassing group-IV compounds, binary configurations of group III-V elements, metal chalcogenides, complex oxides, and more.[274,275] However, among this vast array, only graphene,[276] silicene, germanene,[277] a few graphynes (sp-sp2 carbon allotropes),[278] and selected other systems[279,280,281] have been projected to exhibit Dirac-like properties, as outlined in Figure 6.

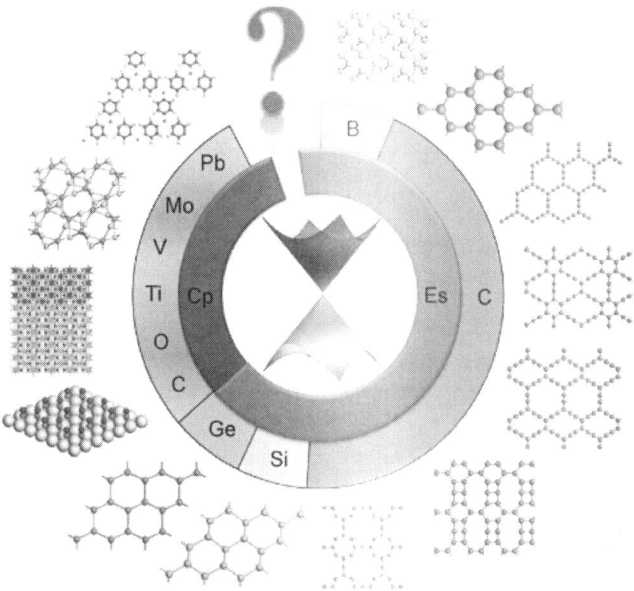


Figure 6. 2D materials featuring Dirac cones comprise a diverse range of elemental substances (Es) and compounds (Cp), incorporating elements such as B, C, O, Si, Ge, Mo, V, Pb, and others. Reproduced with permission from the reference.[284]

2D materials hosting Dirac and Weyl fermions have garnered significant interest in the field of condensed matter physics due to their unique electronic properties and potential applications. These materials exhibit distinct behavior related to the presence of Dirac and Weyl fermions, which are massless particles with specific chiralities. Graphene, a one-atom-thick honeycomb structure of carbon atoms, is a well-known 2D material hosting

Dirac fermions.[282] The Dirac-cone structure in graphene gives rise to massless fermions, leading to novel phenomena like half-integer/fractional quantum Hall effects and high carrier mobility.[283] Besides graphene, various other 2D materials have been identified as hosting Dirac fermions, including silicene, germanene, graphynes, boron and carbon sheets, transition-metal oxides, organic crystals, $MoS_2$ allotropes, and artificial lattices. These materials exhibit unique structural and electronic properties that enable the emergence of Dirac points.[284]

Symmetry-protected 2D Dirac semimetals have been studied for their distinct behavior compared to graphene and 3D Dirac semimetals. These materials feature Dirac cones at high-symmetry points that are not gapped by spin-orbit interactions, showcasing behavior unique to Weyl semimetal phases.[285] The presence of non-symmorphic space group symmetries plays a crucial role in tuning these systems to the boundary between topological and trivial insulators. Breaking these symmetries allows access to trivial and topological insulators as well as Weyl semimetal phases in 2D materials.[286] Additionally, heterostructures composed of different 2D materials offer intriguing opportunities for engineering Dirac and Weyl fermions.[287] By stacking layers with distinct electronic properties, researchers can manipulate and tailor the emergent electronic states, paving the way for designing novel quantum devices.

The study of Dirac and Weyl fermions in 2D materials not only deepens our understanding of fundamental physics but also holds promise for applications in fields such as spintronics, valleytronics, and quantum computing. As research in this area progresses, further insights into the behavior of these exotic particles and their practical implications are expected to unfold. The exploration of 2D materials hosting Dirac and Weyl fermions opens up avenues for studying exotic quantum phenomena and developing novel electronic devices with enhanced functionalities. These materials represent a frontier in condensed matter physics research, offering insights into the behavior of fundamental particles at the nanoscale level.

**Exciton & Multibody Physics:**

An exciton is a quasi-particle composed of a bound electron-hole pair, where an electron is excited into a higher energy state, creating a positively charged space known as a hole. In some cases, the electron and hole exist in a bound state, forming an exciton. This bound state is maintained by Coulombic interactions, with the strength of this bond quantified by the exciton binding energy.[288] Excitons play a crucial role in determining the behavior of semiconductor materials and are essential for understanding the potential applications of these materials in optoelectronic devices like photovoltaics, LEDs, and lasers.[289]

In 2D materials, excitons are formed due to the reduced dielectric screening and confinement in low dimensions. The unique properties of 2D materials, such as graphene and TMDs, lead to the creation of stable excitons with relatively high binding energies even at room temperature. These materials provide an ideal platform for studying excitonic effects and their impact on material properties, offering opportunities for developing advanced optoelectronic devices.[290]

Excitons play a pivotal role in the electronic and optical response of 2D materials, influencing their fundamental properties and potential applications in various optoelectronic devices. In 2D materials, excitons are responsible for the following key aspects:

- Optical Response: Excitons are crucial for the optical response of 2D materials, governing phenomena like absorption and emission of light. The presence of excitons leads to distinct optical properties, such as sharp absorption peaks and efficient light emission, which are essential for applications in photovoltaics, LEDs, and lasers.[291]

- Enhanced Light-Matter Interactions: Excitons in 2D materials exhibit strong light-matter interactions due to their large oscillator strength and enhanced binding energies. This enhanced interaction results in efficient recombination and emission processes, making excitons a central component in shaping the optical behavior of 2D materials.[291]

- Tunability: The nature of excitons in 2D materials allows for easy tunability through external stimuli like electric fields or mechanical strain. This tunability enables the manipulation of excitonic properties, leading to the development of novel photonic devices with tailored functionalities.[291]

- Device Applications: Understanding and controlling excitons in 2D materials is essential for unlocking their potential in future optoelectronic devices. Excitons can shape the design and performance of devices such as excitonic light-emitting diodes, lasers, optical modulators, and other optoelectronic components, showcasing the critical role they play in advancing 2D optoelectronics.[291]

Some experimental techniques used to study excitons in 2D materials include:

- Transient Microscopy: This technique involves using pulses of light to excite the material and observe the location of excitons, providing insights into their motion and behavior within the 2D material.[292]
- Time-Resolved Luminescence: By imaging electron-hole pairs with a streak camera, researchers can track exciton transport at different temperatures, allowing for the observation of exciton movement and dynamics in 2D materials. [292]
- Optical Transmission Spectra: Researchers can measure the optical transmission spectra of thin-film samples of 2D materials to observe features related to the formation of excitons. This technique helps in identifying excitonic transitions and studying their properties.[293]

These experimental methods play a crucial role in understanding the behavior, dynamics, and properties of excitons in 2D materials, providing valuable insights into their fundamental physics and potential applications in optoelectronic devices.

**TMD Materials as host of excitons:**

Monolayer (1L) TMDs represent direct bandgap semiconductors characterized by remarkable quantum confinement, broken inversion symmetry, and reduced dielectric screening, leading to the formation of robustly bound electron-hole pairs i.e. excitons.[294,295] Consequently, light-matter interactions within these systems are predominantly governed by excitonic effects, which have been harnessed for the development of devices based on excitons.[296,297,298,299] Beyond these two-body complexes, 1L-TMDs also manifest higher-order excitations, including three-body complexes termed charged excitons or trions,[300] and four-body complexes [301] known as biexcitons, among others [302,303] as shown in Fig 7. These advanced excitations exhibit discernible characteristics such as excitation power dependence,[304,305] distinct lifetimes, [301] spatial distribution,[306,307] and possess specific spin-valley configurations, [308,309,310] rendering them valuable for innovative quantum optoelectronic applications. However, the external manipulation of these excitations in 1L-TMDs remains a significant challenge, thereby limiting their practical utility.

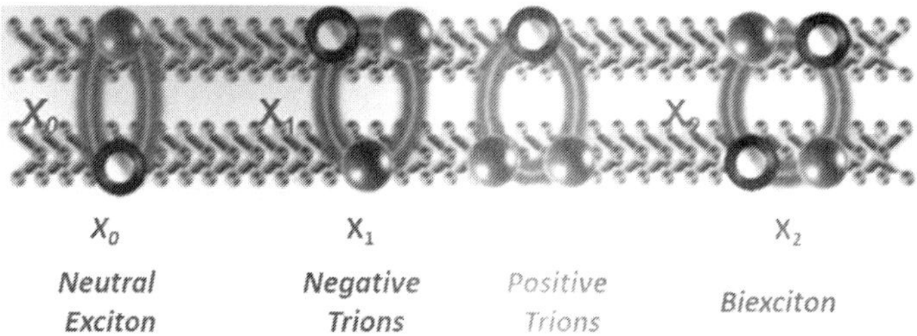


Figure 7. Schematics showing different types of excitons generated in a heterostructure of 2D materials. Reproduced with permission.[311]

Bilayer (BL) systems, composed of two monolayers of either identical or different TMDs, have garnered significant attention for the creation of innovative quantum materials, owing to their intriguing evolving characteristics.[312,313,314,315] The pivotal factor in designing these materials lies in the interlayer coupling, which is dependent on the stacking angle and serves as an external parameter to modulate their optical and electronic behaviors.[316,317,318] Recently, there have been demonstrations[319] of electrically controlled excitonic switching in BL-TMDs,[320] facilitated by the emergence of interlayer excitonic entities.[321]

In summary, excitons significantly influence the electronic and optical response of 2D materials by dictating their light-matter interactions, optical properties, tunability, and device applications. Their unique characteristics make them indispensable for harnessing the full potential of 2D materials in cutting-edge optoelectronic technologies. Excitons in 2D materials can be manipulated and tuned through various methods, such as electric field gating, strain engineering, and stacking different layers to form van der Waals heterostructures. Understanding excitonic properties in 2D materials is essential for developing next-generation optoelectronic devices, such as photodetectors, solar cells, and light-emitting diodes, with enhanced performance and efficiency.

# CHAPTER 6

# APPLICATIONS IN NANOELECTRONICS

The applications of 2D materials in nanoelectronics have garnered significant attention due to their unique properties, such as high carrier mobility and mechanical flexibility. These materials, like graphene and TMDs, hold promise for the development of next-generation electronic devices, including transistors, sensors, and flexible electronics. Their atomically thin nature allows for the creation of ultra-compact and energy-efficient components, making them a focal point of research and innovation in the field of nanoelectronics. Moreover, 2D quantum materials have shown promise in applications beyond traditional field-effect transistors (FETs), including thin-film transistors (TFTs), flexible electronics, and sensors. Their atomically thin nature, coupled with their mechanical flexibility and optical transparency, makes them suitable for integration into flexible and wearable electronic devices, opening up new avenues for futuristic electronic technologies. Overall, the applications of 2D quantum materials in nanoelectronics encompass a broad spectrum of electronic devices and technologies, ranging from transistors and sensors to spin-based devices and quantum computers. As research in this field continues to advance, the integration of 2D quantum materials into electronic devices is expected to drive innovation and reshape the landscape of nanoelectronics in the years to come.

## 6.1 Field Effect Transistors

The application of 2D materials in FETs has emerged as a promising avenue for advancing nanoelectronics. 2D materials, such as graphene, TMDs, and phosphorene, offer unique properties that are advantageous for FETs. These materials can serve as semiconducting gates in FET structures, providing inherent over-voltage protection and boosted ON-current, as demonstrated in a study on 2D/3D heterostructures-based FETs.[322] Additionally, the use of 2D materials in nanoscale tunnelling FETs has shown higher on-state current and lower sub-threshold swing, highlighting their potential for enhancing FET performance.[323] The exceptional electrostatic control, high electrical conductivity, and tunable electronic properties of 2D materials make them suitable for FET design, with specific energy bandgaps and naturally confined charge carriers within an ultra-thin body.[323] Furthermore, the integration of 2D materials in FETs has been explored for various applications, including sensing, digital and analogue electronics, memory, neuromorphic computing, and sensing devices, presenting significant opportunities for the future of nanoelectronics.[324] The development of FETs using heterogeneously stacked 2D materials for all components has also been demonstrated, underscoring the potential of 2D materials in enabling next-generation FET technologies.[325] Overall, the application of 2D materials in FETs holds great promise for advancing the performance and capabilities of nanoelectronic devices, with ongoing research focusing on addressing key challenges and realizing their full potential in practical applications.

### 6.1.1 Graphene based FET

Graphene has found significant applications in FETs, particularly in the development of biosensors and electronic devices. Graphene-based field-effect transistors (GFETs) are a class of conductance biosensors that detect targets by measuring the conductivity offset.[326] GFETs typically consist of a graphene channel between two electrodes with a gate contact to modulate the electronic response of the device.[326] The unique properties of graphene, such as its high carrier mobility, mechanical strength, and large surface area, make it an excellent material for FETs, especially in the realm of biological and chemical sensors.[327] The construction of a GFET is similar to that of a conventional FET, with a source, drain, and a top or back gate. However, unlike silicon-based transistors, the GFET has a thin graphene channel, usually tens of microns in size, between the source and the drain metal electrodes. The gate controls the behavior of the graphene channel, allowing for the modulation of electronic properties.

One of the key advantages of using graphene in FETs is its ability to enable label-free, selective, and ultrasensitive detection of biological and chemical species. The 2D nature of graphene, along with its high surface-to-volume ratio, allows for the functionalization of the channel surface and binding of receptor molecules for specific target detection. When a target molecule binds to the receptor on the graphene surface, it induces a change in the electric field across the FET, leading to a change in the electronic conductivity in the channel and the overall device response.

In the realm of electronic devices, GFETs offer benefits over traditional transistors due to the inherent qualities of graphene, such as high carrier mobility and mechanical flexibility. These properties make GFETs suitable for a range of technologies, including biological and chemical sensors, as well as for applications in communications, energy, and other industries. In conclusion, the application of graphene in field-effect transistors, particularly in the form of GFETs, has shown great promise in enabling the development of ultrasensitive biosensors and high-performance electronic devices. The unique properties of graphene make it an attractive material for FETs, and ongoing research and development in this field are expected to further enhance the capabilities and applications of graphene-based transistors.

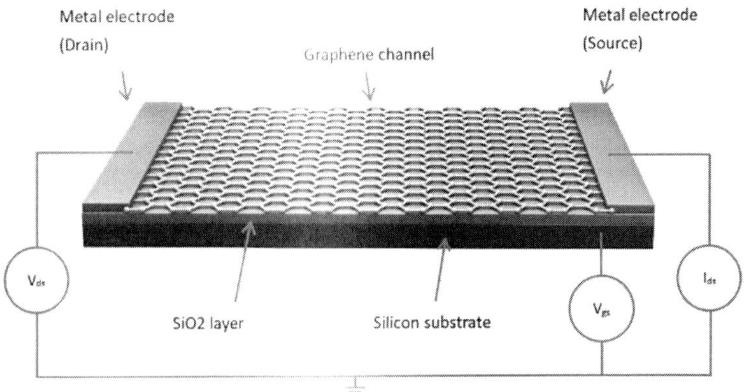


Figure 1. Schematic of Graphene based FET: graphene is used as a channel material. Reproduced with permission.[328]

### 6.1.2 TMD based FET

TMDs like $MoS_2$, $WSe_2$ and $MoSe_2$ have been extensively studied for their potential applications in FETs. TMDs have unique properties that make them suitable for FETs, such as high carrier mobility,[329] large surface area,[330] and tunable bandgap.[331] TMD-based FETs have shown improved performance in terms of low power consumption, high on/off ratio, and high trans- conductance, making them suitable for various applications, including biosensors, digital and analog electronics, and memory devices. TMDs have also been used to improve the performance of devices like metal-oxide semiconductor field-effect transistors (MOSFETs) and tunnelling field-effect transistors (TFETs). The use of TMDs in FET-based biosensors has shown great potential for ultrasensitive detection of biological and chemical species, with high selectivity and specificity. The properties of TMDs such as high surface area, high electrical conductivity, and tunable electrical/optical properties make them suitable for various sensing applications. Overall, TMDs hold great promise for advancing the performance and capabilities of nanoelectronic devices, with ongoing research focusing on addressing key challenges and realizing their full potential in practical applications. The device cross-section is schematically shown in Fig. 2a, and Fig. 2b–d displays optical images of the $WSe_2$, $MoSe_2$ and $MoS_2$ FETs. Measured transfer and output characteristics of micrometre-scale FETs with $WSe_2$, $MoSe_2$ and $MoS_2$ are presented in Fig. 2e–j, respectively. All display n-type behavior, because electron conduction is typically dominant for evaporated Au contacts to these materials.

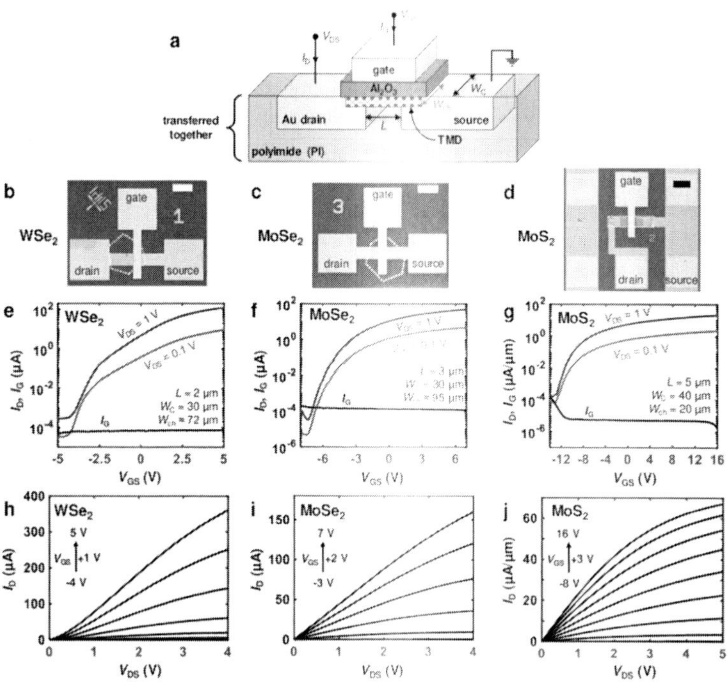


Figure 2. | Flexible FETs with TMDs. (a) Schematic cross-section, (b–d) Optical microscope images of FETs with (b) $WSe_2$ (c) $MoSe_2$ and (d) $MoS_2$. The white dotted lines mark the outline of the 2D materials. Scale bars, 50 μm. (e–j) Measured transfer and output characteristics of (e,h) $WSe_2$, (f,i) $MoSe_2$ and (g,j) $MoS_2$. Red and blue lines in e–g represent drain current, ID. The gate current, IG is often negligible, although for some devices it can limit the on/off ratio. Reproduced with permission.[332]

### 6.1.3 Thin Film Transistors

The application of 2D materials in thin-film transistors (TFTs) has been a topic of significant research in recent years. 2D materials, such as graphene, TMDs, and phosphorene offer unique properties that are advantageous for TFTs. These materials can serve as semiconducting gates in TFT structures, providing inherent over-voltage protection and boosted ON-current, as demonstrated in a study on 2D/3D heterostructures-based TFTs.[333] Additionally, the use of 2D materials in nanoscale tunnelling FETs has shown higher on-state current and lower sub-threshold swing, highlighting their potential for enhancing TFT performance.[323] The exceptional electrostatic control, high electrical conductivity, and tunable electronic properties of 2D materials make them suitable for TFT design, with specific energy bandgaps and naturally confined charge carriers within an ultra-thin body.[323] Furthermore, the integration of 2D materials in TFTs has been explored for various applications, including sensing, digital and analogue electronics, memory, neuromorphic computing, and sensing devices, presenting significant opportunities for the future of nanoelectronics. The development of TFTs using heterogeneously stacked 2D materials for all components has also been demonstrated, underscoring the potential of 2D materials in enabling next-generation TFT technologies.[334] Overall, the application of 2D materials in TFTs holds great promise for advancing the performance and capabilities of nanoelectronic devices, with ongoing research focusing on addressing key challenges and realizing their full potential in practical applications.

#### 6.1.3.1 Graphene Thin Film Transistors

Graphene has shown significant potential in the application of TFTs as evidenced by several research papers from reputed international journals. When integrated into TFTs, graphene serves various purposes, such as a channel material for radio frequency (RF) applications, a tunable contact for high-performance TFTs, and a conducting channel layer for TFTs based on graphene/semiconductor heterojunctions.

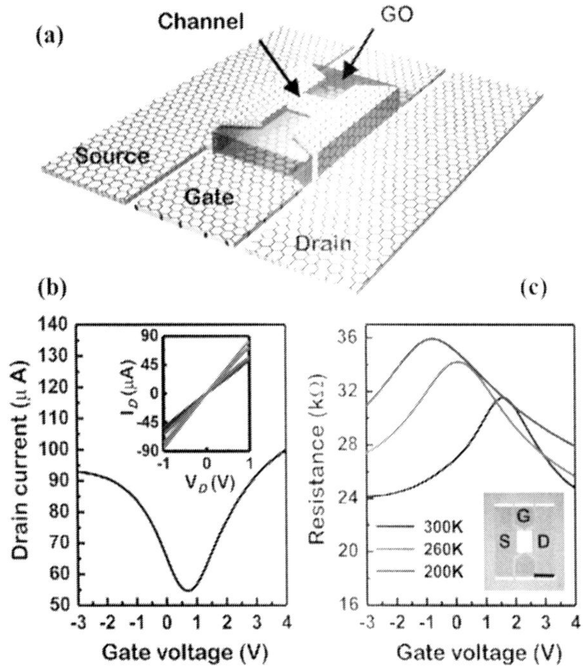


Figure 3. (a) Schematic illustration of bottom-gated graphene/GO transistor. The graphene channel which was monolithically patterned with source and drain electrodes is above the GO dielectric. (b) Typical transfer characteristic of graphene/GO transistors, indicating the hole and electron mobilities are 300 and 250 $cm^2/(V \cdot s)$ at $V_D = -0.1$ V, respectively. (c) Resistance versus applied gate voltage at different temperatures. Inset shows the microscope image of the real device that has a channel length (Lc) of 10 μm and width (Wc) of 25 μm, respectively (scale bar: 100 μm). Reproduced with permission.[335]

In a study published in a reputable journal, it was found that graphene thin film is not suitable for logic applications but can serve as the channel material for RF applications.[336] Another research paper demonstrated the use of graphene as a tunable contact for amorphous indium gallium zinc oxide (IGZO) thin film, leading to high-performance TFTs.[337] Additionally, TFTs based on two-dimensional graphene and graphene/semiconductor heterojunctions have been investigated, showing promising results such as high electron mobility and consistent performance even at high gate voltages.[338] Furthermore, a study on graphene-based vertical thin film transistors (VFETs) highlighted the potential of graphene as the channel material in TFTs, particularly in the context of vertical field-effect transistors.[339] Moreover, the suitability of graphene as a component of flexible/wearable electronics was demonstrated in a study on all-parylene flexible wafer-scale graphene thin film transistors, emphasizing its potential for applications in flexible and wearable devices.[340] In summary, the application of graphene in thin film transistors has been the subject of extensive research, with findings indicating its potential for various TFT configurations and applications, including RF, high-performance, and flexible and wearable electronics.

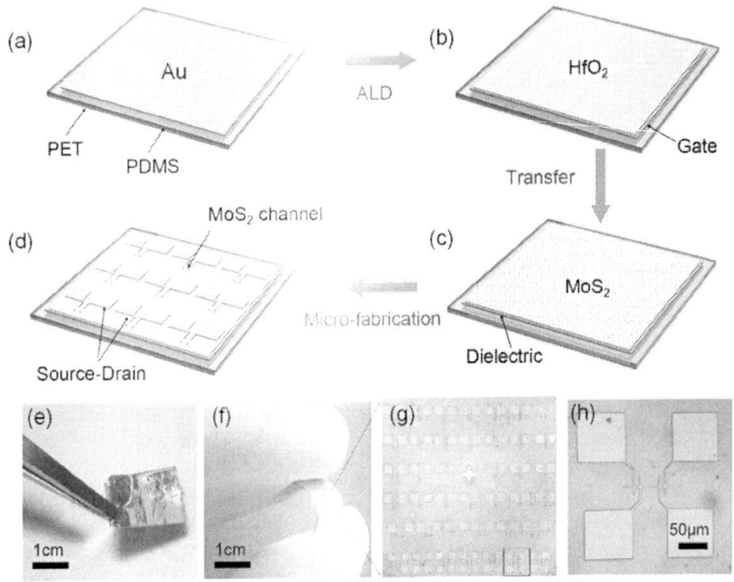


Figure 4. Schematic depiction of the $MoS_2$ TFT device fabrication process. (a) Deposition of Au as back gate on flexible PET substrate precoated PDMS. (b) Atomic layer deposition of 30 nm $HfO_2$ layer on the substrate. (c) Transferring $MoS_2$ film on the high-*k* $HfO_2$ layer. (d) Fabrication of two terminal $MoS_2$ device. (e) Optical image of picking up floating $MoS_2$ from DI (deionization) water using PET substrate. (f) Optical image of the integrated devices on flexible substrate. (g-h) The zoom-in image is the optical image of $MoS_2$ TFTs array on flexible PET substrate and one as-made device with the channel length ≈ 85 μm and width ≈ 2.5 μm, respectively. Reproduced with permission.[341]

### 6.1.3.2 TMD based Thin Film Transistors

The application of 2D materials, particularly TMDs, in TFTs has garnered significant attention due to their unique properties, making them suitable for various electronic applications. TMDs, such as $MoS_2$ and $WSe_2$, have been identified as natural candidates for the fabrication of TFTs, which are extensively used in liquid crystal displays, light-emitting diodes, and other electronic devices.[342] TMDs possess sizable band gaps and impressive carrier mobilities, enabling high on/off ratios, which are highly desirable features of TFTs.[339] The inherent n-typing and p-typing of TMDs enable digital logic circuitry in limited cases, further expanding their potential applications.[339] The device fabrication process by Zhao *et al.*, as depicted in Figure 4 (a–d), commenced with the utilization of 300 μm thick polyethylene terephthalate (PET) films as flexible substrates.[338] These films underwent spin coating with a 1–2 μm thick layer of polydimethylsiloxane (PDMS), followed by baking at 80 °C for 3 hours to achieve surface smoothness. Gold (30 nm) was deposited on the prepared substrates as gate electrodes. Due to the inert nature of the gold surface, direct deposition of high-k material posed challenges. To address this, a 1-2 nm thick titanium layer was deposited on the gold surface, followed by an oxidation process in an oxygen atmosphere at 200 °C for 1 hour. Subsequently, a 30 nm thick $HfO_2$ layer was deposited via atomic layer deposition (ALD) to serve as the gate dielectric layer. The channel material, a centimetre-scale continuous monolayer of $MoS_2$ film grown on a $SiO_2$ substrate through CVD, was transferred onto the $HfO_2$ layer using the PMMA-assisted wet chemical etching method, as illustrated in Figure 4 (e). The growth conditions were controlled to achieve a $MoS_2$ grain size of around 1 μm. The optical image of integrated devices on the flexible substrate is presented in Figure 4 (f), with a zoomed-in view depicting a typical as-made device featuring a channel length of ≈ 85 μm and a width of ≈ 2.5 μm. Research has shown that TMD-based TFTs offer several advantages, including low power consumption, reduced leakage current, and minimized short-channel effects, making them suitable for low-power applications.[343] Additionally, TMDs have been demonstrated to exhibit outstanding electrostatic control, further enhancing their potential for use in TFT devices.[340]

Furthermore, TMDs have been explored for their applications in TFT-based biosensors, leveraging their unique electronic properties for ultrasensitive detection of biological and chemical species.[344] The development of TFTs using TMDs, particularly $MoS_2$, has been the subject of extensive research, with findings indicating their potential for various TFT configurations and applications, including high-performance and flexible/wearable electronics.[345] TMDs have shown great promise for advancing the performance and capabilities of TFTs, with ongoing research focusing on addressing key challenges and realizing their full potential in practical applications.

# 6.2 Flexible Electronics

2D materials have emerged as promising candidates for flexible electronics due to their unique mechanical, electrical, and optical properties. These materials, such as graphene, TMDs, and phosphorene, have been identified as natural candidates for the fabrication of flexible electronic devices, which are extensively used in various applications, including wearable electronics, sensors, and displays. The exceptional mechanical flexibility, high electrical conductivity, and tunable electronic properties of 2D materials make them suitable for flexible electronics design, with specific energy bandgaps and naturally confined charge carriers within an ultra-thin body. The development of 2D material-based flexible electronic devices has been the subject of extensive research, with findings indicating their potential for various configurations and applications, including high-performance and wearable electronics. While early progress and many important innovations have already been achieved, the field of flexible electronics has many challenges before it becomes part of our daily life, representing a huge opportunity for scientific research and development to rapidly advance the field. [346, 347]

## 6.2.1 Graphene in Flexible Electronics

Graphene has demonstrated significant potential in the field of flexible electronics, offering a wide range of applications due to its unique properties. The 2D nature of graphene, along with its exceptional mechanical flexibility and high electrical conductivity, makes it an ideal candidate for the development of flexible and wearable electronic devices. Research has shown that graphene can be used in various applications, including transparent electrodes, RF transistors, memory devices, and energy storage, among others.[348] Additionally, graphene has been identified as a key component in the production of flexible and wearable electronics, such as smartwatches, fitness trackers, and other portable devices.[349] One of the key advantages of graphene in flexible electronics is its ability to improve the electrical circuits required for wearable and flexible devices, as demonstrated by researchers at Caltech.[345] Furthermore, graphene's potential in the development of flexible electronics has been highlighted in a study on all-parylene flexible wafer-scale graphene thin film transistors, emphasizing its suitability for flexible and wearable applications.[350]

Moreover, graphene's role in the creation of the next generation of electronics, including faster transistors, semiconductors, and bendable phones, has been recognized, indicating its potential to revolutionize the electronics industry. The unique properties of graphene, such as its mechanical flexibility and high conductivity, make it an attractive material for the development of flexible and wearable electronics, offering a wide range of potential applications in various electronic devices.

### 6.2.2 TMD in Flexible Electronics

The application of 2D TMDs in flexible electronics has gained significant attention due to their unique properties, making them suitable for various electronic applications. TMDs, such as $MoS_2$ and $WSe_2$, have been identified as promising materials for the fabrication of flexible electronic devices, including transistors, energy storage devices, and sensors.[351] TMDs exhibit tunable properties as a function of layers, such as semiconducting/metallic transition, direct/indirect band transition, strengthened flexibility, and increased transparency with the decreasing layer number, allowing them to be promising materials for flexible devices.

Research has shown that TMD-based flexible devices offer several advantages, including high electronic performance, high flexibility, and stretch ability, making them suitable for various applications, including wearable electronics, sensors, and displays. TMDs have also been demonstrated to exhibit high electronic performance as well as high flexibility, affording them a significant advantage over traditional semiconductors for flexible nanoelectronics.[352] Besides, recent studies have highlighted the potential of TMD-based flexible electronics, including the development of bendable TMD transistors with high electronic performance on soft substrates, demonstrating their potential for use in flexible and stretchable electronic devices. The practical channel thickness of TMD-based thin film transistors (TFTs) can be significantly less than that of thin films of traditional semiconductors, further enhancing their potential for flexible electronics.[353] In summary, TMDs have shown great promise for advancing the performance and capabilities of flexible electronics, with ongoing research focusing on addressing key challenges and realizing their full potential in practical applications.[354]

The following figure demonstrates the usages of numerous 2D materials in flexible electronics and the output characteristics of those devices fabricated from them.

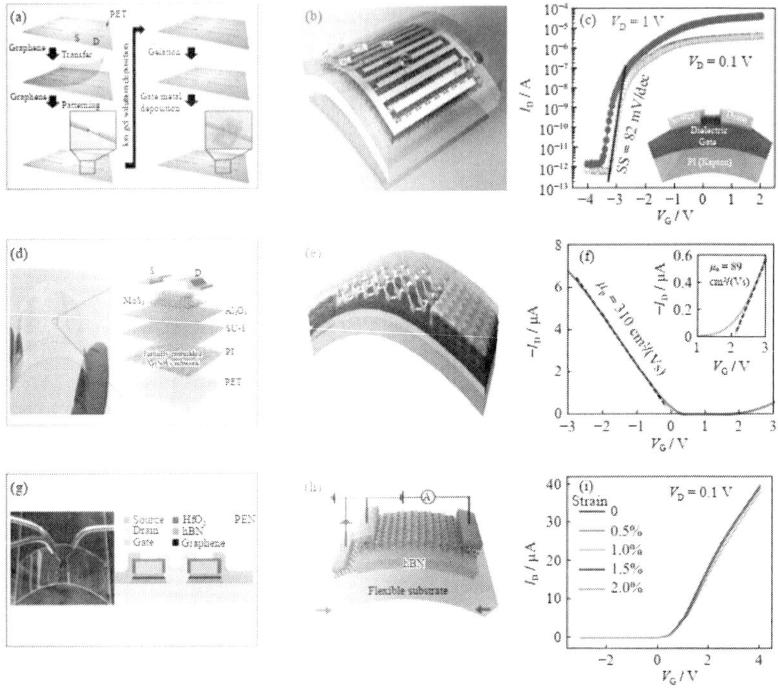


Figure 5. (a) Fabrication process flowchart of graphene FET array on a flexible substrate.[355] (b) Schematic diagram of a waterproof stretchable multi-finger embedded-gate graphene FET.[356] (c) Transfer curves of multilayer $MoS_2$ FETs at different source drain voltages flow through the device ($V_D$). The $I_{ON}/I_{OFF}$ ratio exceeds $10^7$, and the subthreshold swing is approximately 82 mV/dec.[357] (d) Picture of the flexible $MoS_2$ thin film transistor and the decomposition structure.[358] (e) Sketch map of the flexible BP FET on a polyimide substrate. (f) Linear graph of transfer curve at −10 mV bias.[359] (g) Flexible graphene FETs encapsulated with BN and Polyethylene naphthalate (PEN).[360] (h) Schematic diagram of the heterostructure InSe FET formed on a flexible substrate with Cr/Au as contact electrode. (i) Transfer curves of InSe FET under different bending conditions at $V_D$ = 0.1 V. $I_D$, $V_G$, $\mu_n$, and $\mu_p$ represent the source drain current flow through the device, the gate voltage applied on the device, the electron and the hole mobility, respectively.[361] Reproduced with permission.[363]

Graphene, owing to its small effective mass (m*), demonstrates the highest FET mobility among 2D VdW materials.[362, 363] Kim *et al.* fabricated a flexible graphene FET on a polymer substrate using a solution-processed ion gel as a gate dielectric, exhibiting mechanical flexibility; however, both electron and hole mobilities were limited (Fig. 5a).[352] Lee *et al.*[364] employed a low-temperature printing approach to create stretchable, transparent, and monolithic graphene FETs, demonstrating high electron mobility (1131 ± 96 $cm^2 \cdot V^{-1} \cdot s^{-1}$) and stable electrical characteristics, utilizing CVD grown graphene for both channel and contact electrodes. Lee *et al.* utilized $Al_2O_3$ as a dielectric layer and a multi-finger embedded gate on flexible polyimide (PI) substrates (Fig. 5(b)), achieving a high-performance graphene FET with a mobility of 3900 $cm^2 \cdot V^{-1} \cdot s^{-1}$ and stratified mechanical flexibility.[353]

While graphene exhibits high mobility, its unsuitability for digital transistors due to the absence of a bandgap has led to increased interest in 2D layered materials, particularly TMDs.[365,366] $MoS_2$, known for its large bandgap and environmental stability, serves as an excellent channel material in electrical devices.[367] Pu *et al.*[368] reported early work on $MoS_2$ thin-film FETs with ion gel gate dielectrics, exhibiting an electron mobility of 12.5 $cm^2 \cdot V^{-1} \cdot s^{-1}$ and an $I_{ON}/I_{OFF}$ ratio of $10^5$. Additionally, multilayer $MoS_2$ FETs on PI substrate with conventional high solid-state dielectrics achieved an $I_{ON}/I_{OFF}$ ratio as high as $10^7$ (Fig. 5(c)), a SS of approximately 82 mV·dec−1, and a stable mobility of 30 $cm^2 \cdot V^{-1} \cdot s^{-1}$ even with a bending radius of 1 mm.[354] Recent studies on $MoS_2$ FETs with a laser-welded 2D random network of Ag nanowires embedded in a PI substrate and organic material (SU-8) with atomic layer deposition (ALD) $Al_2O_3$ gate insulator demonstrated a field-effect mobility as high as 141 $cm^2 \cdot V^{-1} \cdot s^{-1}$ (Fig. 5(d)), an $I_{ON}/I_{OFF}$ ratio of 5 × 105, and stable electrical performance under bending radii of 10 and 5 mm.[355]

Other 2D semiconductor materials are also under development. Flexible multilayer $MoSe_2$ channel and organic SU-8 solution gate dielectric FETs exhibited a high mobility of 120 $cm^2 \cdot V^{-1} \cdot s^{-1}$, an $I_{ON}/I_{OFF}$ ratio of $10^4$, and a bending radius of 5 mm on a PI/PET substrate.[369] Zhu *et al.*[356] fabricated the first flexible black phosphorus (BP) FETs with ALD $Al_2O_3$ bottom gate insulator and conformal encapsulation layer on a PI substrate (Fig. 5(e)), displaying an excellent hole mobility of approximately 310 $cm^2 \cdot V^{-1} \cdot s^{-1}$. The mechanical robustness of the BP/PI flexible FET demonstrated a large uniaxial tensile strain exceeding 2% under 5000 bending cycles (Fig. 3(f)).

2D VdW HSs, including graphene, TMDs, and *h*-BN are easily constructed on flexible substrates owing to their dangling-bond-free surface, ultrathin atomic layer thickness plane, and rich material system of 2D layered materials.[370] Researchers are particularly interested in *h*-BN due to its high phonon energies, atomically smooth surface, high dielectric breakdown field (approximately 7.9 $mV \cdot cm^{-2}$), dielectric constant (approximately 3–4), and high thermal conductivity, rendering it an ideal dielectric for flexible electronics. Petrone *et al.*[357] fabricated a flexible graphene FET with a channel length of 2 μm (Fig. 5(g)) exhibiting the highest carrier mobility of approximately 10000 $cm^2 \cdot V^{-1} \cdot s^{-1}$, exceptional current saturation characteristics, a low output resistance of approximately 2000 Ω, and excellent mechanical flexibility with a strain limitation of approximately 1%, attributed to *h*-BN limiting scattering from charged surface states and impurities. Wu *et al.*[358] developed a flexible FET based on an indium selenide (InSe)/*h*-BN/graphite heterostructure (Fig. 5(h)). The constructed flexible FET displayed a high mobility of 1146 $cm^2 \cdot V^{-1} \cdot s^{-1}$ and a high $I_{ON}/I_{OFF}$ ratio of $10^{10}$ at room temperature. The flexible InSe/*h*-BN/graphite heterostructure on a polyethylene naphthalate (PEN) matrix substrate exhibited a high mobility of 536 $cm^2 \cdot V^{-1} \cdot s^{-1}$ without bending, and the transfer curve changed only slightly under large strain (Fig. 5(i)).

# CHAPTER 7

# OPTOELECTRONICS DEVICES

The rapid advancement of digital technology has led to a growing demand for electronic devices with high performance. In response to this demand, scientists have been exploring new materials to complement existing electronics. 2D materials have emerged as a proven choice in numerous applications in photonics, particularly in optoelectronic devices, such as optical fibers, photovoltaic cells, sensing, quantum computing, and security.[371] Optoelectronics, which involves the study of optical power from light-detecting devices, functions as an electrical-to-optical or optical-to-electrical transducer.[372] The unique properties of 2D materials, such as graphene, BP, and TMDs, make them promising candidates for a wide range of optoelectronic applications.[373] Their excellent tunable band structure, ultra-fast carrier mobility, and ultra-high nonlinear co-efficiency can be utilized in various applications in optoelectronic devices.[371] The vdW forces of 2D materials allow them to directly combine with other structures, making them suitable for large-scale production.[371] Recent research has demonstrated the potential of 2D materials in optoelectronic devices, including photodetectors, LEDs, optical switches, and optical sensors.[372, 374] The growth of optoelectronic devices based on traditional materials such as III-V and Si has been well-established. However, the lower efficiency and lack of highly efficient optical properties of these materials have led to the exploration of 2D materials for optoelectronic applications.[375] The following sections summarize the application of 2D materials in various optoelectronic devices.

## 7.1 Photodetectors

Photodetectors, or photodiodes, are common optoelectronic devices converting incident light or electromagnetic radiation into electrical signals, primarily based on semiconductor PN junctions. Silicon photodetectors, being CMOS compatible and cost-effective, play a vital role in modern communication networks. However, the challenges arise in achieving high bandwidths (>100 GHz) and photodetector responsivities (>1 A/W) with lower power consumption due to intrinsic limitations in pure silicon, such as its indirect bandgap (1.14 eV) and restricted carrier drift velocity (approximately 1000 $cm^2/(V.s)$). The utilization of 2D materials in photodetectors has emerged as a promising avenue, driven by their unique electronic properties and structural characteristics. These materials, such as graphene, TMDs, and BP, exhibit exceptional sensitivity to light across various spectral ranges. The exploration of 2D materials in photodetector applications holds significant potential for advancing optoelectronic technologies.

### 7.1.1 Graphene based Photodetectors

Graphene has garnered significant attention for its potential applications in photodetectors, owing to its unique properties that make it well-suited for high-performance optoelectronic devices. Graphene exhibits exceptional properties such as high carrier mobility, broadband light absorption, and ultrafast response, making it an ideal candidate for photodetection across a wide spectrum, from visible to infrared light.[376] Extensively researched for its unique absorption spectrum covering terahertz, far-infrared, mid-infrared, near-infrared, short-wave infrared, ultraviolet, and visible light, graphene emerges as a promising primary material for ultra-broadband photodetectors.[377] As a gapless semimetal, graphene surpasses silicon in interacting with long-wavelength light, including microwaves, enabling photodetection across a broader spectrum. Its high carrier mobility facilitates ultrafast photon-to-electrical-current conversion, and its CMOS compatibility allows seamless integration with silicon photonics, while the material's high thermal conductivity ensures energy-efficient operation. One of the key advantages of graphene in photodetectors is its high carrier mobility, which enables the efficient and rapid transport of photo-excited carriers, leading to a fast response time and high sensitivity. Additionally, graphene's unique band structure allows for the detection of light over a broad range of wavelengths, making it suitable for various applications in photodetection, including telecommunications, imaging, and sensing.[378]

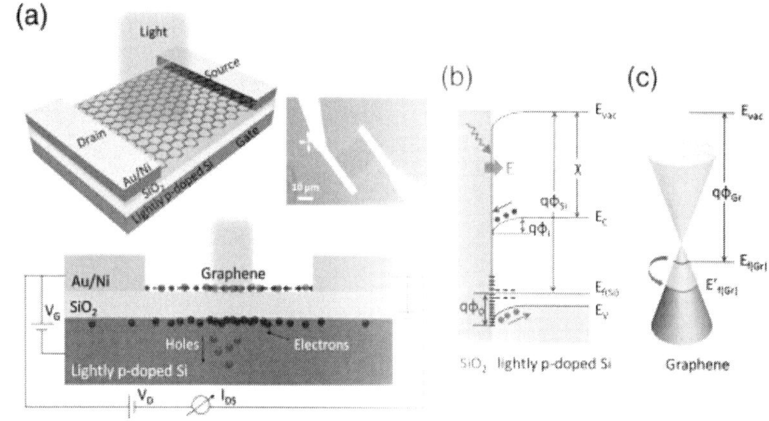


Figure 1. Graphene photodetector based on interfacial gating. (a) Schematic diagram and optical image of the graphene photodetector on lightly p-doped $SiO_2$/Si substrate; (b-c) energy band diagrams of the lightly p-doped $SiO_2$/Si interface with positive localized states ($q\varphi_0$) and its effect on graphene, respectively.The accumulation of photogenerated electrons (blue points) at the interface results in additional negative voltage under light illumination, lowers the Fermi level ($E_{f(Gr)}$) to its new position ($E'_{f(Gr)}$), and results in a light-induced p-type doping in graphene. Reproduced with permission.[378]

In Figure 1(a), the device schematic and an optical microscopy image are presented, utilizing a lightly p-doped silicon wafer (1–10 Ω cm) and a 300 nm thick $SiO_2$ layer as the gate electrode and dielectric, respectively. The working principle involves the energy band diagram of the oxide–silicon interface, creating localized positive charge states (qφ0) inducing a negative depletion layer (−) and forming a built-in electric field (E).[379] Photogenerated electron–hole pairs in lightly p-doped Si separate, causing holes to diffuse into bulk Si and electrons to accumulate at the $SiO_2$/Si interface [Figure. 1(b)]. This additional negative voltage effectively gates the graphene channel through capacitive coupling, lowering the Fermi level (Ef(Gr)) to its new position (E'f(Gr)), resulting in increased hole density and high positive photocurrents in graphene [Figure. 1(c)].

The photoresponse characteristics at $V_D = 1$ V and $V_G = 0$ V are examined with a 514 nm wavelength laser focused on the device (spot size ~1 μm). Figure 2(a) illustrates positive photocurrents observed with increasing laser power, and the corresponding photocurrent dependence on light power is depicted in Fig. 2(b). The microamp-scale photocurrents, measurable without amplification, exhibit saturation with rising light power due to the accumulation of photogenerated electrons at the $SiO_2$/silicon interface altering the electric field balance. Responsivity (R = Iph/P) calculations, presented in Fig. 2(c), reveal a remarkable responsivity of ~1000 $AW^{-1}$ at ~0.6 nW incident light power. Figure 2 (d) exhibits the spectral photocurrent response from visible to near-infrared (~0.05 μW light power), revealing a flat excitation wavelength dependence in the visible regime and a sharp decline beyond ~1050 nm. This aligns with the proposed photoresponse mechanism, highlighting lightly doped silicon as the light-absorbing medium with a photon response range from ~200 to 1100 nm. The potential extension of the photoresponse to longer wavelengths, such as mid-IR photodetection, is suggested through the substitution of silicon with semiconductors like HgCdTe with an adjustable bandgap (0.7 to 25 μm).[380]

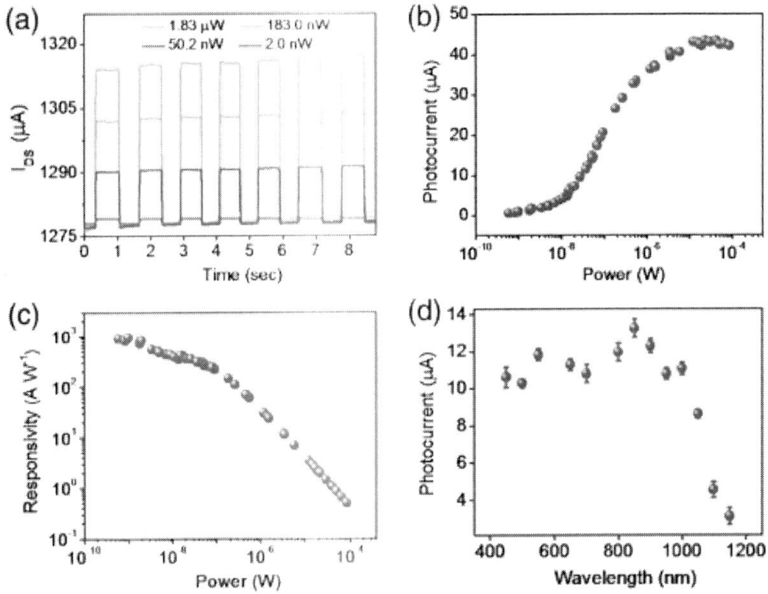


Figure 2. Graphene Photoresponse characteristics as a function of light power and wavelength. (a) Photoswitching characteristics of the graphene photodetector under varying light power; (b-c) photocurrent and responsivity at $V_D$ = 1 V and $V_G$ = 0 V of the device as a function of the light power. The laser wavelength is 514 nm. (d) The spectral photocurrent response of the device at ~0.05 μW light power from 450 to 1150 nm. Reproduced with permission.[378]

Several reports have demonstrated that graphene-based photodetectors exhibit high responsivity, low noise, and fast response times, making them promising for high-performance optoelectronic applications.[381] Furthermore, the integration of graphene into photodetectors has been shown to enhance their performance, enabling the development of next-generation photodetection devices with improved sensitivity and speed.

### 7.1.2 TMD as Photodetectors

The TMDs have emerged as promising materials for photodetectors, owing to their unique properties that make them well-suited for high-performance optoelectronic devices. TMDs exhibit excellent electronic and optoelectronic properties, making them suitable for various applications in photodetection, including photovoltaic cells, phototransistors, and photoconductive devices.[382] The sandwich structure of TMDs leads to excellent electronic and optoelectronic properties, making them ideal building blocks for photodetectors.[383]

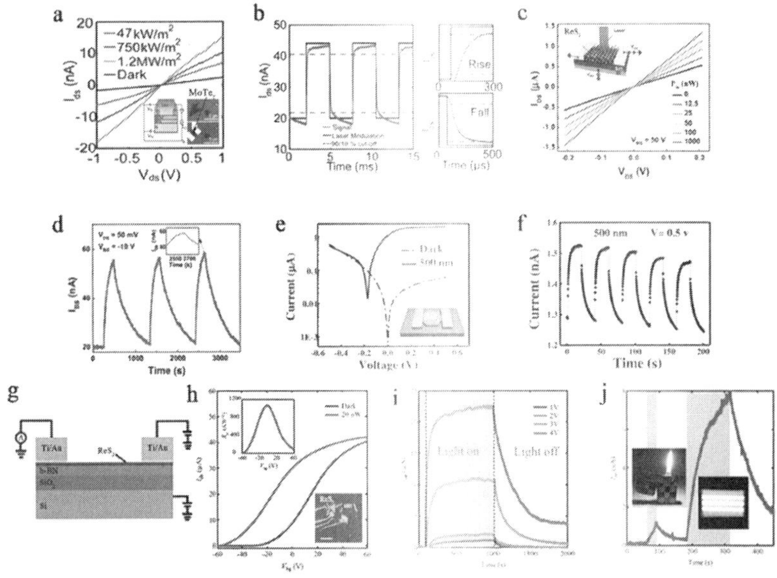


Figure 3. (a) The curves of $V_{ds}$-$I_{ds}$ fixed at $V_g = 0$ V under a different power of 685 nm laser. Inset: The schematic and two optical images of $MoTe_2$ few layer FET. (b) Photo-response occurred in $MoTe_2$ device at $V_{ds} = 5$ V, $V_g = 0$ V under a laser of 178 Hz and 20 μW. Zoomed region shows a rise time and a fall time.[384] c) Output characteristics of the curve of the $ReS_2$ back-gate device under different incident laser powers. Inset: The schematic structure of $ReS_2$ photodetector. d) Photoresponse occurred with time-dependent $I_{ds}$ of the device under an intermittent

laser illumination. Inset: Enlarged image of the photoresponse peak over a small range.[385] e) The characteristic I–V curves were illuminated by light of the single $ReS_2$ device. Inset: The schematic image of the device. f) Photoresponse occurred in $ReS_2$ polycrystalline bilayer film with photocurrent under 500 nm incident light.[386] g) The schematic of the few-layer $ReS_2$ phototransistors. h) Linear transfer curves in the dark state and under illumination. Inset: Photoresponsivity as a function of $V_{bg}$ at the top and optical microscopy image and AFM image of the same device at the bottom. i) Photoswitching behaviors with high photoresponsivity under various $V_{ds}$. j) Weak signal detection in a five-layer $ReS_2$ phototransistor, fixing $V_{bg}$ at -50 V and $V_{ds}$ at 2.0 V.[387] Reproduced with permission.[383]

$MOTe_2$, a group VI TMD, offers a layer-dependent bandgap, enabling a tunable detection range. Zhang *et al.* predicted a theoretical ultrahigh mobility for $MoTe_2$, reaching up to 2526 $cm^2\ V^{-1}\ s^{-1}$, several times higher than $MoS_2$.[388] Enhanced mobility accelerates the conversion of absorbed photons into electrical signals, improving responsivity and reducing photoresponse time. Octon *et al.* observed a rapid photoresponse (≈160 μs) in few-layered $MoTe_2$ (6.5 nm) photodetectors under 685 nm laser, yielding a high responsivity of 6 A $W^{-1}$ (Figure 3 a, b).[384] However, restricted mobility arises due to high contact resistance between the electrode and $MoTe_2$. Yin *et al.* investigated electrode materials' contact resistance, revealing a Schottky barrier in Au-contacted devices.[389] Strikingly, introduction of a tunnelling mechanism led to enhanced electron injection, achieving the highest electron mobility of ≈ 25.2 $cm^2\ V^{-1}\ s^{-1}$ and a photoresponsivity of 2560 A $W^{-1}$ under 473 nm laser. In contrast, few-layer $ReX_2$, a direct bandgap semiconductor with weak interlayer coupling, eliminates the need for a single-layer structure, reducing cost and preparation difficulty. Multilayer $ReX_2$ enhances light absorbance in device construction. With a distinct bandgap, $ReS_2$-based photodetectors generally exhibit high responsivity (>10 A $W^{-1}$) and external quantum efficiency (EQE) (>1000%), suitable for detecting extremely weak signals. Zhang *et al.* fabricated back-gate photodetectors with a maximum gate-tunable responsivity of 16.14 A $W^{-1}$ and EQE of 3168% using few-layer $ReS_2$ (Figure 3 c, d).[385] Though responsivity is comparable to graphene and $MoS_2$, its applications are limited due to a slow response time (≈500 s) and a small on/off ratio of 2.8. In 2016, Hafeez *et al.* constructed a fast-response-time photodetector based on $ReS_2$ flakes, exhibiting a response time of 2 ms, high responsivity (604 A $W^{-1}$), and EQE of $1.50 \times 10^5$% under 500 nm light of 3.11 mW $cm^{-2}$ (Figure 3 e, f).[386] Additionally, Liu *et al.* demonstrated an ultrahigh photoresponsivity of 88600 A $W^{-1}$ and EQE of $2 \times 10^7$% at a wavelength of 532 nm in few-layer $ReS_2$ phototransistors (Figure 3 g, h).[387] This high responsivity surpasses monolayer $MoS_2$ by two orders of magnitude, making it suitable for weak signal detection (Figure 3 i, j). The elevated photoresponsivity is attributed to increased photon absorption and a photogain mechanism involving trap states in $ReS_2$, with a trap state density estimated at $1.96 \times 10^{13}$ $cm^2$ through temperature-dependent field-effect mobility studies.

Another research has demonstrated that TMD-based photodetectors exhibit high responsivity, low noise, and fast response times, making them promising for high-performance opto-electronic applications.[390] Furthermore, the integration of TMDs into photodetectors has been shown to enhance their performance, enabling the development of next-generation photodetection devices with improved sensitivity and speed.[382] Moreover, TMDs have been explored for their applications in flexible photodetectors, leveraging their unique electronic properties for ultrasensitive detection of light.[391] The development of photodetectors using TMDs, particularly $MoS_2$, has been the subject of extensive research, with findings indicating their potential for various photodetection configurations and applications, including high-performance and flexible/wearable electronics.[392] In summary, TMDs have shown great promise for advancing the performance and capabilities of photodetectors, with ongoing research focusing on addressing key challenges and realizing their full potential in practical applications.

## 7.2 Solar Cells

In a solar cell, the electric field arises from the potential energy barrier at the interface of dissimilar materials, facilitating the separation of electrons and holes generated by light absorption. Junctions in solar cells are categorized into homojunctions, heterojunctions, and Schottky junctions based on the materials at the interface.[393] In a conventional silicon solar cell, a homojunction (p–n junction) forms between identical materials with equal bandgaps but different doping. Conversely, a heterojunction occurs between two different semiconductors with varying energy bandgaps. Isotype heterojunctions involve similar dopant types, resulting in n–n or p–p heterojunctions, while anisotype heterojunctions (p–n heterojunctions) involve different dopant types.[394] When a metal contacts a semiconductor, a Schottky junction or metal–semiconductor junction forms, inducing band bending akin to a p–n junction for charge equilibrium. Consequently, a Schottky junction solar cell is established when semi-metallic graphene, characterized by a zero bandgap, interfaces with silicon semiconductors. Graphene and other 2D materials, such as TMDs, have been identified as promising materials for the fabrication of solar cells, offering advantages such as high carrier mobility, broadband light absorption, and ultrafast response.[395] Research has demonstrated that graphene-based solar cells exhibit high efficiency, low noise, and fast response times, making them promising for high-performance optoelectronic applications.[393] Furthermore, the integration of TMDs into solar cells has been shown to enhance their performance, enabling the development of next-generation photovoltaic devices with improved sensitivity and speed. Moreover, recent studies have highlighted the potential of TMD-based solar cells, leveraging their unique electronic properties for ultrasensitive detection of light.[396] The development of solar cells using TMDs, particularly $MoS_2$, has been the subject of extensive research, with findings indicating their potential for various photovoltaic configurations and applications, including high-performance and flexible/wearable electronics.[397]

### 7.2.1 Graphene based Solar Cells

Graphene was discovered in 2004,[398] but the first graphene solar cell on n-type Si was reported in 2010 by Li *et al.*[399] A schematic depiction of this graphene–Si Schottky junction solar cell is depicted in Figure 4 (a). A cross-sectional view and an optical photograph are presented in Figure 4 (b - c). The graphene sheets (GS), grown via CVD on nickel foils were wet transferred onto prepatterned $Si/SiO_2$ substrates with an effective area ranging from 0.1 to 0.5 $cm^2$. The GS forms a conformal coating on the exposed n-Si substrate, establishing the Schottky junction. Gold (Au) electrodes were employed to establish contact with the graphene sheet.

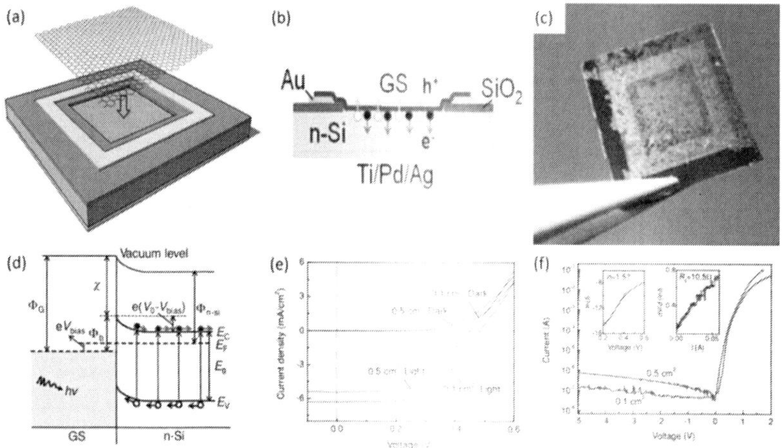


Figure 4. (a) Schematic illustration of the device configuration. b) Cross-sectional view: photogenerated holes ($h^+$) and electrons ($e^-$) are driven into the GS and n-Si, respectively, by the built-in electric field. c) Photograph of a GS/n-Si Schottky cell with a 0.1 $cm^2$ junction area. d) Energy diagram of the forward-biased GS/n-Si Schottky junction upon illumination. e) Light *J-V* curves of the cells illuminated with simulated AM 1.5 global light. f) Semi-logarithmic-scale dark *I-V* curves of two GS/n-Si cells of different junction areas. The ideality factor ($\eta$) and the series resistance ($R_s$) of the 0.1 $cm^2$ cell extrapolated from the linear regimes in the insets are 1.57 and 10.5 $\Omega$, respectively. The shunt resistance is up to 45 M$\Omega$, which is estimated from reverse bias *I-V* sweep.[399] Reproduced with permission.[395]

Figure 4 (d) illustrates the energy band diagram of the graphene/Si solar cell under illumination. The difference in work function between graphene ($\Phi_G$) and silicon ($\Phi_{n\text{-}Si}$) results in the depletion of electrons from n-Si to graphene, creating a positive charge on the semiconductor side. This leads to the upward bending of energy bands in silicon at the interface, establishing a built-in electric field that acts as a barrier for majority charge carriers' transport.[400] In contrast to conventional p–n junction solar cells, the Schottky junction solar cell doesn't require a high-temperature diffusion step for fabrication. However, the Schottky barrier height is typically lower than the built-in potential in a p–n junction silicon solar cell. When incident photons exceed the silicon bandgap, electron–hole pairs are generated and separated at the Schottky junction.[401] In the case of incident light on graphene, electron–hole pairs are generated in graphene, with holes collected by graphene and electrons crossing the Schottky barrier to be collected by n-Si.[401] Figure 4 (e) displays the *J-V* characteristics of the solar cell in dark and illuminated conditions. From the illuminated solar-cell characteristics, four important solar cell parameters are deduced.[393] From the characteristics of the illuminated solar cell, key parameters are derived. Firstly, the short-circuit current ($I_{sc}$) represents the current when the solar cell is short-circuited i.e., the voltage across the solar cell is zero. Secondly, the open-circuit voltage ($V_{oc}$) is obtained when the cell is open-circuited i.e., when the current through the cell is zero. However, at both of these operating points, the power from the solar cell is zero. So the fill factor (FF) which is the third important metric, is defined as

$$FF = \frac{P_{max}}{V_{OC}I_{SC}} = \frac{V_{mp}I_{mp}}{V_{OC}I_{SC}} \tag{1}$$

Where $P_{max}$ is the maximum power extracted from the solar cell, and $V_{mp}$ and $I_{mp}$ are the voltage and current at the maximum power point, respectively. Lastly, the solar cell's power conversion efficiency (PCE), which is the fraction of incident power that is converted to electrical energy, is defined as

$$PCE = \frac{P_{max}}{P_{inc}} = \frac{V_{mp}I_{mp}}{P_{inc}} = \frac{V_{OC}I_{SC}FF}{P_{inc}} \tag{2}$$

Where $P_{inc}$ is the power of the incident light. Graphene/silicon solar cells show $V_{oc}$ of 0.42 - 0.48 V, $J_{sc}$ of 4 - 6.5 mA $cm^{-2}$, and FF of 45 - 56%, yielding an overall solar-energy conversion efficiency (η) of 1 - 1.7% as shown in Figure 4 (e). Dark-current voltage characteristics reveal a rectification factor of $10^4$ - $10^6$, an extrapolated diode ideality factor of 1.57, and a series resistance of 10.5 Ω as seen in Figure 4 (f). Various strategies have been employed to enhance graphene-Si solar cell performance, focusing on optimizing light absorption and maximizing the separation of electron–hole pairs for improved charge carrier flow in the external circuit. In a graphene/Si heterojunction solar cell, graphene plays a dual role in achieving enhanced performance.[395]

### 7.2.2 TMD based Solar Cells

The introduction of 2D TMDs holds promise in enhancing solar cell performance through their unique electronic and optical properties. These atomically thin materials, such as $MoS_2$ and $WSe_2$, offer opportunities for improved light absorption and charge carrier mobility, paving the way for more efficient and compact solar cell designs. Figure 5 illustrates the calculated absorbance for monolayers of three TMDs - $MoS_2$, $MoSe_2$, and $WS_2$ along with graphene, showcasing the potential of a single TMD monolayer, with sub-nanometre thickness, to absorb sunlight more efficiently than 50 nm of Si or 15 nm of GaAs.[402] This capability enables the generation of electric currents as high as 4.5 mA cm-2 theoretically.[402]

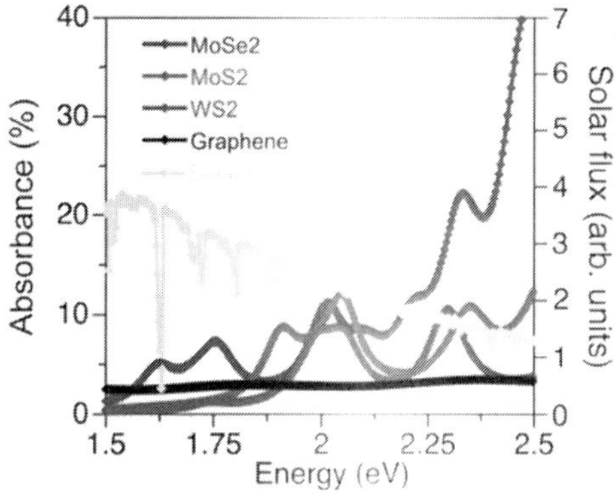


Figure 5. Absorbance of three TMDC monolayers and graphene, overlapped with the incident AM1.5G solar flux. [402] Reproduced with permission.[395]

The absorbance (A) of graphene and other TMDCs are calculated using

$$A = 1 - e^{-\alpha L} \tag{3}$$

Where the absorption coefficient $\alpha$ is obtained from the literature for the materials, and converted to the absorbance A, for a flat layer of thickness L = 1 nm.[402,403,404, 405]

The absorbed photon flux can be translated to the equivalent short-circuit current in the ideal case by

$$J_{\mathrm{abs}} = e \int_{E_{\mathrm{g}}}^{\infty} A(E)\, J_{\mathrm{ph}}\,(E)\, \mathrm{d}E \tag{4}$$

Where $E$ is the photonic energy, $J_{ph}(E)$ is the incident photon flux, $E_g$ is the optical bandgap of the absorbing material, and $e$ is the elementary charge. With a mere thickness of 0.65 nm, monolayer $MoS_2$ efficiently absorbs 5-10% of incident light. The work functions of TMDs exhibit considerable variation based on the thickness of the transition metal; for monolayers, $WS_2 \approx 4.6$ eV, $WSe_2 \approx 4.3$ eV, and $MoS_2 \approx 5.1$ eV, while for bulk counterparts, $WS_2 \approx 4.2$ eV, $WSe_2 \approx 3.9$ eV, and $MoS_2 \approx 4.5$ eV).[406]

In a Si/$MoS_2$ p-n junction solar cell, the direct bandgap (1.9 eV) of monolayer $MoS_2$ facilitates the formation of a Schottky junction with p-Si, resulting in a solar cell with a power conversion efficiency of 5.23% and a

$V_{oc}$ of 0.41 V.[403] The device structure (Figure 6a) illustrates a $MoS_2$/p-Si heterojunction solar cell, and the energy band diagram (Figure 6b) reveals the type-II hetero-junction formed between $MoS_2$ and p-Si, achieving efficiencies of 5.23% (Figure 6c).[407] The introduction of an $Al_2O_3$ passivation layer enhances the built-in potential at the n-$MoS_2$/p-Si interface.[408]

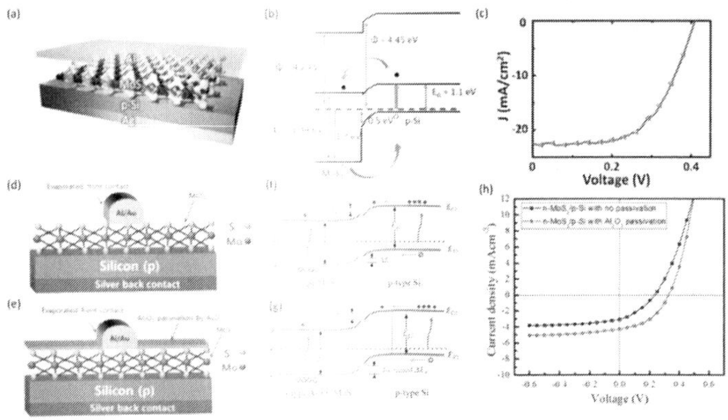


Figure 6. (a) Device structure of $MoS_2$/p-Si heterojunction solar cell. (b) Band diagram of $MoS_2$/p-Si heterojunction obtained from UPS measurements. $\Phi$ is the work function, and $E_G$ is energy bandgap. (c) *J-V* characteristics. Reproduced with permission.[407] (d, e) Schematic illustrations of n-$MoS_2$/p-Si solar cells: (d) No-passivation layer, and (e) 10 nm thick $Al_2O_3$ passivation layer employed on $MoS_2$ surface. (f, g) Energy band diagrams of $MoS_2$/p-Si p-n junction (f), and (g) $MoS_2$/p-Si p-n junction with $Al_2O_3$ passivation at the $MoS_2$ surface (g); $\Delta E_C = E_{C2} - E_{C1}$ and $\Delta E_V = E_{V1} - E_{V2}$ have increased with $Al_2O_3$ passivation; $E_g$ is the energy bandgap, $E_F$ is the Fermi energy level, $E_V$ is the top of the valence band, and $E_C$ is bottom of the conduction band. h) Photovoltaic characteristics of the solar-cell devices with/without $Al_2O_3$ passivation layer at the $MoS_2$ surface. Reproduced with permission.[408]

The photovoltaic performance of $MoS_2$/Si solar cells resembling bulk improves with $Al_2O_3$-based passivation at the $MoS_2$ surface.[408] Schematics of the device with and without the $Al_2O_3$ passivation layer are shown in Figure 6 d,e, while passivation enhances the built-in field at the $MoS_2$/Si interface (Figure 6 f,g). This passivation, achieved by atomic-layer-deposited $Al_2O_3$, reduces the surface trap density of $MoS_2$, resulting in a ominously increased power conversion efficiency (PCE) of 5.6% compared to 2.21% for $MoS_2$/p-Si solar cells with no passivation as seen in Figure 6 (h).[408] Additionally, a $MoS_2$/ITO Schottky-junction solar cell, formed by transferring a CVD-grown $MoS_2$ nanomembrane onto an ITO-coated glass substrate, reported efficiencies of 0.7% and 1.8% for $MoS_2$ thicknesses of 110 nm and 220 nm, respectively.[409] In a vertically stacked indium tin oxide electrode/multilayer $MoS_2$/ITO structure, a p-n junction was formed within the $MoS_2$ layers through plasma-induced p-doping, increasing the PCE to 2.8% under AM1.5G solar illumination.[410]

### 7.2.3 2D heterojunction based Solar Cells

The concept of an "all-2D" solar cell, while seemingly less effective in light absorption due to thinner active materials, surprisingly holds potential to outperform conventional Si and GaAs-based photovoltaics. TMDs, with visible to near-infrared bandgaps, offer advantages as direct-bandgap semiconductors at monolayer thickness, enabling high radiative efficiency. Despite monolayer TMDs absorbing only 10% of the solar spectrum, their higher absorption coefficient per unit thickness, surpassing Si, GaAs, and perovskites makes them suitable for ultrathin, high absorbing photovoltaic devices. Light-trapping designs are crucial for reaching high efficiency in "all-2D" solar cells. [411, 412] Seamless heterojunctions formed by stacking various 2D materials offer effective carrier extraction,[413,414,415] with reduced thickness and sharp interfaces minimizing electron-hole recombination. However, the presence of a built-in potential at heterointerfaces induces a high electric field, leading to interlayer tunnelling recombination, an aspect yet unexplored in solar-cell performance.[413, 416] While 2D/2D interfaces exhibit reduced interface trap densities, the scalability of all-2D solar cells is constrained by the less accessibility of large area 2D materials.[417]

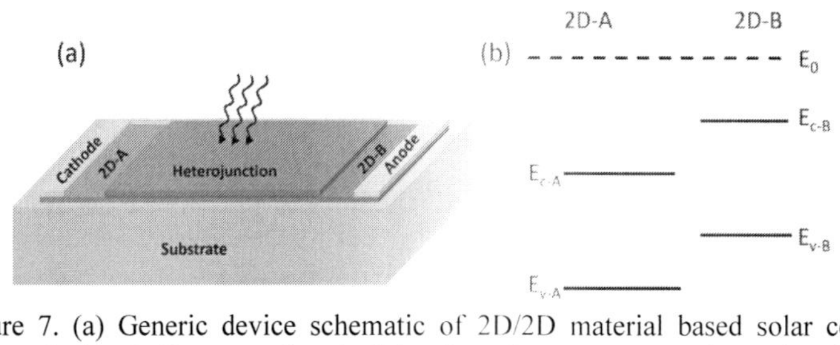


Figure 7. (a) Generic device schematic of 2D/2D material based solar cell. b) Schematic band diagram of a dual bandgap heterostructure. Reproduced with permission.[395]

In Figure 7 (a - b), a schematic of a generic 2D/2D heterojunction solar cell and the band diagram of the type-II heterojunction along a vertical section illustrate the benefits of a vertical type-II heterostructure, allowing efficient charge separation. This contrasts with a longer p–n junction, which, when scaled up, can decrease charge separation efficiency. In the vertical heterojunction device, excitons can diffuse through the 2D/2D interface for dissociation, especially when they have a long lifetime. An effective strategy to enhance the efficiency of 2D/2D solar cells involves employing multiple materials with distinct bandgaps, leading to a multi-bandgap vdW solar cell.

While the formation of a vertical 2D/2D heterojunction holds promise, its actual implementation is restricted to a limited number of chalcogenides, and alloy structures at the interface hinder the realization of a seamless vdW heterojunction. Novoselov and Neto have highlighted the significance of assembling 2D materials into 3D heterostructures for efficient light collection and charge transfer essential for the photovoltaic effect.[418] In their recent publication, they expanded the range of applications for such heterostructures beyond demonstrating the photovoltaic effect.[419] Various fabrication processes, including vdW epitaxy, direct growth, and large-scale CVD growth, are employed to realize thin vdW heterostructures such as graphene/*h*-BN,[420,421,422] $MoS_2$/graphene,[423,424] $MoS_2$/*h*-BN,[425] $WS_2$/*h*-BN,[426] $WS_2$/$MoS_2$,[427] and $MoS_2$/$WSe_2$ heterojunctions.[428]

Highly crystalline vertically stacked bilayers and in-plane interconnected $WS_2$/$MoS_2$ hetero-structures can be achieved under different temperatures. Vertically stacked bilayers form at 850 °C, while in-plane lateral heterojunctions are observed at 650 °C (Figure 8 a, b).[427] The lateral heterostructure, consisting of monolayer materials, exhibits good rectification and a photovoltaic effect with $V_{oc}$ of 0.12 V and $J_{sc}$ of 5.7 pA (Figure 8 c). Additionally, a two-step CVD growth process for $WSe_2$/$MoSe_2$ heterostructures has been reported, where $MoSe_2$ is synthesized first, followed by the epitaxial growth of $WSe_2$ on the edge and top surface of $MoSe_2$. The two-step method, avoiding cross-contamination, offers spatial and size control of each 2D component,[429] resulting in type-I and type-II $WSe_2$/$MoSe_2$ heterostructures with efficiencies of 0.12% and $V_{oc}$ of 55 mV for a laser illumination of 619 mW cm$^{-2}$ at 514 nm (Figure 8 d, e). The obtained heterostructure exhibits rectification, photodetecting properties, and photovoltaic performance.

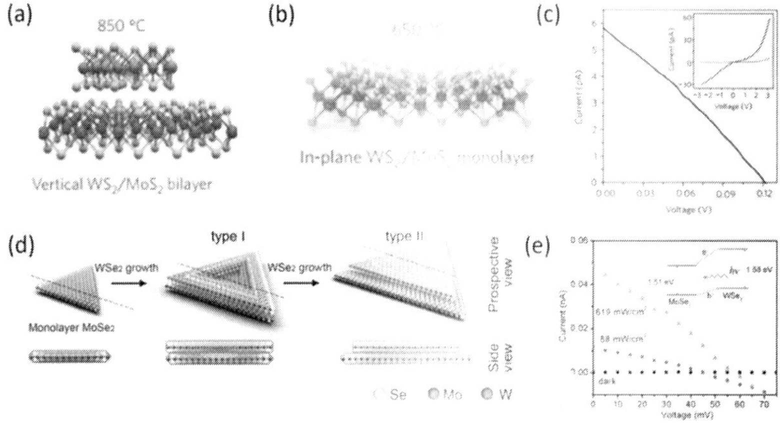

Figure 8. (a) Schematic image of the vertically stacked $WS_2/MoS_2$ heterostructures synthesized at 850 °C, showing the bilayer feature. b) Schematic image of the $WS_2/MoS_2$ in-plane heterojunctions grown at 650 °C. c) Photovoltaic effect of the in-plane heterojunction. The inset shows the typical *I-V* curve of the junction with (black) and without (red) illumination, showing the p-n junction behavior.[427] d) Schematics of the $MoSe_2$-templated $WSe_2$ growth. $WSe_2$ always grows from the edge regions of the $MoSe_2$ layer, forming both a vertical bilayer heterostructure on the $MoSe_2$ surface and a seamless in-plane heterostructure. The $MoSe_2$ template is outlined by white lines, where the in-plane $MoSe_2/WSe_2$ interface would form. The type-I and type-II heterostructures are obtained after different $WSe_2$ growth times by using monolayer $MoSe_2$ as a template. e) The photovoltaic effect *I-V* curve of the $MoSe_2/WSe_2$ heterojunction. The inset shows the schematic diagram of $MoSe_2/WSe_2$ heterojunction band structure and the photoexcitation process in the $MoSe_2/WSe_2$ heterojunction.[429] Reproduced with permission.[395]

A solar cell featuring a lateral monolayer $MoS_2$/monolayer $WSe_2$ interface achieved a power conversion efficiency of 2.56% under AM1.5G illumination,[430] leveraging the exposure of the full depletion area to light for excellent omnidirectional light-harvesting characteristics. Rivera *et al.*[431] explored the interlayer exciton in a $MoSe_2/WSe_2$ vertical heterostructure using PL spectroscopy, suggesting applications in lasers, light-emitting diodes, and photovoltaic devices. Additionally, a high-quality n-$MoS_2$/p-$MoS_2$ bilayer junction device demonstrated the feasibility of converting n-type $MoS_2$ nanosheets to p-type via UV-ozone plasma treatment for photovoltaic and other optoelectronic applications.[432] The $MoS_2/WSe_2$ junction's gate-tunable rectifying electrical characteristics and photovoltaic response were investigated by optimizing the number of layers, band alignment, and interface lattice mismatching.[433] These low-cost, large-area heterostructures are fully compatible with various flexible and transparent substrates, holding potential for next-generation flexible electronics.[434]

## 7.3 Light Emitting Diode (LED)

Highly intriguing optical and electrical properties of 2D materials,[435,436] particularly TMDs and phosphorene, have garnered significant attention, featuring tunable band structures,[437,438] direct to indirect band transitions,[439,440,441] large exciton binding energies,[442,443,444] strong light matter interactions,[445,446] and robust second harmonic generations.[447,448] These characteristics have spurred extensive studies into advanced optoelectronic devices like light emitting diodes (LEDs) in parallel to photodetectors. As TMDs and phosphorene yield promising research outcomes, organic-inorganic halide perovskites have emerged as compelling materials for high-performance LEDs, driven by their distinctive structure and exceptional optoelectronic assets.

### 7.3.1 TMD based LEDs

The bulk TMDs exhibit an indirect bandgap, but this limitation transitions to a direct bandgap in single layers, meeting a fundamental requirement for efficient light emission applications.[403] The diverse TMDs, featuring various bandgaps and work functions, not only allow tunable emissions across a broad wavelength range but also offer opportunities for bandgap engineering in heterostructures.[449] Light emission from TMDs is predominantly governed by exciton species,[450,451] observable at room temperature due to quantum confinement and strong Coulomb interactions. Different exciton species, including excitons, trions, biexcitons, and localized excitons, can be generated optically or electrically in TMDs, presenting manipulation possibilities through chemical doping,[452] electrostatic gating,[453,454] and excitation power adjustments.[455] LEDs, relying on electroluminescence (EL) in semiconductors, convert electrical energy into optical energy through the radiative recombination of electrons and holes. DC-driven LEDs can be categorized based on the presence of a p-n junction, while AC-driven LEDs share a similar structure to a capacitor.

In semiconductors, when a valence electron gains sufficient energy, it transitions to the conduction band, leaving behind a hole in the valence band, forming a conductive electron free to move within the crystal lattice. This electron and the hole can be attracted to each other by the Coulomb force, creating an exciton, an electrically neutral quasiparticle studied extensively in III-V semiconductor systems.[456,457] The exciton binding energy, representing the minimum energy required to separate electron-hole pairs, is low in III-V semiconductors like GaAs, limiting observations to low temperatures. However, the 2D nature of monolayer TMDs enhances Coulomb interaction, resulting in tightly bound excitons. $MoS_2$ has a predicted exciton binding energy between 0.5 and 1 eV,[458,459] measured at 240 meV.[460] Monolayer $WS_2$,[461] $WSe_2$,[442] $MoSe_2$,[462] and $MoTe_2$[454] exhibit even larger exciton binding energies of 0.71 eV, 0.37 eV, 0.55 eV, and 0.58 eV, respectively compared to typical III-V semiconductors (InP: 3.71 meV, GaAs: 4.76 meV).[463] These substantial binding energies enable the observation of exciton dynamics at room temperature, and the strong Coulomb interaction in TMDs allows higher-order exciton quasiparticles, such as trions and biexcitons, to be studied. Trions, charged quasiparticles resulting from exciton binding, can be influenced by electrostatic gating and chemical doping, offering a platform for tunability. Additionally, monolayer $MoSe_2$ has demonstrated a biexciton binding energy of ~20 meV,[464] and charged biexcitons, a five-particle system, have been observed in *h*-BN encapsulated monolayer $WSe_2$, showcasing 2D TMDs as an ideal material system for studying exciton dynamics and many-body interactions.

The p-n junction is a pivotal component in optoelectronic applications, notably photodetectors and LEDs. Due to the ultra-thin nature of TMDs, these materials are highly responsive to the surrounding dielectric environment, enabling electrostatic gating control for monolayer TMDs doping. A common technique involves forming a p-n junction by applying two split gates beneath the monolayer TMDs, as depicted in Figs. 4(a) and 4(b).

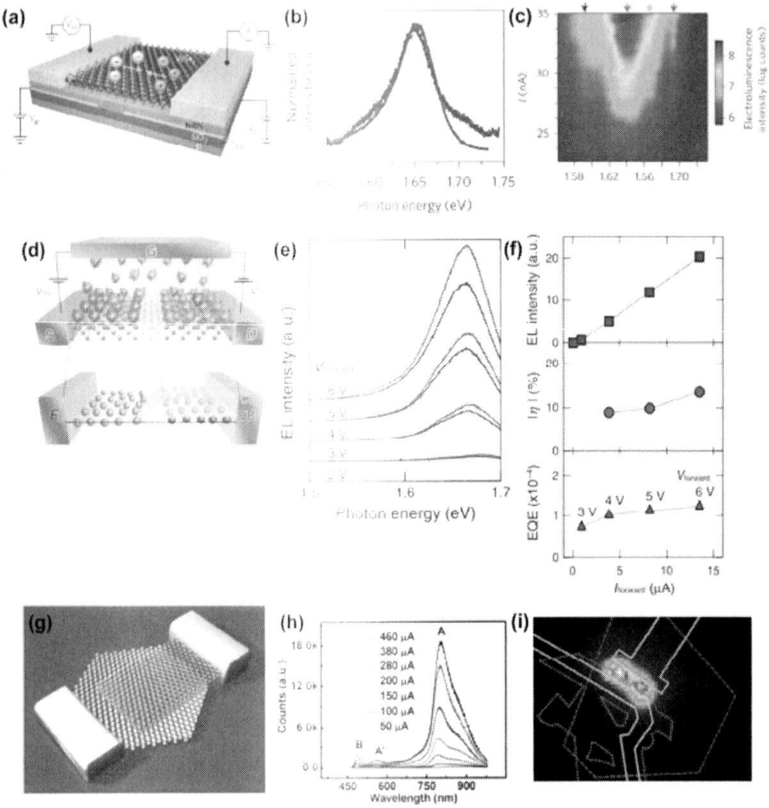


Figure 9. LEDs based on TMDs p-n junction. (a) Schematic plot of multiple monolayer $WSe_2$ p-n junction devices with palladium back gates ($V_{g1}$ and $V_{g2}$) and source (S) and drain (D) contacts. The source–drain voltage ($V_{sd}$) is applied to one contact and the current (A) is read out of the other. During EL in the $WSe_2$, electrons (blue) and holes (yellow) move towards each other (arrows) and recombine. The back gates are separated from the $WSe_2$ by *h*-BN. The device sits on a layer of silicon dioxide on a silicon substrate. (b) EL spectrum (blue) generated by a current of 5 nA closely resembles the photoluminescence spectrum (red) at 300 K. (c) EL intensity plot as a function of bias current and photon energy. From left to right, the arrows indicate the impurity-bound exciton ($X^I$), the charged excitons ($X^-$ then $X^+$) and the neutral exciton ($X^0$).[465] (d) The upper panel is device structure of TMD electric-double-layer transistor (EDLT) under ambipolar charge accumulation. The bottom panel shows schematic band structure of EDLT induced p-i-n junction under equilibrium. (e) Voltage dependence of EL spectra of the device. (f) Upper: current dependence of total EL intensity extracted from (e). Middle: current dependence of EL polarization extracted from (e). Bottom: current dependence of external quantum efficiency extracted from (e).[466] (g) Schematic illustration of the $WSe_2/MoS_2$

vertical heterojunction p-n diode. (h) The EL spectra of a ML- $WSe_2/MoS_2$ heterojunction at different injection current. (i) The false colour EL image of the heterojunction device under an injection current of 100 μA. The purple dashed line outlines the ML-$WSe_2$, the blue dotted line outlines the $MoS_2$ and the golden solid line outlines the gold electrodes.[467] Reproduced with permission.[468]

Monolayer $WSe_2$ LED devices fabricated using this method demonstrate a total photon emission rate of ~16 million per second at the largest applied current of 35 nA, equivalent to one photon per 10^4 injected electron–hole pairs.[465] While the overall quantum efficiency is modest at 0.01%, the emission wavelength can be finely tuned across various exciton regimes, as illustrated in Fig. 4(c). This splitting gate concept has gained popularity in the realm of TMD-based LEDs. Ambipolar control using local gates allows monolayer $WSe_2$ devices to achieve both PN and NP configurations, exhibiting diode ideal factors (n = 1.9) and a notable rectification factor of $10^5$.[468] Despite a modest estimated EL quantum efficiency of ~ 1%, these devices demonstrate versatile functionalities. For light emission in monolayer TMDs, the ambipolar FET presents an alternative means to form a p-n junction within 2D materials.[469] By applying gate voltage, effective neutral doping of the TMDs occurs. Consequently, TMD-based FET devices operate in the ambipolar injection regime, allowing simultaneous injection of holes and electrons at the two opposite contacts. This achievement is exemplified by the light-emitting transistor based on monolayer $WSe_2$ devices, as illustrated in Fig. 4(d).[466] Notably, circularly polarized EL emanates from p-i-n junctions electrostatically formed in transistor channels (Fig. 4(e)). The unique property of this phenomenon is qualitatively explained by the in-plane electric field controlling the hole-electron overlap. The EQE exhibits variation from 0.002% to 0.06%, depending on samples (Fig. 4(f)). However, the efficiency of light emission from lateral p-n junction devices is restricted by the narrow one-dimensional junction interface, resulting in low LED efficiency. Therefore, the preference leans toward vertically stacking TMDs to form p-n junctions, offering an increased optical active area from a one-dimensional (1D) line to a 2D overlapping area. $MoS_2$ and $WSe_2$ vertical p-n junction LEDs are depicted in Fig. 4(g), where the p-n junction area covers the entire $MoS_2/WSe_2$ overlapping region.[467] The EL of this device reveals prominent band edge excitonic emission and notably enhanced hot-electron luminescence (Fig. 4(h)). However, the emitted light is localized at the overlapping area near the electrodes rather than covering the entire overlapping region, as illustrated in Fig. 4(i). This phenomenon can be attributed to the current injection in such p-n junction, limited by lateral contacts and lateral carrier transport due to the depletion of ultrathin TMDs.

To realize large and uniform area LEDs using this vertical p-n junction structure, achieving vertical current injection is crucial. Graphene, as a uniform conductor, can be employed in designing vertical p-n junction TMD LEDs. However, directly fabricating vertical contacts on both top and bottom TMDs could result in significant current leakage through ultrathin 2D materials. Using a quantum well structure and inserting a thin dielectric material between graphene and TMDs will substantially minimize current leakage, thereby increasing the efficiency of LEDs.

### 7.3.2 LEDs from 2D Perovskite Material

Perovskite, currently a focal point in academia and industry, holds promise for the next generation of optoelectronic devices. In contrast to its conventional three-dimensional counterpart, 2D perovskite films are increasingly intriguing due to their unique structural characteristics and distinct performance. While 3D organometal halide perovskites have been widely applied in optoelectronic applications for their exceptional electrical and optical properties, such as a broad and strong absorption spectrum, long charge diffusion length, high carrier mobility, and extended carrier lifetime, their bulk crystal lattice is susceptible to damage when exposed to high temperatures, high humidity, and ultraviolet irradiation.[470, 471] These conditions accelerate perovskite decomposition, generating reactants that rapidly degrade device performance.[470] Perovskite-based applications, including photodetectors, solar cells, FETs, and LEDs, demand high-performance devices. The performance of these devices is significantly influenced by perovskite material properties.[472] Therefore, the development of more appealing 2D films with enhanced humidity resistance, smaller grain size, and improved surface coverage is essential.[471] This approach offers considerable stability against humidity and temperature, making these films more suitable for use in optoelectronic applications.[470]

In recent times, considerable attention has been directed towards 2D and quasi-2D perovskites as light-emitting materials, aiming to enhance the luminescent efficiency of perovskite LEDs. These materials exhibit outstanding performance, marked by effective exciton recombination, high quantum yield, and improved thermal stability. The notable advantages position 2D perovskites with significant potential for various optoelectronic applications. Emphasizing this potential, Yang *et al.* highlight a remarkable enhancement in the external quantum efficiency of halide perovskite LEDs, reaching up to 14.36% when employing quasi-2D perovskites as emitters.[473] In contemporary research, organic-inorganic hybrid perovskites have garnered significant attention for their extensive exploration in LEDs.[474] This interest stems from their noteworthy PL efficiency and the capability to generate a broad spectrum of colours, spanning from visible to near-infrared.[475] A substantial focus has been placed on enhancing the performance of perovskite LEDs through the incorporation of 2D perovskite emitter layers.[476] Layered 2D perovskites exhibit favourable film processability coupled with excellent optical properties, making them promising candidates for LED fabrication. Recent publications have reported encouraging results in the realm of 2D or quasi-2D perovskite-based LEDs. [477,478]

In conclusion, the application of 2D materials in optoelectronic devices has emerged as a transformative field, showcasing remarkable progress and opening new frontiers in device design and performance. The unique structural and electronic properties of 2D materials, such as graphene and TMDs, have paved the way for advancements in diverse optoelectronic applications. In photodetectors, 2D materials exhibit exceptional sensitivity and responsivity, with the ability to detect light across a broad spectrum. Their tunable bandgaps and efficient charge transport properties make them ideal candidates for efficient solar cells. Additionally, the incorporation of 2D materials in LEDs and lasers has enabled precise control over light emission properties, including wavelength and intensity, offering enhanced performance and energy efficiency. The versatility of 2D materials in forming heterojunctions and stacking configurations has further fuelled innovation in the development of novel device architectures. Despite challenges such as scalability and fabrication techniques, ongoing research continues to address these issues, unlocking the full potential of 2D materials in optoelectronic devices. As the field evolves, the integration of 2D materials promises to revolutionize optoelectronics, leading to the creation of more efficient, compact, and flexible devices. The journey from fundamental research to practical applications underscores the significant impact of 2D materials in shaping the future landscape of optoelectronic technologies.

# CHAPTER 8

# ENERGY STORAGE AND CONVERSION

The depletion of fossil fuels and the challenges posed by climate change represent significant issues facing modern society. Embracing renewable energy conversion and storage technologies offers a clean and sustainable solution to reduce dependence on fossil fuels and mitigate environmental concerns. Essential to enabling a wide array of energy conversion and generation processes is the strategic design and advancement of innovative, cost-effective materials. In recent years, the field of energy storage and conversion has undergone a paradigm shift with the emergence of 2D quantum materials.

2D structures, also known as nanosheets or thin films, are characterized by their atomic-scale thickness and significantly larger lateral dimensions, often with a ratio between the in-plane *a-b* axes and the perpendicular *c*-axis exceeding 100.[479] When layered materials are reduced to single- or few-layer thicknesses, they exhibit unique electronic properties that differ dramatically from their bulk 3D counterparts, owing to quantum confinement effects.[480] These effects include the opening of band gaps and the ease of atomic doping, rendering such materials particularly intriguing for catalyst design. Each 2D material including graphene, TMDs, *h*-BN, black phosphorus, etc. exhibit distinct quantum phenomena and offers specific advantages for energy-related applications. Graphene, for instance, boasts exceptional electronic conductivity and mechanical strength, making it an ideal candidate for various energy storage devices such as supercapacitors and batteries. TMDs, on the other hand, exhibit tunable bandgaps and strong light-matter interactions, rendering them suitable for photovoltaic and photocatalytic applications. Additionally, their inherently large specific surface area positions them as promising contenders for uses in supercapacitors and as innovative battery electrode materials. For instance, graphene boasts a specific surface area of approximately 2600 $m^2/g$, while a 10-layer-thick flake offers around 260 $m^2/g$, placing 2D materials within a specific surface area range comparable to activated carbons.

In a nutshell, the unique quantum properties of 2D materials arise from their ultrathin nature, which leads to quantum confinement effects, large surface-to-volume ratios, and enhanced quantum coherence. These features enable efficient charge and energy transport, catalytic activity, and electrochemical reactions, thereby enhancing the performance of energy storage and conversion devices. Moreover, the atomically precise control over the structure and composition of 2D materials allows for tailored design and optimization of their properties for specific energy applications. This chapter aims to provide a comprehensive overview of the role of 2D quantum materials in energy storage and conversion technologies. We explore their fundamental properties and recent advances in device fabrication and performance. Additionally, we discuss key challenges and future directions in harnessing the potential of these materials for addressing global energy needs. From batteries and supercapacitors to hydrogen evolution and fuel cells, the integration of 2D quantum materials promises to drive innovation and unlock new possibilities in the quest for sustainable energy solutions.

## 8.1 Supercapacitors

Energy storage plays a pivotal role in advancing renewable energy technologies. [481, 482, 483, 484] Ultracapacitors, also known as supercapacitors or electrochemical capacitors, are increasingly prominent as short-term, high-power-density storage solutions. They fill the gap between applications requiring electrolytic capacitors (e.g., AC to DC conversion)[485,486] and those necessitating batteries.[487] Supercapacitors boast significantly higher capacitance compared to conventional electronic circuitry or power supplies for such circuits. With over 40 years of commercialization, these devices find widespread use in various applications, including consumer and commercial transportation vehicles like trains and automobiles,[488 489] as well as in handheld power tools and lighting products.

There are two primary categories of these capacitors: (1) Electrical Double Layer Capacitors (EDLCs), and (2) Pseudocapacitors (PSCs).[490] EDLCs operate by forming high-density ion double layers at electrodes when a polarizing potential is applied. This mode of energy storage is well-established in colloid and surface chemistry, where ions surround electrodes subjected to a potential. PSCs also incorporate or exhibit a capacitance component similar to that of EDLCs. Additionally, they employ faradaic redox processes to store extra energy, often the predominant component, by capturing reduced or oxidized ions.[491] They store energy through ions' electro-sorption to surface sites (which involves potential-dependent coupling between double-layer charging and faradaic surface processes[492]) and by intercalating ions via activated or facile diffusion into layered 2D materials. The distinction between supercapacitors and batteries can blur when relying on faradaic redox chemistry to provide energy, as is also the case with batteries. PSCs serve as an intermediate device between EDLCs and batteries. Progress in PSC technology is anticipated to yield benefits in battery technology as well.

### 8.1.1 2D Materials in Supercapacitors

EDLCs closely resemble electrolytic capacitors as they rely on a dielectric medium facilitating ion diffusion and migration, often displaying polarity. Electrode materials must possess reasonable electrical conductivity, with the overall capacitance directly proportional to the total electrode surface area within the EDLC. Typical values of 20.0 $\mu Fcm^{-2}$ are achieved for activated carbons and related nanocarbons, a benchmark also observed for graphene.[493] Should both sides of single-layer graphene be efficiently utilized, its exceptionally high specific surface area (2630.0 $m^2 g^{-1}$) suggests an upper limit of approximately 550.0 F $g^{-1}$ for a graphene-based EDLC. This assessment serves as the upper boundary for EDLC charge storage established on 2D materials.[494]

When referring to graphene in supercapacitor formulations, it typically indicates reduced graphene oxide (rGO), obtained through the reduction of graphene oxide (GO). While some electrodes incorporate GO, their electrical conductivity tends to be limited. Traditional methods for preparing rGO supercapacitors, such as gas-phase hydrazine reduction[495] and THF solution borane reduction[496] of GO to rGO (resulting in B-doped rGO), have demonstrated specific capacitances of up to 205.0 F $g^{-1}$ at discharge rates of 0.5 A $g^{-1}$. Alternatively, microwave expansion and ammonia reduction, producing N-doped rGO, have shown capacities of 261.0 F $g^{-1}$ at 50.0 A $g^{-1}$.[488,489] This enhanced performance is attributed to factors including low oxygen content, high specific surface area (1000 $m^2$ $g^{-1}$), and good conductivity (15.0 S $cm^{-1}$).

A notable application showcasing the fabrication of EDLCs using GO-based films involved the creation of rGO on a commercial DVD drive through laser scribing to photoreduce GO to rGO.[494] By employing an organic electrolyte (Et4NBF4/$CH_3CN$) and an ionic liquid electrolyte (EMIMBF4), these EDLCs achieved potential windows of 3.0-4.0 V and exhibited a specific capacitance of approximately 270.0 F $g^{-1}$, retaining over 100.0 F $g^{-1}$ when operated at 1.0 kA $g^{-1}$. Utilizing a flexible PET substrate, these devices demonstrated robust EDLC operation even under severe bending, indicating their potential in printed electronics applications. An extension of this research led to the development of planar interdigitated electrode supercapacitors boasting high power density (200.0 W/$cm^3$) and exceptional storage and discharge characteristics (ranging from 2.3 mF $cm^{-2}$ @ 16.8 A $cm^{-2}$ to 1.4 mF $cm^{-2}$ @ 18.4 A $cm^{-2}$; approximately 1100.0 A $g^{-1}$).[497] A PVA/H2SO4 electrolyte-gel served as the solid-state electrolyte, applied as a thin coating over these interdigitated rGO electrodes. Alternatively, employing a fumed silica/ionic liquid (1-butyl-3-methylimidazolium (CF3SO2)2N-, BMIm (CF3SO2)2N) ionogel as the electrolyte coating expanded the potential window from 1.0 V to 2.5 V, while maintaining similar storage and discharge performance as reported for the PVA/$H_2SO_4$ electrolyte-gel. This microminiaturization led to a substantial 100-fold increase in specific surface area performance, elevating it from $2 \times 10^{-5}$ F $cm^{-2}$ to $2 \times 10^{-3}$ F$cm^{-2}$. Organic conjugated materials, particularly conjugated polymers, exhibit intriguing pseudocapacitive properties attributable to their redox reactions. Numerous conjugated polymers such as PAn, polypyrrole and polythiophene, serve as viable electrode materials in supercapacitors.[498] Hence, the fabrication of composites comprising graphene and conjugated polymers presents a promising avenue for designing high-performance electrode materials that exploit pseudocapacitance in supercapacitors. Geng *et al.*[499] synthesized nanostructured PAn Fibers within macroporous graphene thin films (MGTFs) deposited on flexible graphene papers (GPs) via in situ electropolymerization. As depicted in Figure 1 a, b, the resulting PAn@MGTF@GPs hybrids integrate the macroscale pores provided by graphene sheets with the nanoscale pores contributed by PAn Fibers. These structural characteristics enable the utilization of both electrical double-layer capacitance from graphene and pseudocapacitance from PAn. Figure 1 c presents a comparison of the cyclic voltammetry (CV) curves obtained at a scan rate of 10 mV $s^{-1}$ in a 1 M H2SO4 electrolyte for PAn@MGTF@GPs and control samples lacking either PAn (referred to as MGTF@GPs) or macroporous graphene (designated as PAn@GPs). Notably, PAn@MGTF@GPs exhibit superior capacitive performance, with a stable area-specific capacitance of 538 mF $cm^{-3}$, surpassing that of the control samples.

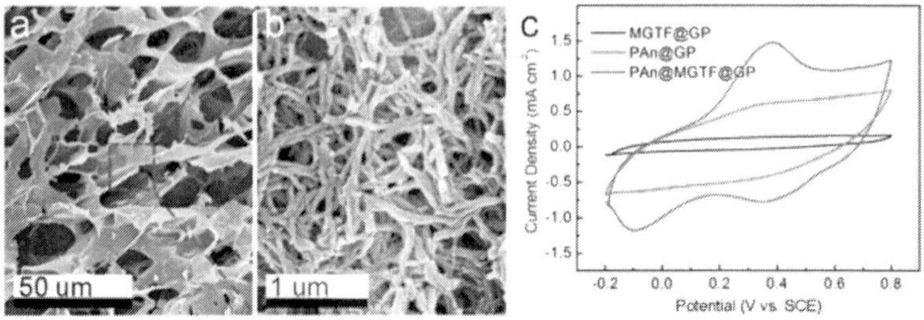


Figure 1. (a, b) Scanning electron microscopy images of the PAn@MGTF@GPs prepared by electrodeposition of polyaniline (PAn) into MGTF@GPs. (c) CV curves recorded for MGTF@GP, PAn@GP, and PAn@MGTF@GP electrodes. Reproduced with permission.[537] Copyright 2016 American Chemical Society.

Titanium carbides, known as $Ti_3C_2T_x$ ($T_x$ represents surface functional halides, hydroxides, or oxygens[500]), belong to the MXene family and possess sufficiently high conductivity to serve as electrodes. They can be polarized adequately to form EDLCs.[501502] $Ti_3C_2T_x$, derived from $Ti_3AlC_2$ through HF etching, constitutes a multi-nanosheet material capable of intercalating various cations to create EDLCs.[503,504] When prepared with carbon black to enhance intersheet conductivity and PVDF fibers as binders,[503] $Ti_3C_2T_x$ electrodes achieved specific capacitances of up to 130.0 F $g^{-1}$ in various alkali hydroxide and sulphate aqueous electrolytes, particularly at low scan rates.[504] Additionally, a $Ti_3C_2T_x$ paper electrode, fabricated via vacuum filtration of a nanosheet suspension, exhibited a specific capacitance of approximately 200.0 F $g^{-1}$ in aqueous KOH and other electrolytes. Despite the intrinsic dependence of high discharge capability on inter-sheet cationic intercalation, an apparent paradox was resolved by considering the effective inter-sheet ionic diffusion, likely facilitated by solvation structures between the sheets.[504] Recent advancements have led to the development of transparent capacitor symmetric electrodes based on $Ti_3C_2T_x$, featuring flexible electrode structures with an area-capacitance of 1.6 mF $cm^{-2}$ and remarkable cycle life exceeding 20,000 cycles.[505]

The same MXene material ($Ti_3C_2T_x$) exhibited even higher volumetric capacitance in an acidic electrolyte ($H_2SO_4$), reaching up to 900 F·$cm^{-3}$, with long-lasting stability observed over 10,000 life cycles, as depicted in Figure 2 a,b.[506] The theoretical underpinning for this amazing capacitance was elucidated using density functional theory, coupled with an ion-solvation model, to predict the redox mechanism of MXene electrodes in 1 M $H_2SO_4$ electrolyte. According to this model, the charge storage mechanism was proposed to follow the reaction:

$$Ti_3C_2O_2^q + 2H^+(aq) \rightarrow Ti_3C_2O_{2-2x}(OH)_{2x}^q + 2(1-x)H^+(aq)$$

Where q represents the net charge and x denotes the surface hydrogen coverage ranging from 0 to 1.[507] In contrast, nonaqueous (organic) electrolytes offer a higher operating potential window due to the absence of water molecules in the electrolysis reactions. MXenes composited with carbon nanotubes (MXene/CNT) were evaluated as active electrodes in a 1 M 1-ethyl-3-methylimidazolium bis(trifluoromethylsulfonyl)imide ionic electrolyte solution (Figure 2 c), demonstrating a volumetric capacitance of 245 $F \cdot cm^{-3}$ within a potential window of 1.8 V.[508] Lin *et al.* reported an increase in the voltage window to 3 V for MXenes with the same ionic electrolyte.[509] The large electronegativity of MXenes could enhance the communal interaction of imidazolium cations, thereby facilitating higher potential windows.[508,509] Despite the appreciable capacitance values observed in nonaqueous electrolyte solutions, a three- to four-fold decrease compared to acidic/aqueous electrolyte solutions was noted. This reduction can be attributed to the absence of water molecules, protons, and other constituents in nonaqueous electrolyte solutions, leading to decreased activation of the electrode surface in supercapacitor applications.[510] The drawback associated with nonaqueous electrolytes poses limitations on the utilization of MXenes in supercapacitors. Hence, careful selection of electrolytes is crucial to optimize the practical applications of MXenes.

Significant applications of TMDs in thin film and flexible electronics include those involving $MoS_2$[511] and $VS_2$[512]. The use of organolithium chemistry has facilitated the exfoliation of $MoS_2$ nanosheets in both aqueous salt solutions and organic solvents. These nanosheets exhibit a substantial negative electrokinetic charge, which is believed to facilitate extensive alkali cation intercalation, resulting in a moderate capacitance in the range of approximately 100.0 F $g^{-1}$.[511] Two key aspects emerge from these findings: Firstly, the $MoS_2$ nanosheets obtained are predominantly of the 1T phase, known for their electrical conductivity (ranging from 10.0 to 100.0 S $cm^{-1}$). Secondly, in $CH_3CN$, the electrochemical window accessible is approximately 3.0-3.5 V. Detailed mechanistic discussions and studies suggest that these $MoS_2$ materials also function as PSCs, involving faradaic processes.[513] Flexible thin film coatings on polyimide substrates have been successfully demonstrated. Similarly, flexible thin-film interdigitated single-layer capacitors composed of $VS_2$ were coated onto cellulose membranes, enabling transfer to various flexible and rigid substrates.[512] Gold current collectors were subsequently deposited, followed by the deposition of a dielectric film comprising a PVA/ionic liquid gel. The obtained devices exhibited a moderate volume capacitance of around 317.0 F $cm^{-3}$ and an area capacitance of about 4760.0 Mf $cm^{-2}$, suggesting potential practical applications for such materials.

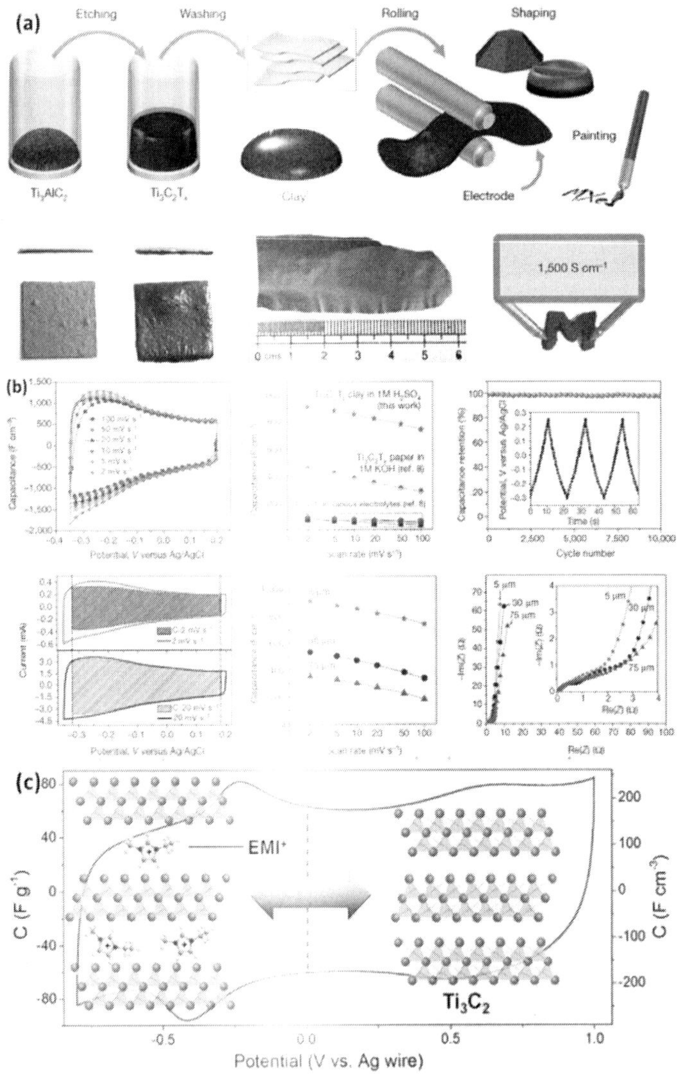


Figure 2. (a) Schematic of MXene clay synthesis and electrode preparation. (b) Electro-chemical performance of rolled, fee-standing electrodes (MXenes) in an acidic electrolyte. (c) Graphical representation of $Ti_3C_2T_x$ in a nonaqueous electrolyte solution. Reproduced with permission.[508]

The solvothermal synthesis of β-NiS hierarchical flower-like structures was conducted (Figure 3) utilizing diethanolamine as both the coordination agent and the solvent.[514] These synthesized materials were engineered as electrodes for supercapattery applications and evaluated in a 2 M KOH electrolyte, exhibiting a maximum specific capacitance of approximately 857 $F \cdot g^{-1}$. However, the cycling stability of the materials proved inadequate, with a specific capacitance retention of merely 40% after 1000 cycles, attributed to the phase transformation of NiS to $Ni(OH)_2$, an irreversible electrode reaction, and alterations in the morphology of the materials. Another approach involved the in situ growth of β-NiS on a $Ni_3S_2$ porous network synthesized via solvothermal methods, targeting supercapattery applications.[515] Electrodes derived from this material showcased a maximum specific capacitance of 1158 $F \cdot g^{-1}$ and an enhanced specific capacitance retention of 97.4% after 2000 charge-discharge cycles in a 6 M KOH electrolyte. Assembling an asymmetric supercapacitor with this composite electrode as the positive electrode and activated carbon as the negative electrode yielded a notable energy density of 55.1 $Wh \cdot kg^{-1}$, highlighting its viability for commercial supercapattery electrode utilization. Additionally, various NiS composites such as NiS/rGO,[516] NiS/N-doped carbon,[517] NiS/$Ni_3S_4$,[518] NiS/CoO,[519] and NiS/NiO,[520] have been synthesized utilizing diverse techniques and applied in supercapattery applications.

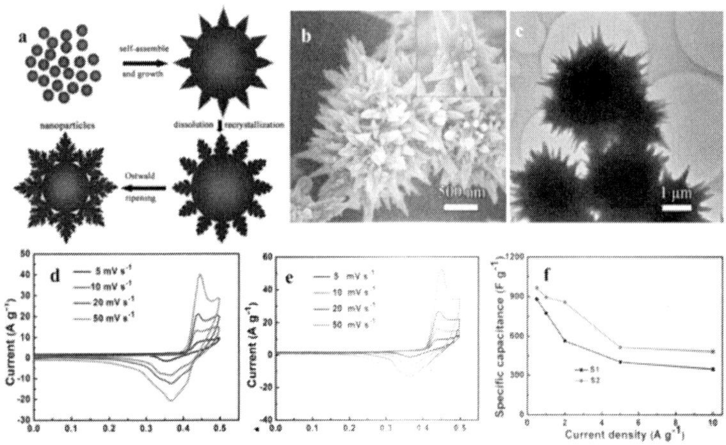

Figure 3. (a) Schematic of hierarchical structure development of β-NiS flower-like structures. (b) SEM and (c) TEM images of the NiS structure. (d, e) CV plots of NiS electrodes calcined at different temperatures. (f) Variation of specific capacitance of the NiS structure at different current densities. Reproduced with permission.[508]

$NiCo_2O_4$, a ternary metal cobaltite, underwent examination as a prospective electrode material for Supercapattery. By replacing oxygen with Sulfur, the chemical bonds of $NiCo_2O_4$ were elongated, aiming to engineer a more flexible structure conducive to enhanced ion transport and electrochemical storage capabilities.[521] This substitution led to a reduction in the material's optical band and a significant increase in electrical conductivity, which escalated by two orders of magnitude compared to pristine $NiCo_2O_4$.[522] A comparative investigation contrasted the direct growth of $NiCo_2S_4$ nanotubes on nickel foam with $NiCo_2O_4$ nanowires synthesized via the hydrothermal method. Results favoured $NiCo_2S_4$ nanotubes, exhibiting a superior performance with a specific capacitance of 1936 $F \cdot g^{-1}$ compared to 1370 $F \cdot g^{-1}$ for $NiCo_2O_4$ nanowires when employed as supercapattery electrodes.[523] Furthermore, CuS2, another chalcogenide, served as the sulfur source in a one-step hydrothermal synthesis yielding mesoporous $NiCo_2S_4$ nanosheets designed for supercapattery applications.[524] These nanosheets were employed as positive electrodes in a $NiCo_2S_4$//AC asymmetric supercapacitor configuration utilizing a 3 M KOH electrolyte. This assembly achieved a peak energy density of approximately 26 $kW \cdot kg^{-1}$ within a voltage window of 1.6 V, while retaining 86% of its initial specific capacitance following 4000 cycles. Additionally, hollow mesoporous $NiCo_2S_4$ ellipsoids, fabricated through calcination and chemical transformation, were subjected to electrochemical characterization as supercapattery electrodes, demonstrating a capacitance of 495 $F \cdot g^{-1}$.[525]

PSCs offer superior energy storage densities compared to EDLCs,[526] albeit with power density constraints imposed by electrical conductivity. Recent advancements in PSC technology often involve the integration of conductive EDLC nanosheets as substrates for novel inorganic 2D materials, combining electrosorption, ion intercalation, and redox chemistry. Among the most effective PSCs are those incorporating rGO, demonstrating specific capacitances ranging from 361.0 $F \cdot g^{-1}$ at 2.0 $A \cdot g^{-1}$ to 3344.0 $F \cdot g^{-1}$ at 1.25 $A \cdot g^{-1}$.[527,528] Hybrids such as $MnO_2/Ni(OH)_2$ on rGO exhibit remarkable charge storage capacities of 2628.0 $F \cdot g^{-1}$ at 3.0 $A \cdot g^{-1}$ and 1328.0 $F \cdot g^{-1}$ at 20.0 $A \cdot g^{-1}$.[529] PSCs are poised to offer substantially higher energy storage capabilities compared to EDLCs, bridging the technological gap between traditional capacitors and batteries. Moreover, PSCs are being explored in flexible electronics, with $VOPO_4$/rGO compositions on PET reaching 8.4 mF $cm^{-2}$ @ 20.0 μA $cm^{-2}$.[530] Several notable electrode materials have emerged, including $Co_3O_4$/Ti with a specific capacitance of 1033.0 $F \cdot g^{-1}$ at 2.5 $A \cdot g^{-1}$,[531] $(NiCo)_3BO_3/Ni(OH)_2$/Ni foam with capacities ranging from 1982.0 $F \cdot g^{-1}$ @ 0.5 $A \cdot g^{-1}$ to 1445.0 $F \cdot g^{-1}$ at 10.0 $A \cdot g^{-1}$,[532] $Ni_3S_2$/rGO heterostructures delivering 1886.0 $F \cdot g^{-1}$ at 1.0 $A \cdot g^{-1}$ to 1621.0 $F \cdot g^{-1}$ at 20.0 $A \cdot g^{-1}$,[533] and $Ni_3S_2$/Ni foam with capacities ranging from 2900.0 $F \cdot g^{-1}$ at 2.0 $A \cdot g^{-1}$ to 1200.0 $F \cdot g^{-1}$ at 50.0 $A \cdot g^{-1}$.[534] While MXenes show promise, their structures pose challenges in facilitating facile faradaic redox transformations, limiting their applicability in PSCs.[506,535]

## 8.2. Batteries

In contrast to the high-power density of supercapacitors, batteries are known for their high energy densities. Rechargeable batteries, in particular, offer resource conservation benefits along with low costs, sustainability, and minimal environmental impact compared to primary batteries. Hence, rechargeable batteries hold immense promise in various applications such as portable electronic devices, electric vehicles, and stationary battery energy storage systems for smart grids. Among rechargeable batteries, lithium-ion batteries (LIBs) emerged as the most successful version post their commercialization by Sony in 1991. As illustrated in Figure 4, LIBs store energy through reversible Li-ion intercalation and deintercalation into and from cathodes, accompanied by similar processes on anodes. The voltage output of LIBs is determined by the difference in the electrochemical potentials of cathode and anode materials. Energy densities are correlated with the specific capacities of cathode and anode materials, while their rate performances indicate the intercalation and deintercalation rates of Li ions. Commonly used cathode and anode materials in commercial LIBs include $LiCoO_2$, $LiFePO_4$, and graphite, respectively. However, their low theoretical specific capacities (273, 170, and 372 mA h $g^{-1}$ for $LiCoO_2$, $LiFePO_4$, and graphite) limit current LIBs in terms of energy densities, cycle lives, and charge rates. Consequently, extensive research efforts have focused on designing novel cathode materials with larger specific capacities (>200 mA h g-1) or higher electrochemical potentials (>4 V vs. Li/Li+). Promising cathode materials include nickel-rich layered oxides ($LiNi_{1-x}M_xO_2$, M = Co, Mn, Al) and lithium-rich layered oxides ($Li_{1+x}M_{1-x}O_2$, M = Mn, Ni, Co). Additionally, numerous 2D nanomaterials have demonstrated potential as anode materials due to their unique structural attributes, including ultra-thin thickness, large lateral size, and high specific surface area. Leveraging the electrochemistry in LIBs, 2D nanomaterials are expected to enhance electrode reactions and reduce ion diffusion lengths in electrodes,[536] thereby enhancing battery performance.

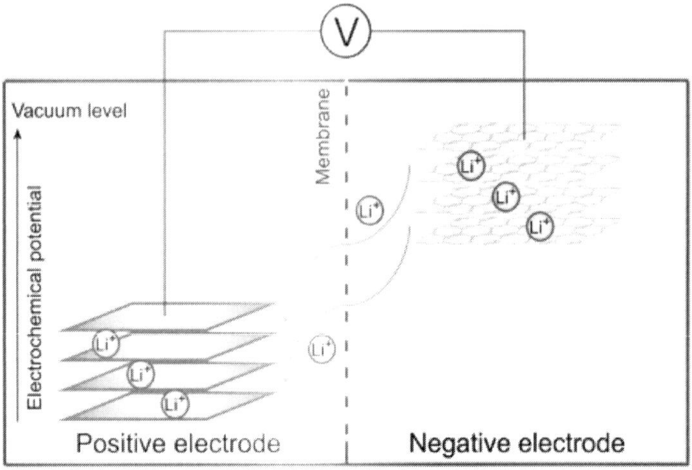


Figure 4. Schematic representation and operating principle of lithium-ion batteries. Reproduced with permission.[537] Copyright 2020 American Chemical Society.

In comparison to traditional lead-acid and nickel-cadmium batteries, LIBs offer higher energy density, longer cycle life, and reduced weight, making them more compact. However, concerns regarding cost, safety, and the availability of lithium may hinder widespread adoption for large-scale applications. Sodium-ion batteries (SIBs) present a promising alternative due to the abundance, high theoretical capacity, and lower cost of sodium. Despite these advantages, the larger ionic radius of $Na^+$ (1.02 Å vs. 0.76 Å) and it's lower reduction potential (0.3 V higher than that of lithium) result in slower ion transport and lower gravimetric and volumetric energy densities.[538] 2D nanosheets with expansive surface areas offer plentiful active sites (such as interlayer regions, surface/interlayer interfaces, and edges) for cation storage, leading to higher specific capacity compared to 0D and 1D materials in many instances. Moreover, the surface adsorption processes of 2D nanosheets typically involve low kinetic barriers. Additionally, their atomic-thin nature provides excellent mechanical flexibility, making them suitable for the development of thin, flexible, and stretchable batteries, as well as roll-to-roll coating processes. Indeed, 2D nanosheets exhibit promising performance in rechargeable batteries, demonstrating improved rate capability and higher cycling stability compared to bulk counterparts.[539,540]

### 8.2.1. 2D Materials in Batteries

Few-layer metal dichalcogenides, including $MoS_2$, $MoSe_2$, $WS_2$, $WSe_2$, $TiS_2$, $ReS_2$, and $VS_2$, exhibit promising potential as alternative electrode materials for LIBs.[541,542] Among these, 2D $MoS_2$, a typical TMD material, demonstrates an enhanced lithium storage capacity, both theoretically (ranging from 669.0 to 1675.0 mAh g−1) and experimentally (up to approximately 900.0 mAh g−1), in comparison to graphite.[543] $MoS_2$ nanosheets exhibit higher coulombic efficiency as an anode material compared to graphene oxide and its derivatives in LIBs. However, $MoS_2$ faces challenges related to its low electronic conductivity, leading to poor cycling stability, and a polysulfide shuttling effect that accelerates electrode degradation. These challenges can be mitigated by integrating $MoS_2$ nanosheets with graphene. Composite materials such as $MoS_2$/rGO have demonstrated first-cycle discharge capacities of up to 2200.0 mAh g−1, with minimal capacity degradation observed after more than 100 cycles and high coulombic efficiency. The reversible capacities of TMD-graphene composites can be adjusted within the range of 600.0 to 1546.0 mAh g−1, depending on factors such as the TMD component, content, and fabrication methods. Heterostructures like the 2D/2D $MoS_2$/$WS_2$ exhibit enhanced lithium-ion binding and mobility.[544] Additionally, $VS_2$, a metallic TMD, possesses low diffusion barriers, with a monolayer of $VS_2$ predicted to adsorb three layers of lithium stably, resulting in a theoretical capacity of 1397.0 mAh g−1.[545] Predictions suggest that monolayers of group IV monochalcogenides (SiS, SiSe, GeS, GeSe, SnS, and SnSe) could serve as potential candidate anode materials for LIBs. These materials are characterized by low diffusion barriers, moderate to high specific capacities, low open-circuit voltages, small volume changes, and favourable electrical conductivities.[546]

Nanosheets composed of transition metal oxides, including $Co_3O_4$,[547] NiO,[548] $FeO_x$,[549] $MnO_2$ (based on conversion reactions),[550] $VO_x$,[551] $TiO_2$ (based on intercalation-de-intercalation),[552] and $SnO_2$ (based on conversion and alloying reactions),[553] have garnered considerable attention as potential anode materials in LIBs. Their appeal lies in their high theoretical capacities, cost-effectiveness, low environmental impact, and abundance in nature. Among these metal oxides, $Co_3O_4$ nanosheets have demonstrated notable rate capability and cycling stability. However, VOx, $TiO_2$, and $SnO_2$ nanosheets exhibit lower theoretical capacities. Additionally, $SnO_2$ and FeOx nanosheets undergo significant volume changes during cycling, which can lead to anode pulverization and capacity degradation. Transition metal oxide $MO_2$ (M = Mn, Co, and Ni) nanosheets have also been identified as potential cathode materials for LIBs,[554] with $MnO_2$ showing the highest specific density and smallest volume expansion (<0.35%). Other binary metal oxide nanosheets reported as electrode materials for rechargeable batteries include $Cr_2O_3$,[555] CuO,[556] $MoO_2$,[557] and ZnO.[558] The incorporation of holey 2D morphologies appears to enhance battery performance by preserving structures and facilitating ion transport. This concept has been demonstrated with 2D pierced $ZnMn_2O_4$ nanosheets, which maintained a stable specific capacity of 480.0 mAh $g^{-1}$ with an average Coulombic efficiency of 99.8% after 1000 cycles at 800.0 mA g-1.[559] However, one major challenge when using metal oxide nanosheets as electrode materials in LIBs is their inherently low conductivity. To address this issue, researchers have explored combining oxide nanosheets with conductive graphene materials.

Unadorned MXenes exhibit metallic conductivity and possess low barriers for lithium migration, showcasing promising potential for lithium storage. The surface terminations of MXenes play a crucial role in determining initial coulombic efficiency, electronic conductivity, and Li-ion mobility. MXenes terminated with OH or F groups often exhibit semiconductor behavior with narrow band gaps. However, the presence of surface $F^-$ and $OH^-$ groups may hinder the transport of $Li^+$ ions, resulting in a substantial initial irreversible capacity and restricted reversible capacity. [560,561] Strategies involving the surface modification of MXenes with metal ions,[562] metal oxides,[563] or in-situ generated $TiO_2$ [564] offer pathways to enhance $Li^+$ ion storage. A novel MXene incorporating hafnium ($Hf_3C_2T_z$) was recently introduced, demonstrating a notable reversible volumetric capacity of 1567.0 mAh $cm^{-3}$ for LIBs at a current density of 200.0 mA g-1 after 200 cycles, while maintaining volume stability.[565]

2D transition metal borides (e.g., $Mo_2B_2$, $Fe_2B_2$) have been demonstrated to possess a uniformly low diffusion energy barrier and a high capacity for lithium atom storage, suggesting their potential as electrode materials in lithium-ion batteries (LIBs). The theoretical specific capacities of 2D $Mo_2B_2$ and $Fe_2B_2$ were estimated to be approximately 444.0 and 665.0 mAh $g^{-1}$, respectively.[566] On the other hand, 2D metal nitrides (such as GaN) are also being explored as promising electrode materials for electrochemical storage devices due to their excellent electrical conductivity, high theoretical capacity, and low, consistently stable potentials that closely resemble those of lithium metal.[567]

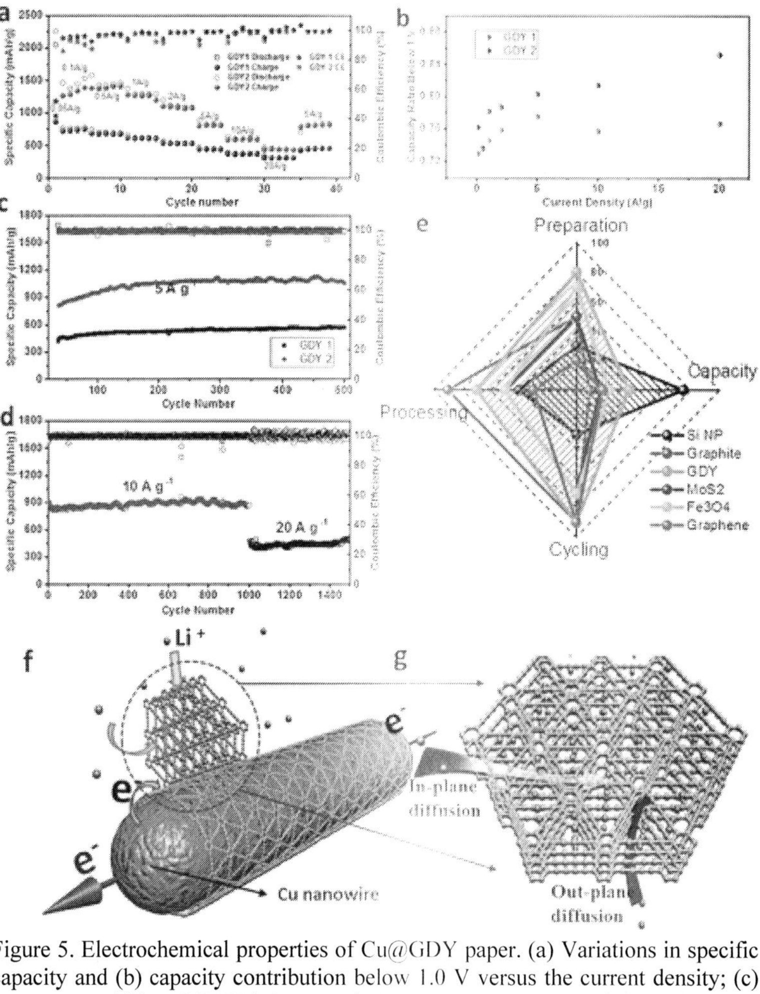


Figure 5. Electrochemical properties of Cu@GDY paper. (a) Variations in specific capacity and (b) capacity contribution below 1.0 V versus the current density; (c) long-term stability of GDY 1 (unsmooth GDY coats on Cu nanowires) and GDY 2 at 5.0 A g−1 and (d) GDY 2 at 10.0 and 20.0 A g−1. (e) Overall comparisons with other anodes. (f) and (g) Schematic representation of a possible mechanism for the high rate performance. Reproduced with permission.[570]

Graphdiyne (GDY), a novel artificial carbon allotrope, features a 2D framework with nano-pores arranged in-plane.[568] Computational simulations suggest that GDY serves as an effective intercalation electrode, offering a theoretical storage capacity of 740.0 mAh g-1 by forming C3Li, which traps Li+ ions within its in-plane nanopores and minimizes volume changes.[569]

Experimental electrochemical assessments demonstrated that GDY nanosheets (with an average thickness of approximately 1.9 nm) grown on Cu nanowires (referred to as GDY 2 in Fig. 5) exhibited a capacity of 1388.0 mAh g-1 at 0.1 A g-1 after activation, surpassing the performance of numerous graphene-based materials and transition metal oxides.[570] This value approaches double the theoretical capacity of GDY (740.0 mAh $g^{-1}$), suggesting a model of lithium ion storage on both sides. GDY2 demonstrated outstanding rate capability, retaining reversible capacities of 824.0, 596.0, and 449.8 mAh $g^{-1}$ at 5.0, 10.0, and 20.0 A $g^{-1}$, respectively (Fig. 5a). The primary contributions to capacity from GDY were predominantly observed below 1.0 V (vs. Li/ Li+) (Fig. 5b). Moreover, GDY 2 exhibited robust capacity retention even after 500 cycles (Fig. 5c and d). Comparative analyses with several well-studied anodes, as depicted in Fig. 5e, indicate that GDY holds promise in achieving a balance among its performance, fabrication, and processing. The exceptional Li-ion storage properties of GDY are attributed to its 3D transfer channels facilitated by tubular GDY and ultrathin GDY nanosheets grown on the Cu nanowires, effectively shortening ion and electron transfer distances (Fig. 5f and g).

Black phosphorus (BP), the most stable allotrope of phosphorus from a thermodynamic standpoint, boasts considerable electrical conductivity, approximately 300.0 S $m^{-1}$, and offers a high theoretical specific capacity of 2596.0 mAh $g^{-1}$ by absorbing lithium to form $Li_3P$. However, BP faces challenges such as low reversible capacity, rapid capacity degradation due to substantial volume expansion of 300.0% and structural cracking during lithiation.[571] While prone to oxidation in the presence of air, BP can be dispersed and stabilized in oxygen-free water.[572] Notably, single- and double-layered BP can maintain their layered structures with minimal volume changes (0.2%) during lithiation and delithiation.[573] Theoretical capacities for single-layer and double-layer BP are estimated at 432.79 and 324.59 mAh g-1, respectively.[574] Particularly noteworthy is the rapid diffusion of lithium along the zigzag direction of BP, with diffusivity ranging from 102 to 104 times faster than that of graphene and MoS2. Additionally, phosphorene exhibits a shallow lithium migration barrier (0.08 eV) but a relatively high average voltage of 2.9 V. Hybridizing phosphorene with graphene[575,576] or *h*-BN [577] can further enhance lithium binding energy, air stability, and retention capability.

Utilizing graphene, whether as a cathode or an anode, facilitates swift electron and ion transportation within electrodes, thereby enabling rapid charging and discharging processes. When employed as an anode material for LIBs, graphene exhibits a theoretical specific capacity of 744.0 mAh g-1, with lithium ions binding on both sides of graphene to form $Li_2C_6$.[578] RGO has been shown to achieve initial discharge capacities ranging from 540.0 to 1600.0 mAh g-1 at 50.0 mA g-1, a range dependent on factors such as specific surface area, interlayer spacing, presence of disordered structures, defects, edges, chemical functionalities, and textural properties.[579] Defects and edges are conducive to lithiation and delithiation reactions, acting as storage sites for lithium ions and thereby enabling a specific capacity higher than the material's theoretical value.

Introducing heteroatoms (N, B, S, P, and F) into graphene to modify its atomic charge density, spin density, and interlayer distance can expose more active sites, facilitating rapid surface adsorption and diffusion of lithium ions as well as electron transport. This enhancement leads to improved battery capacities, surpassing 1000.0 mAh $g^{-1}$ at 50.0 mA $g^{-1}$.[580] While N, S, and P substitutions generate electron-rich structures, B doping induces an electron-deficient lattice.[581] B-doped graphene exhibits the lowest potential barrier for lithium diffusion among these four types of doped graphene, boasting a very high theoretical capacity of 2271.0 mAh g-1 for lithium storage.[582] Monolayer BC2 demonstrates a greater number of empty states above the Fermi level, lower diffusion energy barriers, and open circuit voltage compared to other BCx graphene-like carbon sheets, resulting in enhanced Li kinetics.[583] Furthermore, codoping graphene with multiple heteroatoms, such as B/N,[584] P/N,[585] S/N,[586] F/N,[587] and Ge/N,[588] to create synergistic effects can boost capacity reversibility and rate performance.

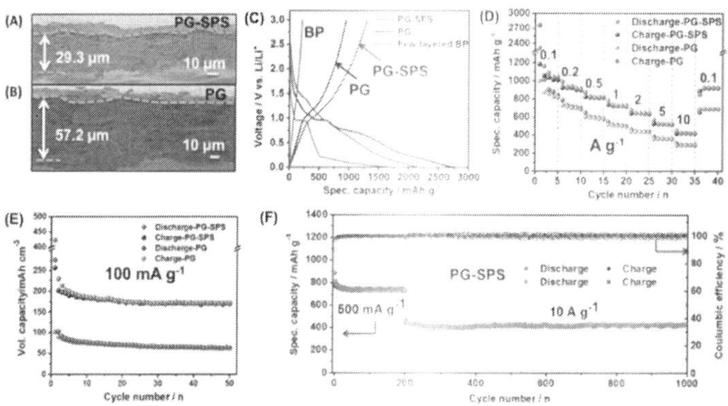


Figure 6. Cross-sectional SEM images of (A) PG-SPS and (B) PG electrodes on Cu foil with a same mass loading (0.65 mg $cm^{-2}$). (C) The first galvanic charge/discharge measurements of few-layered BP, PG, and PG-SPS at 100.0 mA $g^{-1}$. (D) Rate capabilities of PG and PG-SPS electrodes at various current densities.

(E) Cycling life of PG and PG-SPS electrodes based on volumetric capacities at 100.0 mA $g^{-1}$. (F) Long-term cycle life of PG-SPS electrode at 500.0 mA $g^{-1}$ (200 cycles) and 10.0 A$g^{-1}$. Spec. capacity and Vol. capacity stand for specific capacity and volumetric capacity, respectively. Reproduced with permission.[576]

Incorporating (doped) graphene with electrochemically active materials can enhance the rate capability and cycling stability of LIBs. For instance, a 2D/2D $MoS_2$-S doped graphene composite demonstrated a reversible specific capacity exceeding 1500.0 mAh $g^{-1}$ at 0.1 A $g^{-1}$.[589] Recently, Yan and colleagues reported an improvement in first-cycle coulombic efficiency (60.2%) for a phosphorene–graphene composite fabricated via a spark plasma sintering process (PG-SPS), surpassing phosphorene (11.5%) and a loosely stacked phosphorene–graphene composite (PG: 34.3%). This composite also exhibited high gravimetric capacity (1306.7 mAh $g^{-1}$) and volumetric capacity (256.4 mAh $cm^{-3}$), excellent rate capabilities (415.0 mAh $g^{-1}$ at 10.0 A $g^{-1}$), and long-term cycling stability (91.9% retention after 800 cycles at 10 A $g^{-1}$), as illustrated in Figure 6.[576]

While strides have been made in utilizing 2D nanosheets as electrode materials for LIBs, several challenging issues persist for their integration into large-scale energy storage systems. A significant obstacle is the high first-cycle irreversible capacity and inadequate coulombic efficiency stemming from the formation of solid electrolyte interface layers. Safety concerns arise from the degradation of electrolytes, resulting in the production of lithium carbonate and organic components, particularly accelerated when employing 2D anodes. Moreover, the utilization of 2D nanosheets contributes to low packing density, limiting the volumetric energy densities of such systems, thereby posing challenges for portable electronic applications.

## 8.3 Hydrogen Production

Hydrogen ($H_2$) stands out as an optimal energy carrier due to its eco-friendliness - its oxidation by-product being $H_2O$, in contrast to CO, $CO_2$, and pollutants like sulphur and nitrogen oxides emitted by petroleum-based fuels. Additionally, hydrogen boasts a high energy capacity of 143.0 mJ $kg^{-1}$.[590] Recently, 2D materials have garnered increasing attention as electrocatalysts and photocatalysts for hydrogen evolution reactions (HER) owing to their advantages over conventional metal-based catalysts, such as being more cost-effective, abundantly available in the Earth's crust, and environmentally sustainable. The HER ( $2H^+ + 2e^- \rightarrow H_2$, in acidic electrolyte; $2H_2O + 2e^- \rightarrow H_2 + 2OH^-$, in alkaline electrolyte) unfolds at the cathode of an electrolyzer, generating high-purity hydrogen fuel from water. The process commences with the adsorption of a proton from acidic solutions onto the electrode surface (Volmer step), where it undergoes reduction to form $H_{ad}$. In alkaline media, water dissociation supplies a proton for adsorption ($H_2O + e^- \rightarrow H_{ad}^{+} + OH^- + e^- \rightarrow H_{ad} + OH^-$) that transforms into a hydrogen atom, $H_{ad}$. This $H_{ad}$ then either combines with adjacent adsorbed hydrogen atoms to yield molecular hydrogen (Tafel step, $2H_{ad} \rightarrow H_2$) or directly reacts with a hydrated proton from the acidic solution, along with an electron, to produce hydrogen gas (Heyrovsky step, $H_3O^+ + H_{ad} + e^- \rightarrow H_2 + H_2O$).[591] The rate-determining step following hydrogen adsorption can be discerned by the Tafel slope ($b$) of polarization curves,[591] correlating overpotential ($\eta$) to the logarithm of current density ($j$) through the Tafel equation ($\eta = b \log j + a$). A Tafel slope close to 40.0 mV $dec^{-1}$ indicates that the Heyrovsky step governs the rate, whereas a slope approaching 30.0 mV $dec^{-1}$ suggests the Tafel step is rate-limiting. In alkaline conditions, the HER's apparent activity is constrained by a high energy barrier for hydrogen adsorption, attributable to the additional water dissociation process or changes in entropic barriers across the water double layer.[592]

The Gibbs free energy of hydrogen adsorption ($\Delta G_{H^*}$) serves as a pivotal descriptor for acidic HER activity. Optimal HER catalysis hinges on hydrogen binding neither too strongly (signified by a highly negative $\Delta G_{H^*}$) nor too weakly (indicated by a highly positive $\Delta G_{H^*}$) to the electrode surface. Positioned at the peak of HER activity in a volcano plot, Pt boasts a $\Delta G_{H^*}$ close to zero (around 0.09 eV), showcasing exceptional HER performance characterized by a sizable exchange current density and a diminutive Tafel slope (30.0 mV $dec^{-1}$). Apparent alkaline HER activity is governed by three crucial factors: suitable energies for $H_{ad}$ and/or $OH_{ad}$ (neither excessively strong nor weak), along with a low barrier to water dissociation.

## 8.3.1 2D Materials for HER

A diverse array of 2D transition metal-based materials have emerged as compelling alternatives to noble metal catalysts for the HER. These encompass transition metals (Ni),[593] mixed metal chalcogenides (TMDs, GaS, SnS, $Ni_3S_2$, $Co_9S_8$, $NiCo_2S_4$, Fe-Ni sulphide, InSe, and IVA mono-chalcogenides),[594,595,596,597,598,599,600] oxides (like $WO_3$ and $\alpha$-$MoO_{3-x}$),[601602] hydroxides (such as NiCoFe hydroxide, $Ni(OH)_2$, and $Co(OH)_2$),[603,604,605] F-rich MXenes,[606] carbides (like WC,[607] TaC,[608] NbC),[609] nitrides (including

$NiMoN_x$,[610] $Cu_1Ni_2$-N,[611] and $Mo_5N_6$)[612], phosphides (like FeP, CoP, and NiP2),[613,614,615] mixed phosphosulfides like $Ni_{0.9}Fe_{0.1}PS_3$,[616] and trichalcogenide phosphites (like $MnPX_3$, where X = S or Se).[617] In particular, WO3 nanosheets rich in oxygen vacancies have demonstrated a low overpotential of 38.0 mV at -10.0 mA cm-2 and a Tafel slope of 38.0 mV dec-1, placing them as a potential alternative to Pt for HER electrocatalysis.[618] Moreover, few-layer pnictogens like As, Sb, and Bi have shown activity in HER, with Sb surpassing the others in terms of onset potential[619] (-0.68 V for Sb, -0.91 V for Bi, and -0.98 V for As vs. reversible hydrogen electrode (RHE)). However, the electrocatalytic performance of these new 2D monoelements falls short of commercial Pt/C. While borides have demonstrated high HER catalytic activity,[620] exploration of boride nanosheets in HER has been limited, possibly due to challenges in their synthesis.

Metal chalcogenides typically demonstrate superior electrical conductivity, mechanical and thermal stability, and enhanced electrochemical activity compared to their corresponding metal oxide counterparts. Nanosheets composed of TMDs like $MoS_2$, $WS_2$, etc. as well as $TiS_2$ and $VSe_2$, exhibit huge potential in HER electrocatalysis.[621,622,623,624] The catalytic efficacy of six TMD nanosheets ($MoS_2$, $MoSe_2$, $MoTe_2$, $WS_2$, $WSe_2$, and $WTe_2$) was observed to follow a hierarchy, with selenides showing superior performance over sulphides and tellurides.[594] Among these, $MoSe_2$ demonstrated the highest performance. Furthermore, metallic $VS_2$ nanosheets exhibited superior performance compared to the aforementioned TMDs, approaching the efficiency of Pt. They displayed a current density of -10.0 mA $cm^{-2}$ at a potential approximately 30.0 mV more negative than Pt, with a Tafel slope of 34.0 mV $dec^{-1}$ compared to Pt's 30.0 mV $dec^{-1}$.[598]

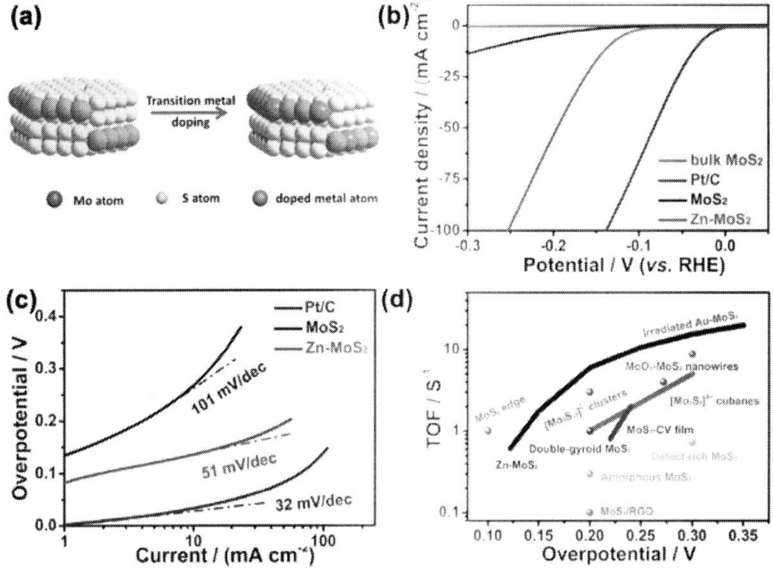


Figure 7. (a) Schematic diagram of doping $MoS_2$ material with other transition metals. (b) Polarization curves of HER for various catalysts. (c) Tafel plots of the corresponding electro-catalysts derived from the early stages of HER polarization curves. (d) TOF values of the Zn-$MoS_2$ catalyst (black line) and other $MoS_2$-based catalysts reported recently. Reproduced with permission.[625]

Innovative trials and DFT calculations have shed light on the crucial role of exposed edges of TMDs in the HER. However, the precise nature of the active sites in TMDs remains a topic of debate. It has been proposed that unsaturated sulfur edge atoms in 2H-$MoS_2$ trigger the HER[626] primarily through a Volmer-Heyrovsky mechanism. This mechanism involves the adsorption of H+ on sulfur edges, followed by the diffusion of $H^+$ from sulfur to molybdenum to form molybdenum hydride (MoeH), ultimately resulting in the generation of $H_2$ from MoeH and $H_3O^+$.[627] Both metal and chalcogen edges of $MoSe_2$ and $WS_2$ have been suggested as potential sites for HER activation, with only selenium edges being active sites for $WSe_2$.[628] Defects, pore structure, and conductivity of TMDs also significantly influence HER efficiency. [629,630] The metastable octahedral 1T phase, in contrast to the thermodynamically stable trigonal prismatic 2H phase, exhibits inherently lower charge-transfer resistance but higher active site density due to strained lattice distortion, thereby facilitating faster HER kinetics.[631,632] Both edge and basal sites of 1T-$MoS_2$ are catalytically active for HER, although this phase is not ultimately stable and reverts back to the

2H phase over time or under heat. Covalent chemical functionalization with phenyl rings to form an S-C bond can kinetically stabilize the 1T phase,[633] resulting in superior stability and long-term performance under HER conditions. Remarkably high intrinsic HER activity has been achieved on porous 1T-$MoS_2$ nanosheets, with an overpotential of 153.0 mV at a current density of -10.0 mA $cm^{-2}$ and a Tafel slope of 43.0 mV $dec^{-1}$.[632] The activity of TMD nanosheets can be tailored by (1) applying strain [623] to induce fluctuations in the lattice spacing, (2) doping the edge/basal plane with heteroatoms like Cu, Zn, Mn, Ru, Rh, and Ta to modify hydrogen binding on the surface,[634] (3) engineering defects (disorder) to provide a high density of low-coordinated active sites that facilitate favourable hydrogen binding,[635,636] (4) manipulating phase and nanostructure,[631] and (5) modulating interfaces by hybridization with a support such as graphene or reduced graphene oxide,[637,638] N-doped graphene,[639] metallic substrate,[640] or a substrate with HER activity,[641] or by forming heterostructures.[642] While Fe, Co, and Ni have shown marginal enhancement in the activity of $MoS_2$, Zn doping has been found to improve HER by increasing the coverage of H and enhancing adsorption-desorption kinetics thus improving HER with an onset potential of -0.13 V (vs. RHE) and a Tafel slope of 51.0 mV $dec^{-1}$ as shown in Fig. 7.[625] Amorphous metal chalcogenides are thought to have more catalytically active sites due to structural disorder compared to their crystalline counterparts.[643] Proton intercalation has been shown to enhance the HER catalysis of $MoS_2$ nanosheets without generating new active sites.[644] Calculations suggest that the adsorption free energy of hydrogen shifts closer to zero in the presence of protons, and intercalated protons also increase the electric conductance of $MoS_2$ in both vertical and horizontal directions, aiding the HER in many ways.

Porous $NiSe_2$ nanosheets, derived from the conversion of LDH precursors, demonstrate superior performance compared to most reported 2D non-noble metal catalysts in acidic electrolyte. Their onset overpotential is unusually low, as low as 90.0 mV.[645] Overpotentials required to achieve current densities of -10.0, -100.0, and -200.0 mA cm-2 are measured at 135.0, 183.0, and 202.0 mV, respectively. In alkaline media, $Ni_xCo_{3-x}S_4/NiS_2/Ni$ foam on $NiS_2$ nanosheets surpass Pt/C at potentials more negative than -360.0 mV (vs. RHE).[646] This catalyst exhibits stability over operational durations exceeding 50 hours. The observed catalytic activity is attributed to the high electrical conductivity and surface chemical effects of cobalt (Co) and sulfur (S). However, the Tafel slope, at 156.0 mV $dec^{-1}$, is notably higher than the 43.0 mV $dec^{-1}$ exhibited by 20.0 wt% Pt/C. Improved performance has been achieved by combining $MoS_2$ with N-doped graphene, requiring only an overpotential of 56.0 mV to achieve -10.0 mA

$cm^{-2}$ for the HER. This 2D/2D hybrid demonstrates a Tafel slope of 41.3 mV $dec^{-1}$, approaching the commercial 20.0 wt. % Pt/C value of 30.0 mV $dec^{-1}$.[647]

In alkaline environments, core-shell hybrid nanosheets of $Co(OH)_2$@polyaniline supported on Ni foam demonstrate an onset overpotential for the HER of 50.0 mV and a Tafel slope of 91.6 mV $dec^{-1}$. These values outperform many other non-precious catalysts and even Pt/C.[604] Another noteworthy HER catalyst is the quaternary mixed nickel iron phosphosulfide nanosheets ($Ni_{0.9}Fe_{0.1}PS_3$), which exhibit superior performance in alkaline solutions. This catalyst achieves an overpotential of 72.0 mV vs. RHE at a current density of -10.0 mA $cm^{-2}$ and a Tafel slope of 73.0 mV $dec^{-1}$, surpassing the standard expensive Pt/C.[616]

Pristine graphene exhibits a significant positive $\Delta G_{H*}$ (1.85 eV), indicating challenges in H* adsorption. However, doping graphene with *p*-block elements (B, N, P) has been shown to reduce $|\Delta G_{H*}|$, thereby enhancing the activity of the HER.[648] Codoping further decreases $|\Delta G_{H*}|$, resulting in lower HER overpotential and smaller Tafel slopes, similar to the effects observed in the oxygen reduction reaction (ORR)[648] and oxygen evolution reaction (OER).[649] According to a model that considers the most active sites as carbons in non-edge positions, the values of ΔGH* follow the sequence B-G (0.61 eV) < P-G (0.71 eV) < N-G (0.81 eV) < O-G (0.97 eV) < S-G (1.01 eV).[650] Experimental findings have demonstrated that B-G and N-G outperform other singly doped samples in terms of overpotential and exchange current density. [651] However, all singly doped graphene materials exhibit Tafel slopes above 90.0 mV $dec^{-1}$,[652] suggesting a predominant Volmer-Heyrovsky pathway with the Volmer step as the rate-determining step. Codoping with elements such as P/N,[653] S/N,[654] and S/P[655] can further enhance hydrogen adsorption on graphene surfaces, with (S, N)-G > (P, N)-G > (S, P)-G, causing reduced HER overpotential and smaller Tafel slopes.[651 655] Notably, doubly doped (N, B)-G exhibits higher $\Delta G_{H*}$ and greater overpotential than corresponding singly doped graphene materials.[651] Chemically doped graphene[656] edges are proposed to be the primary electro-chemically active sites based on scanning electrochemical cell microscopy observations. Further improvements in HER activity can be achieved through multiple-element doping,[657] introduction of favourable defects[658] and meso-/micropores, or their combinations. A comparative analysis of these doped and codoped graphene materials, along with their current potential performance relative to a commercial Pt/C catalyst, is presented in Figure 8. Tafel plots, charge population analyses of N, S, and P doped graphene, Gibbs free energy profiles calculated using density functional theory (DFT), pore size effects of N/S/P codoped graphene, and the impacts of dopant level and inter-defect distance on N/S/P codoped graphene catalysts are also depicted in Figure 8.[655]

Combining doped graphene with g-$C_3N_4$ exhibits a synergistic effect, enhancing both proton adsorption and reduction kinetics. [659,660,661] These metal-free hybrids can achieve performance levels comparable to state-of-the-art $MoS_2$ catalysts for the HER. Recent DFT calculations have suggested that the p-type dopant boron is likely the most effective doping element in g-$C_3N_4$@MG (where M = B, N, O, F, P, S) for HER. This configuration yields a low overpotential of 0.06 eV and a substantial interfacial adhesion energy of -1.30 eV.[660]

Few-layered BP, characterized by its extensive exposed surface area and high carrier mobility, shows promise as an electrocatalyst for the HER.[662] However, the HER efficiency of BP is restricted by its high hydrogen adsorption energy (1.01 eV). Metal doping presents a solution to this limitation by adjusting the energy associated with hydrogen adsorption and desorption, thereby enhancing the electrocatalytic capabilities of BP.[663] Apart from heteroatom doping, defects, edges, and structural curvature (such as wrinkles and ripples) significantly influence the HER activity of graphene. This influence is demonstrated through DFT calculations of Gibbs free energy profiles for HER at pH = 0, considering pristine graphene and its chemically doped variants (see Fig. 8g).[664] Recent findings support this concept, showing that irregularly shaped and wrinkled single-/few-layer graphene flakes with defective edges (SLG/FLG-DE) can catalyze HER with remarkably low overpotentials of 55.0 and 85.0 mV at -10.0 mA $cm^{-2}$ in 0.5 M $H_2SO_4$ and 1 M KOH solutions, respectively, while maintaining stable HER activity over 200 hours (refer to Fig. 8h and i).[664]

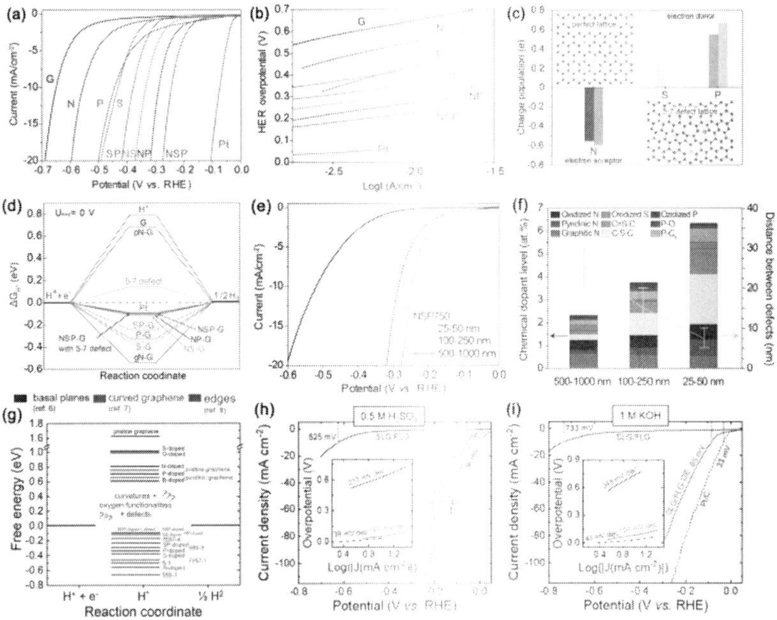

Figure 8. (a) CV curves of pristine and doped nanoporous graphene with various dopants, and commercial Pt. (b) Tafel plots. (c) Charge population analysis of NSP doped graphene on (left bar) perfect graphene lattice and (right bar) 5-7 defect graphene lattice. Inset: proposed doping sites for each element. (d) DFT calculations of Gibbs free energy profiles for Pt, and doped highly curved graphene with a topological defect. (e) Ligament and pore size dependence of HER activities from NSP ternary doped nanoporous-graphene obtained at 750 °C. (f) Size dependence of dopant level and distance between defects in NSP doped graphene with different pore sizes.[655] (g) DFT calculations of Gibbs free energy profiles for HER at pH = 0 for graphene in its pristine and chemically doped basal planes, including curvature effects and edges/defects. H* indicates the adsorbed hydrogen. (h, i) iR-corrected cathodic LSV curves of the SLG/FLG-DE electrodes in acidic and alkaline solutions, respectively. The LSV curves of the SLG/FLG electrodes and Pt/C benchmarks are also shown for comparison.[664] Reproduced with permission.[625]

In summary, noble metal nanosheets containing Ru demonstrate competitive performance with Pt for HER.[665] The most effective non-noble metal electrode identified thus far consists of disordered $MoS_2$ nanosheets combined with reduced N-doped graphene,[639] retaining stability and activity even after 5,000 CV cycles. Among the reported non-metal catalysts, porous layers of graphitic $C_3N_4$ supported on N-doped graphene film exhibit outstanding HER performance, with an onset potential close to that of commercial Pt (8.0 mV vs. 0.0 mV of Pt/C, vs RHE @ -0.5 mA cm$^{-2}$), high exchange current density (0.43 mA $cm^{-2}$), and remarkable durability (minimal activity loss over 5000 cycles).[661] Despite these advancements, significant progress remains elusive in alkaline HER electrocatalysis. Synergistic approaches involving heterostructure formation,[666] pore introduction, amorphous structure development,[667] and electronic modulation (alloying and doping) hold promise for further enhancing HER catalysis.

## 8.4 Fuel Cells

Fuel cells are devices that convert chemical energy into electrical energy through reactions with oxidizing agents like $O_2$.[668] They offer reliable energy production, mainly for renewable sources, due to their high efficiency, zero emissions, and continuous fuel supply. Fuel cell technology is utilized in transportation, stationary power, and portable devices, functioning similarly to batteries but without following traditional combustion cycles.[669] Overcoming the high activation energy barrier of fuel cell electrodes is crucial, which can be achieved by using affordable and stable electrocatalytic materials. Solar and electrical energy can be stored as chemical bonds, enabling water splitting to produce hydrogen for energy generation alongside oxygen reduction.[670] This clean energy production involves electrocatalytic cycles like ORR, OER and HER crucial for fuel cells as well as metal-air batteries but the primary reaction for generating energy in fuel cells is the ORR.[671] Direct methanol/ethanol utilization in the carbon cycle presents a viable solution for energy storage and hydrogen transportation challenges.[672]

Fuel cell-based energy production involves multi-electron charge transfer processes, such as the ORR, which can be classified as a $2e^-$ or $4e^-$ system. These electrons are channelled via the anode to the cathode through an external circuit, generating a current. Water is produced as a by-product from the constituent atoms of $H_2$ and $O_2$. Conversely, $H_2$ (fuel) and $O_2$ are generated through comparatively low energy application. This combination of forward and reverse reactions, involving consecutive hydrogen- and oxygen-associated catalytic processes, forms a cycle known as the water cycle. The increasing hydrogen market has made the electrocatalytic water cycle a significant research focus. The ideal technical water cycle involves water electrolysis via the OER and HER for fuel production, followed by power production through the ORR and HOR. Electrolytic cells facilitate the HER and OER at the cathode and anode, producing $H_2$ gas and $O_2$

molecules, respectively (Fig. 9(a)). The ORR and HOR occur at the cathode and anode for $H_2$ and $O_2$, respectively (Fig. 9(b)).

Significant research efforts have been directed towards achieving high electrocatalytic properties for both the water and carbon cycles. While the kinetics of the $2e^-$ transfer method in half-cell HER and HOR are straightforward, slow rates for multi-electron transfer ORR and OER decrease energy equipment efficiency. Traditional noble metals like Pt, Ir, and Ru are conventionally used as electrocatalysts for these reactions but suffer from poor stability and high cost. Therefore, the development of low-cost, high-performance electrocatalysts is crucial for clean and sustainable energy production.[670] The water cycle, a ground-breaking achievement, was summarized in Grove's 1842 letter "On a gaseous voltaic battery".[673] Two systems, discharging mode (fuel cell) and charging mode (capacitors and batteries) as shown in Fig. 9 (c) and (d), respectively, have been independently examined in alkaline (AEM) and acid media (PEM) in recent decades. Additionally, interest in developing $H_2/O_2$ energy storage systems has led to the combination of both systems, known as the URC, as reported by Mitlitsk *et al.*[674] (Fig. 9(e)). The ORR plays a crucial role in various electrochemical energy conversion and storage technologies. It occurs at the cathode through two pathways, leading to the conversion of $O_2$ to either $OH^-$ or $H_2O$ depending on the type of electrolyte utilized (acidic or alkaline). One pathway involves a "partial" $2e^-$ reduction, resulting in the formation of adsorbed $H_2O_2$ species, while the other is a "direct" $4e^-$ reduction or full reduction pathway. The $4e^-$ reduction pathway is more favourable because the reactivity of $H_2O_2$ is relatively superior to the stability of $H_2O$. In the direct transformation of $O_2$ into $H_2O$, a dissociative method is employed, where $O_2$ is initially adsorbed on the metal/catalyst surface, followed by the breaking of the oxygen-oxygen bond to form adsorbed oxygen atoms. Subsequently, electrons are transferred to the adsorbed oxygen atoms through hydrogen addition, resulting in surface-bound hydroxyl groups. Further reduction and protonation of the hydroxyl group yield an $H_2O$ molecule, which then leaves the metal/catalyst surface. Conversely, the partial reduction of $O_2$ follows an associative mechanism, where the adsorption of $O_2$ on the metal surface does not lead to oxygen cleavage but instead generates $H_2O_2$.

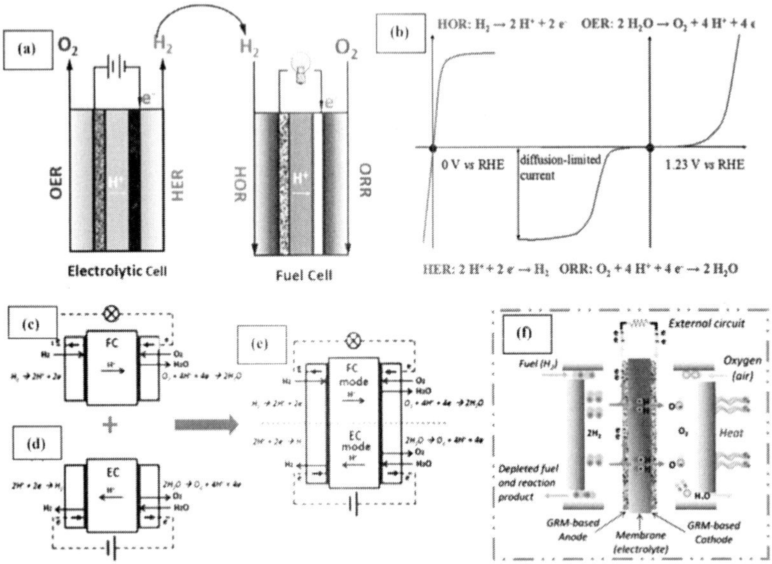


Figure 9. (a) Schematic illustration of an electrolytic/fuel cell. (b) Polarization curves of the $H_2$- (red color curves) and $O_2$-involved reactions (blue color curves), without a proper scale. Innermost polymer electrolyte membrane (PEM) system (c) fuel cell (FC), (d) electrolyzer cell (EC), and (e) unitized regenerative cell (URC). Discharging mode: FC and recharging mode: EC. PEMFCs, where the fuel (e.g. $H_2$) channel starting from one side, is split by the catalyst (e.g. graphene electrode) into $H^+$ ions and $e^-$. (f) More detail explanation of the current generation and combination of hydrogen ions and oxidant at the cathode, generating water and heat.[675] Reproduced with permission.[90]

## 8.4.1 2D Materials for Fuel Cells

The graphene family offers promising, cost-effective alternatives to platinum group metal (PGM) electrocatalysts for the ORR in sustainable energy production. This alternative is gaining traction due to their unique structural features and excellent properties, such as large surface area and high density of active sites. Recent research has shown significant progress in utilizing 2D graphene derivatives to replace conventional PGM-based materials for the ORR. High nitrogen-doped graphene and boron co-doping in nitrogen-graphene have demonstrated notable advancements, achieving performances comparable to benchmark materials. Additionally, introducing defects in graphene has led to ORR results comparable to Pt/C. Although achieving very high nitrogen doping levels in graphene is challenging, it can match the ORR performance of Pt/C. However, codoping with sulfur, iron, and nitrogen has shown limited improvement compared to Pt/C. The chemical stability of graphene is attributed to its $p_z$ atomic orbitals, which hybridize to form stable, delocalized p-bonding orbitals, enabling adsorption through various interactions. Introducing defects by creating pentagons, heptagons, and octagons in graphene, as demonstrated by Yao *et al.* (Fig. 10),[676] further enhances its ORR characteristics. It's crucial to consider the presence of residual nitrogen in graphene sheets, which can significantly affect ORR activity and can be detected through XPS analysis. To mitigate this issue, synthesizing non-doped and edge-doped graphene via argon plasma treatment is being explored for functional materials.

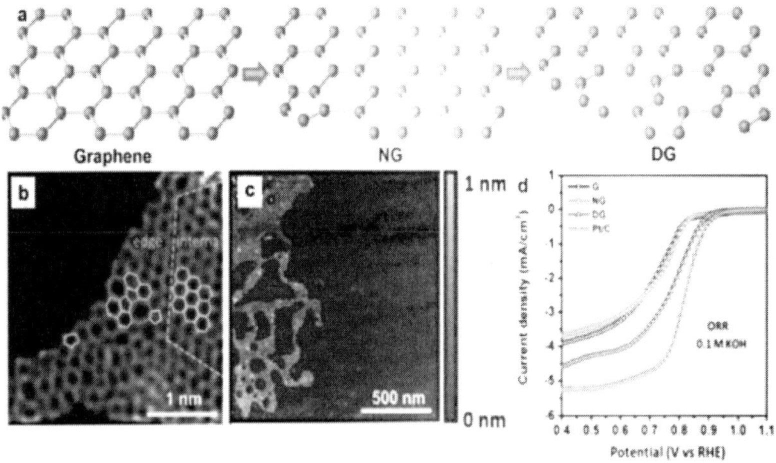


Figure 10. (a) Schematic representation of graphene with defects. (b) HAADF figure of defective graphene with hexagons, pentagons, heptagons, and octagons, shown by orange, green, blue, and red colours, respectively. (c) AFM image of graphene with defects. (d) LSVs for pristine, N-doped and defective graphene for the ORR. Reproduced with permission.[90]

The inherent structure of graphene can be easily modified through heteroatom doping (N, B, S, and P) to create metal-free catalysts for various electrocatalytic applications, including the ORR. Such doping methods can be achieved through in situ techniques like ball milling, CVD, and bottom-up solution-based synthesis, or through post-treatment approaches involving wet chemical methods, heat treatment of GO in specific environments with dopant elements, and plasma or arc-discharge treatments. Chen *et al.*[677] calculated the maximum achievable doping levels of B (10.8 atom %), S (17.6 atom %), and N (29.8 atom %) by fluorinating graphene with $XeF_2$ and annealing it under an $NH_3$

environment, sulfur, or $BCl_3$, respectively, under normal pressure conditions. The nitrogen concentration approached the theoretical maximum value in stable graphene (33.3-37.5 atom %).[678] The synthesized N-doped graphene exhibited catalytic properties approximately 80% that of the benchmark Pt/C in alkaline conditions.

In ORR applications, electron-deficient boron atoms facilitate the chemisorption of negatively polarized oxygen atoms and promote O–O splitting. Recently, Qiao *et al.* demonstrated that B-doped graphene exhibited a lower ORR activation barrier energy, resulting in higher catalytic activity compared to P- and S-doped graphene.[679] A more detailed study suggested that pyridinic nitrogen further enhances the active surface, boosting ORR activity in N-graphene.[680] The significance of N-doping was also explored by Zhang *et al.*, who utilized g-$C_3N_4$ as a template and nitrogen source to synthesize highly porous N-doped graphene, showing ORR catalytic activity similar to benchmark Pt/C in alkaline solution.[681] P-doped graphene showed high ORR performance with a more positive $\eta$ and higher current.[682] Theoretical explanations by Wu *et al.* demonstrated the $4e^-$ pathway for ORR on P-doped graphene.[683] Recent DFT simulations by Yang *et al.* suggested a $2e^-$ pathway followed by a $4e^-$ pathway on P-doped graphene, with OH reduction to $H_2O$ identified as the rate-limiting step. P atoms and nearby C atoms were proposed as suitable adsorption and active sites.[684] Additionally, investigating metallic dopants (Fe, Co, Ni, Cu)[685,686] and metal oxides ($MnO_2$)[687] in graphene is significant for enhancing catalytic activity. Pumera *et al.*[687] doped a small amount of Mn, leading to an 80 mV potential shift in the ORR. Substituting multiple heteroatoms (B/N, S/N, P/N, I/N, B/N/P, N/P/S, N/Fe, N/S/Fe) may further modify graphene's electronic structure.[688]

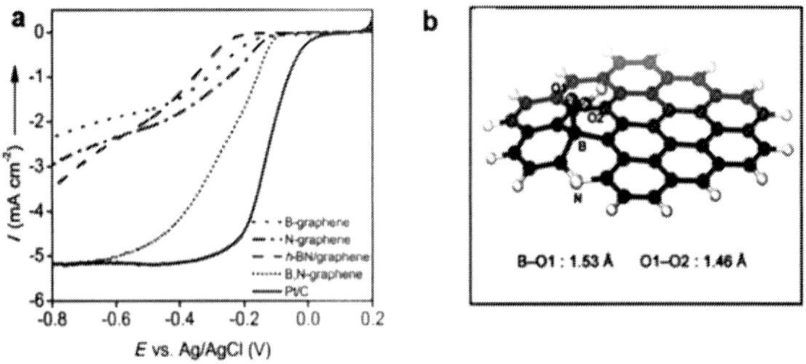


Figure 11. (a) LSVs curves at rotating rate of 1500 rpm, in $O_2$-saturated alkaline solution (0.1 M KOH scan rate: 10 mV $s^{-1}$). (b) $HO_2$ adsorbed on B, N-graphene optimized arrangement structure.[689] Reproduced with permission.[90]

Figure 11 illustrates Co-doped B in N-doped graphene, which Qiao *et al.* indicated as highly beneficial for the ORR in alkaline solutions. In Figure 11(a), the overpotential ($\eta$) of boron and nitrogen-doped graphene is approximately 110 mV with an E1/2 of about 150 mV, while Pt/C exhibits a negative value. Co-doping with S and N further influenced electron donors and acceptors,[690] enhancing active sites.[691] Zhou *et al.* demonstrated a synergistic doping effect using aminothiazole-based co-doped graphene, AT-N/Fe-G,[692] with doping levels of 6.5% for nitrogen, 6.8% for oxygen, 1.5% for sulfur, and 2.6% for iron. These co-doped graphene materials exhibited excellent ORR activity in both alkaline and acidic conditions compared to the benchmark 20% Pt/C. In terms of stability in acidic solutions, the potential of AT-N/Fe-G shifted positively by 12 mV after 10,000 cycles, while the potential of Pt/C shifted by 58 mV within the same period.

One of the most promising applications of the graphene family in conjunction with mayenite electride, C12A7:$e^-$, is in the realm of ORR within fuel cells.[693] In Figure 12(a), the CV curves exhibit cathodic reduction currents displaying evident diffusion-controlled ORR peaks at 0.78 V and 0.8 V (vs. RHE) in $O_2$-saturated 0.1 M KOH solution for tin-doped C12A7:$e^-$ with and without rGO coating, respectively. The diffusion-limited current density increased with higher rotation rates, as commonly observed (Figure 12(b and c)). These findings underscore the inherently high intrinsic activity of the tin-doped C12A7:$e^-$ composite in the ORR, both with and without rGO coating.

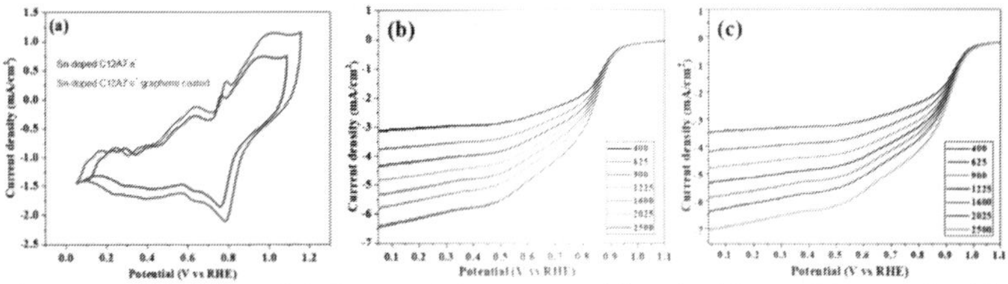


Figure 12. (a) Sn-doped C12A7:$e^-$ CVs curves (with and without rGO coated, with doping level, x = 1) at 50 mV $s^{-1}$ scan rate in $O_2$-saturated 0.1 M KOH electrolyte. Sn-doped C12A7:$e^-$ (x = 1) (b) without rGO coating and (c) with rGO coating in $O_2$-saturated 0.1 M KOH with various rotation rates at a scan rate of 10 mV $s^{-1}$. Reproduced with permission.[90]

Zhao *et al.* conducted theoretical calculations of the ORR performance for N-doped graphene supported by MXenes ($Ti_2C$, $V_2C$, $Nb_2C$, and $Mo_2C$ monolayers) as illustrated in Figures 13(A) and (B).[694] Graphitic sheets hybridized with $V_2C$ (G/$V_2C$) and $Mo_2C$ (G/$Mo_2C$) MXene monolayers exhibited remarkable catalytic activity for the ORR. The hollow-site carbon atoms situated near the pyridinic nitrogen dopants were identified as the most active sites, leading to low overpotentials of 0.36 V (G/$V_2C$) and 0.39 V (G/$Mo_2C$) (Figure 13(C)), which are competitive with the benchmark Pt (0.45 V). The strong electronic coupling between MXene and graphitic materials suggests potential for multifunctional hybrid structures in flexible energy devices. Additionally, Xie *et al.*[695] experimentally evaluated the ORR activity for Pt and $Ti_3C_2T_x$ composites, which remained stable even after 10,000 cycles, maintaining constant half-wave potentials (Figure 13(D)), while Pt/C experienced a 21 mV loss (Figure 13(E)). Introducing alkalization-intercalated MXene ($Ti_3C_2(OH/ONa)_2$) into Ag composites, the urchin-structured MXene-$Ag_{0.9}Ti_{0.1}$ bimetallic nanowire exhibited pronounced electrocatalytic activity for the ORR (Figure 13(F)),[696] with onset and half-wave potentials exceeding those of the benchmark 20 wt% Ag/C and Ag-based catalysts.[694]

MXenes derived from earth-abundant elements present promising prospects in the field of electrocatalysis. Utilizing MXene materials like $Ti_3C_2X_2$ (where X = OH and F) as a support offers two distinct advantages in electrocatalytic applications. Firstly, being titanium-rich materials, they exhibit excellent durability against mechanical, acidic, and oxidative stresses. Secondly, they not only provide effective conductive pathways for electron transfer similar to carbon supports but also effectively mitigate support corrosion, a common issue encountered with precious group metal catalysts (PGMCs) and carbon-based electrocatalysts. For example, while silver demonstrates relatively high ORR performance and robustness in alkaline media, its reported electrocatalytic behavior is often unsatisfactory. Another promising strategy to further enhance catalytic activity involves fabricating more conductive supports combined with effective bimetallic catalysts.

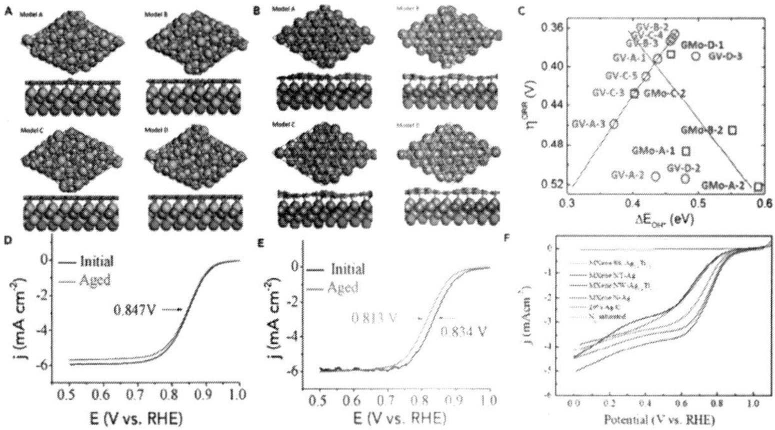

Figure 13. MXenes as catalysts for the ORR. (A-C) Four different models of N-doped graphene on (A) $V_2C$ and (B) $Mo_2C$ MXene monolayer. (C) Volcano plot of ORR overpotential and $\eta^{ORR}$ vs. $OH^*$ binding energy. The black box indicates the dimensions of the supercells. The red numbers label the C sites. C, N, O, V, and Mo atoms are denoted by gray, blue, red, green, and turquoise, respectively. ORR current on (D) $Pt/Ti_3C_2X_2$ and (E) Pt/C catalysts before and after the accelerated durability tests. (F) ORR polarization curves of MXene-$Ag_{0.9}Ti_{0.1}$ nanowire and other control samples in 0.1 M KOH. Reproduced with permission.[90]

## 8.5 Conclusion

In summary, the widespread utilization of 2D materials as electrocatalysts requires careful consideration and several key factors need to be addressed:

1) Understanding the electrocatalytic behavior of 2D materials is essential for their effective design and development. This involves theoretical analysis and in situ characterizations to gain insights into their properties at the atomic level. Such understanding can facilitate the creation of customized nanosheets and precise atomic-level electrocatalysts tailored for specific applications.

2) The scalability of 2D material production is crucial for their practical application. It is imperative to establish methods for the mass production of 2D materials while ensuring control over factors such as the number of layers and flake dimensions, especially for materials like 2D-MXenes. Dimension confinement techniques are particularly important in this regard to achieve consistent and controllable material properties.

3) Controlled modification of 2D materials is necessary to optimize their electro-catalytic performance. This includes techniques such as doping to adjust their electronic structures and surface engineering to introduce functional groups. These modifications can enhance the sustainability and efficiency of 2D materials in various applications.

4) Manufacturing processes should focus on creating heterostructures of nanosheets, allowing for the precise tuning of electronic properties. By engineering heterostructures, it becomes possible to tailor the material properties to meet specific performance requirements, further expanding the potential applications of 2D materials in electrocatalysis and beyond.

# CHAPTER 9

# QUANTUM COMPUTING

The intersection of quantum computing and 2D materials has emerged as a promising frontier in the quest for more efficient and powerful computational technologies. Quantum computing, with its ability to harness the principles of quantum mechanics, promises unprecedented computational capabilities by leveraging quantum bits or qubits. Within this landscape, 2D materials, characterized by their atomically thin structures, have garnered significant attention for their unique electronic and quantum properties.

The applications of 2D materials in quantum computing are manifold, ranging from qubit implementation to enhancing various aspects of quantum information processing (QIP). Graphene, a prototypical 2D material, has been explored for its potential in spin qubits due to its long spin coherence times. TMDs, another class of 2D materials, exhibit strong spin-orbit coupling, opening avenues for spin-valley qubits. These materials offer a rich playground for manipulating quantum states and interactions. Research in this field delves into the development of scalable and controllable qubits using 2D materials, addressing challenges such as decoherence and scalability faced by traditional quantum systems. Additionally, 2D materials provide a platform for the realization of quantum gates and circuits, essential components for building robust quantum computers.

As we embark on this exploration of the synergy between 2D materials and quantum computing, it becomes clear that the unique properties of these materials hold the potential to revolutionize the landscape of information processing. This introduction sets the stage for a detailed examination of the current state of research, challenges, and future prospects in the dynamic field of 2D materials in quantum computing.

## 9.1 Single Quantum Emission

In the era of quantum information, technologies centered on photon qubits are pivotal, utilizing photonic properties like polarization for $|0\rangle$ and $|1\rangle$ states. Essential for quantum tech, the quest for dependable single-photon sources (SPSs) is underway. Simultaneously, the development of accurate single-photon detectors (SPDs) is critical, as most quantum protocols rely on single photons, fostering advancements for efficient detection in incoming light signals. Single quantum emission (SQE) from 2D materials has emerged as a fascinating and promising field at the intersection of quantum optics and materials science. These materials, with their atomically thin structures, exhibit unique quantum properties that make them ideal candidates for generating individual photons with remarkable precision. The ability to produce single photons is crucial for various quantum technologies, including quantum communication, quantum cryptography, and quantum computing. The study of SQE in 2D materials involves exploring the quantum confinement effects and electronic properties of these materials at the nanoscale. Various 2D materials like TMDs and *h*-BN, exhibit remarkable quantum properties that make them promising candidates for single-photon sources. Notably, monolayer TMDs and *h*-BN have garnered significant attention due to their exceptional properties. TMDs, such as $WSe_2$ and $MoS_2$, possess distinct excitonic behaviors,[697] while *h*-BN, with its wide bandgap and excellent optical properties, serves as an efficient emitter.[698]

In TMDs, ($MX_2$, M = transition metal and X = chalcogen), the primary source of quantum emission is associated with exciton formation, representing a bound state of an electron and its corresponding hole. The recombination of this electron-hole pair results in the emission of a single photon. While excitons in TMDs are typically delocalized, the presence of point defects and crystallographic flaws can lead to their localization, enabling localized SQE, particularly at cryogenic temperatures. [699,700]

Conversely, in wide-bandgap *h*-BN, point defects serve as color centres for generating single-photon emissions. Atomic impurities within the solid-state introduce intermediate energy levels situated between the conduction band minimum and valence band maximum of materials with wide energy bandgaps. Excitation of electrons from occupied energy states to higher levels induces spontaneous one-photon emission, even at elevated temperatures of up to 800 K.[701]

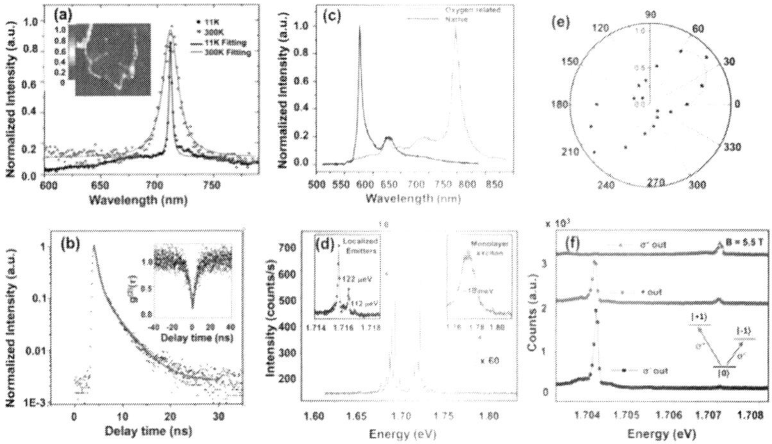


Figure 1. Examples of experimentally measured emission characteristics of *h*-BN and TMD emitters. (a) PL spectra at 11 K and 300 K, of the *h*-BN crystals under a laser excitation of 532 nm, 300 μW before Ar plasma etching. Inset: Confocal PL map showing emitters encircled in red. (b) Experimentally measured lifetime and second-order autocorrelation function $g^{(2)}(\tau)$ (inset) of an *h*-BN emitter. (c) Effect of Ar plasma etching on the PL spectrum of *h*-BN emitters, showing red-shift of emission.[702] (d) PL spectrum of localized emitters from $WSe_2$ monolayer [(left inset) highest peak intensity for a certain SQE; (right inset) the valley exciton emission collected for 60 seconds] are shown. (e) Polar graph of the emission which is linearly polarized at an angle of 49° with respect to the y-axis. (f) Effect of magnetic field on emission polarisation: polarisation-resolved PL spectra shows a change from cross-linear to circular polarization of the emission-doublet on application of magnetic field. The corresponding selection rules are shown using a zero-field energy level diagram [see inset].[703] Reproduced with permission.[712]

Several research groups confirm that localized emitters in *h*-BN and TMDs emit non-classical light in the NIR-visible range.[704,705,706] Bourrellier *et al.*[707] demonstrated UV-range single-photon emission from *h*-BN. Another study reported the first observation of room temperature quantum light from *h*-BN at 998 nm [$g^{(2)}(0) = 0.6$].[708] The emission wavelength corresponds to the peak intensity in the PL spectrum, as shown in Figure 1a. The emission lifetime can be determined from the time-resolved intensity decay curve. Figure 1b illustrates the time-dependent emission intensity decay with $g^{(2)}(\tau)$ (inset) recording $g^{(2)}(0) \sim 0.1$. For quantum technologies, highly pure single photon emission with $g^{(2)}(0)$ values ranging from $g^{(2)}(0) < 0.1$ to $g^{(2)}(0) < 0.001$ is crucial.[709] In fibre optical quantum communications, photon energies in the telecom wavelength range are desirable for low transmission losses. Recent studies reported site-controlled emission in the

1080-1150 nm range from monolayer and few-layer $MoTe_2$ coupled to strain-inducing nanopillars. Quantum emitters in multilayer $MoTe_2$ show emission in the telecom wavelengths (1200-1600 nm).[710] The $g^{(2)}(0) = 0.155$ recorded in this study was measured at 13 K under pulsed laser excitation at a 1540 nm emission wavelength. Temperature-dependent photoluminescence studies reveal broadening of the ZPL linewidth with increasing temperature for *h*-BN fluorescent centres.

Xu *et al.* studied the effect of annealing in air on *h*-BN emitters, showing a red-shift and localization of emitters with higher yield [see Figure 1c].[702] Figure 1d illustrates the PL spectrum arising from $WSe_2$ emitters, where sharp PL emission lines emerge from localized emitters, red-shifted from delocalized valley excitons [right inset].[711] The brightest SQE in this ensemble has a doublet fine structure [left inset], fitted Lorentzian functions. The PL spectrum comprises both localized and delocalized excitons for $WSe_2$, distinguishing it from TMD[711] and *h*-BN[702] emitters. Linearly polarized emission is observed in *h*-BN defects obeying Malus' law [see Figure 1e]. Polarized emission, observed in both *h*-BN and TMD emitters, has applications in certain quantum key distribution protocols. In TMDs, magneto-optical studies[712,713] on single quantum emitters determine the magnitude of Zeeman splitting and effects on polarisation-resolved PL. Most quantum dots in $WSe_2$ reveal cross-linearly polarised doublet peaks in emission spectra even at zero magnetic field. For a non-zero magnetic field applied in the Faraday configuration, the doublet separation increases with the applied field strength, indicating spin-dependent optical transitions.[714] The emission doublet converts from linear to circular polarisation states, as shown in Figure 1f, attributed to the overpowering effect of the Zeeman shift on QD electron-hole exchange energy.[711] The anomalously large Zeeman Effect of those emitters is ascertained by the measurement of relatively large Lande g-factors.[713] Recent studies demonstrate that trions (charged QDs) in TMDs with quenched electron-hole exchange interaction can be optically initialized to a known spin-valley state by exerting magnetic control over the excitation helicity.[715] Strain control and type-II band alignment using heterostructures can moderate the localized creation of such emitters.[716] Magneto-optical studies for QEs in $MoTe_2$ show valley Zeeman splitting with increasing external magnetic field.[710] Some localized emitters exhibit a significant zero-field energy splitting in the range of 1-3.7 meV. Furthermore, at zero external field, some QEs may have a vanishing fine-structure splitting, arising from azimuthally symmetric confinement potentials induced by strain.[717]

## 9.2 Quantum Entanglement

Quantum entanglement (QE) is a fundamental phenomenon in quantum mechanics where two or more particles become correlated and interdependent in such a way that the state of one particle cannot be described independently of the state of the others, regardless of the distance separating them. This concept was famously referred to by Einstein as "spooky action at a distance." The entangled particles exhibit a level of correlation that goes beyond what classical physics can explain. QE is a fundamental phenomenon in quantum physics that plays a crucial role in QIP.[718] It arises between states in an inseparable composite quantum system, connecting two or more states regardless of the distance between them. This non-classical correlation means that acting on one entangled state instantaneously affects the others, even if they are spatially separated. It is a cornerstone of various quantum technologies, including quantum computing, quantum communications, and quantum sensing. One of the core elements of QIP is quantum entanglement, a phenomenon that emerges within an inseparable composite quantum system. This highly non-classical aspect of quantum physics connects two or more states across space-time, making it impossible to manipulate one state without influencing the others. QE holds significant potential for accelerating quantum computing,[719] enhancing the transmission and security of qubits,[720] and improving the resolution of quantum sensing. [721,722]

In the context of quantum computing, entanglement enables the acceleration of computations and assists in securing the transmission of qubits. It also contributes to enhancing the resolution of quantum sensing applications. Quantum teleportation, a key application of entanglement introduced by Bennett *et al.* in 1993, is pivotal in quantum communications.[720] It allows for the transfer of quantum states between distant locations without physically moving the particles involved. This process relies on entangled particles to transmit information securely and efficiently. Entanglement is essential for various quantum protocols and technologies, such as Bell state measurements (BSM) used in quantum communications and quantum teleportation schemes. The fidelity of entanglement, quantified by the overlap between density matrices representing the state and a Bell state, determines the quality of entanglement in a system.[723] Moreover, reliable single-photon detectors are crucial components for measuring entangled photon pairs accurately in BSMs.

Key characteristics of quantum entanglement include:

1) Correlation: Entangled particles share a quantum state, and the measurement of one particle instantaneously influences the state of the other, no matter the spatial separation between them.

2) Non-locality: Changes in the state of one entangled particle can occur instantaneously, seemingly violating the principle of locality, which dictates that information cannot travel faster than the speed of light.

3) Superposition: Entangled particles exist in a quantum superposition, meaning they can be in multiple states simultaneously until a measurement is made, at which point they "collapse" into a definite state.

4) Quantum States: The entangled particles are often described using quantum states such as spin, polarization, or other observable properties.

5) Quantum Entanglement Pairs: Common examples of entangled particle pairs include photons with opposite polarization or electrons with opposite spins.

### 9.2.1 Applications of 2D Materials in Quantum Entanglement

Applications of 2D materials in QE involve leveraging the unique properties of these materials to enhance entanglement generation, manipulation, and detection.

1) Enhancing Entanglement Generation: QE can be generated between 2D materials, such as TMDs, graphene, and silicene.[723] The tunability of the electronic and optical properties of these materials facilitates the design of structures that promote entanglement generation.

2) Manipulating Entanglement: Interface engineering in 2D materials improves the control of entanglement by modifying the electronic and optical properties of the materials. For instance, the band alignment and charge transfer across the interface between different TMD layers can influence the strength and stability of entanglement.[723]

3) Detecting Entanglement: Single-photon detectors play a critical role in measuring entanglement, especially in BSMs. High-performance single-photon detectors are necessary for accurate and efficient entanglement detection. 2D materials like $MoSe_2$ show promise in developing fast and efficient single-photon detectors.[723]

4) Quantum Communication: The unique properties of 2D materials contribute to the development of advanced quantum communication systems. For example, graphene-based qubits could serve as building blocks for topological qubits, which are highly resistant to noise and decoherence.[723] Additionally, 2D materials can facilitate the creation of quantum memories and repeaters, which are vital for long-distance quantum communication networks.

While the bulk of research has primarily focused on the theoretical aspects of entanglement, recent explorations suggest that 2D materials could play a crucial role in practical applications of QE. Here are some potential applications of 2D materials in the realm of QE:

1) Quantum Computing:

   - Incorporating 2D materials in the design of quantum gates and circuits to exploit their unique electronic properties.

   - Leveraging the exceptional quantum coherence of 2D materials for creating entangled qubits.

2) Quantum Sensing:

   - Employing 2D materials to enhance the precision and sensitivity of quantum sensors based on entanglement phenomena.

   - Developing entanglement-based sensing devices using 2D materials for applications in magnetic field measurement, gravimetry, and accelerometry.

3) Quantum Networking:

   - Integration of 2D materials in the development of entanglement-based quantum networks.

- Creating interfaces between quantum systems and 2D materials to facilitate entanglement distribution and manipulation.

4) Quantum Photonic Devices:

- Designing entangled photon sources with improved characteristics using 2D materials.

- Exploring the use of 2D materials in entanglement-based photonic devices for applications in imaging and sensing.

5) Quantum Information Storage:

- Investigating the potential of 2D materials for storing and retrieving quantum entanglement in quantum memory devices.

- Developing memory elements based on entangled states in 2D material structures.

6) Quantum Metrology:

- Utilizing 2D materials to enhance the precision of measurements in quantum metrology through entanglement-based techniques.

- Exploring the use of 2D materials in entangled state generation for improving measurement standards.

7) Quantum Imaging:

- Applying 2D materials in the development of entanglement-based quantum imaging devices.

- Enhancing the resolution and efficiency of quantum imaging systems using the unique properties of 2D materials.

2D materials hold immense potential for advancing quantum entanglement research and applications. Their unique properties allow for the design of structures that generate, manipulate, and detect entanglement effectively. As a result, 2D materials can significantly impact the development of quantum communication systems, quantum sensing, and quantum computing architectures.

## 9.3 Quantum Bits (Qubits)

In classical computing, information is processed using classical bits, which exist in one of two states: 0 or 1. These bits are the fundamental units of classical information and form the basis of traditional computing systems. Classical computers rely on manipulating these bits to perform computations, following classical physics principles.

Now, the landscape transforms when we step into the realm of quantum computing. Quantum bits, or qubits, harness the principles of quantum mechanics to revolutionize information processing. Unlike classical bits, qubits can exist in multiple states simultaneously, thanks to a phenomenon known as superposition. This property enables quantum computers to perform parallel computations, potentially solving certain problems much more efficiently than classical computers.

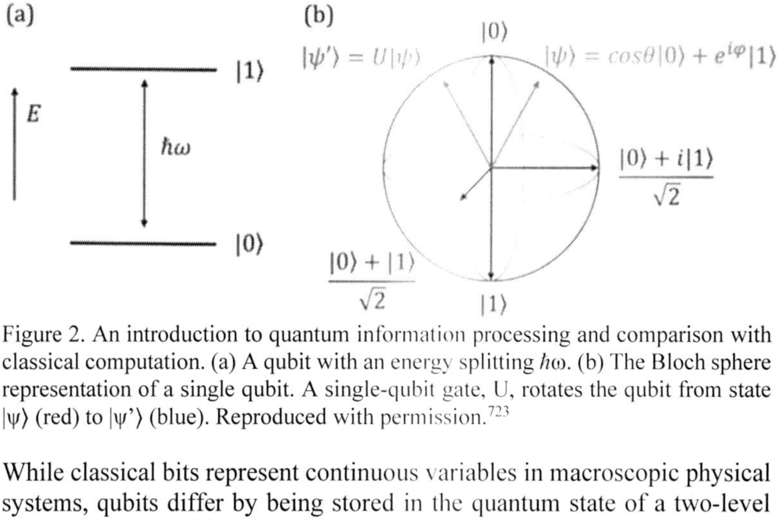


Figure 2. An introduction to quantum information processing and comparison with classical computation. (a) A qubit with an energy splitting $\hbar\omega$. (b) The Bloch sphere representation of a single qubit. A single-qubit gate, U, rotates the qubit from state $|\psi\rangle$ (red) to $|\psi'\rangle$ (blue). Reproduced with permission.[723]

While classical bits represent continuous variables in macroscopic physical systems, qubits differ by being stored in the quantum state of a two-level physical system with a finite energy difference or any linear superposition of these states.[724] These quantum states, $|0\rangle$ and $|1\rangle$ are fundamental in qubit representation as shown in Figure 2 (a). Visualizing a single qubit, the Bloch sphere (Figure 2 (b)) provides a useful tool. Represented as a unit vector on the Bloch sphere, any arbitrary quantum state in a 2D Hilbert space is denoted as $|\psi\rangle = \alpha|0\rangle + \beta|1\rangle$, where $|\alpha|^2 + |\beta|^2 = 1$. In a general N-qubit quantum system, the state exists in a superposition of all possible permutations of the system: While the fundamental unit of quantum information is the qubit, representing a two-level quantum system, the concept can be extended to qutrits (three-level systems) and further to qudits (any d > 2), which encompass many-level systems.[725] The multilevel nature of qudits offers a significantly larger state space for information storage and processing.

Moreover, qubits can be entangled, another quantum phenomenon where the state of one qubit becomes correlated with the state of another, even if they are physically separated. Entanglement allows quantum computers to achieve a level of interconnectedness that classical computers can't replicate, thus facilitates faster and proficient communication of information within a quantum system.

In the context of materials, certain quantum materials exhibit properties that make them suitable for hosting qubits. These materials, often in the form of two-dimensional structures (2D materials), provide a platform for manipulating and controlling qubits due to their unique quantum properties. Quantum materials play a crucial role in providing a stable environment for encoding, manipulating, and preserving the delicate quantum information carried by qubits in the pursuit of building powerful quantum computers. Quantum computers boast a larger computational space per bit: in classical systems, an $N$-bit configuration has a computational space of size $N$, whereas in quantum systems with $N$ qubits, the computational space expands exponentially to $2^N$.

## 9.3.1 2D Materials for Qubits

The following sections will summarize the applications of various 2D materials for qubits. Materials like graphene, TMDs, and *h*-BN exhibit unique electronic and optical properties suitable for hosting qubits. Their atomically thin nature, coupled with the ability to engineer spin and valley degrees of freedom, makes them attractive candidates for realizing qubits with long coherence times and high-fidelity operations. Additionally, heterostructures formed by stacking different 2D materials offer opportunities for tailoring qubit properties and exploring novel quantum phenomena, paving the way for scalable and fault-tolerant quantum computing architectures.

### 9.3.1.1 Graphene

Graphene, an atomically thin sheet of $sp^2$-hybridized carbon, stands out as a promising material for quantum dot spin qubits due to its unique properties, including being a type-I Dirac semimetal with highly tunable electrical characteristics and high mobility. Unlike traditional 2D charge carrier gases, graphene's electrostatic confinement is not limited by physical layer thickness, making it the thinnest 2DEG system.[726] With low intrinsic spin-orbit coupling and negligible hyperfine interaction, graphene presents the potential for long coherence times in spin qubits.[727] To utilize graphene for this purpose, two conditions must be met: opening a bandgap and lifting the valley degeneracy. This can be achieved through semiconducting graphene nanoribbons (GNRs) with armchair boundaries or gate-tunable bandgaps in bilayer graphene (BLG), requiring an out-of-plane electric field[728] and magnetic field,[727] respectively. Additionally, isotopic purification of graphene, akin to silicon, further reduces hyperfine interactions.

Graphene, beyond its spin degree of freedom, introduces a binary valley degree due to broken inversion symmetry, leading to two Kramer pairs: $|\uparrow,K\rangle$, $|\downarrow,K'\rangle$ and $|\uparrow,K'\rangle$, $|\downarrow,K\rangle$. For practical valley qubits, a large, tunable valley splitting is crucial. In graphene quantum dots (GQDs), the electrostatically tunable valley factor ($g_v$) ranges from 20 to 90,[729] providing a valley splitting an order of magnitude greater than Si/SiGe systems.[730] Although graphene faces a challenge with low intrinsic spin-orbit coupling (SOC), recent findings suggest that spin-valley coupling in single-electron GQDs is dominated by Kane-Mele SOC,[731] approximately 60 μeV, hinting at the potential to enhance spin-valley coupling by leveraging the proximity effect on graphene's weak intrinsic SOC.[732]

Figure 3a depicts the device structure for a graphene double quantum dot (DQD) device with a corresponding band structure. A BLG channel is enclosed by layered hexagonal boron nitride (*h*-BN), and side split-gate electrodes laterally confine carriers to a narrow channel, while top finger gate electrodes control QD formation. In a single QD, three finger electrodes are required, while a double DQD device involves two plungers, two side barrier electrodes, and a middle barrier electrode to tune tunnel coupling between dots. The charge stability diagram in Figure 3b demonstrates ambipolar operation, allowing for hole QDs in the valence band.[729] Recent advancements include transport measurements through a BLG DQD at the single-electron level, revealing capacitive interdot coupling effects (Figure 3c,d).[733] Gate-tunable tunnel coupling in a BLG DQD device has also been achieved, a critical step toward various qubit types. Understanding the interplay between spin and valley states in coupled dots is crucial, with ground states transitioning between spin-triplet valley-singlet[734] and spin-singlet valley-triplet[735] states under varying magnetic fields. Additionally, spin and valley blockade, as well as dispersive charge readout, have been demonstrated in graphene DQD systems,[736] paving the way for spin or valley state readout via spin-to-charge conversion.

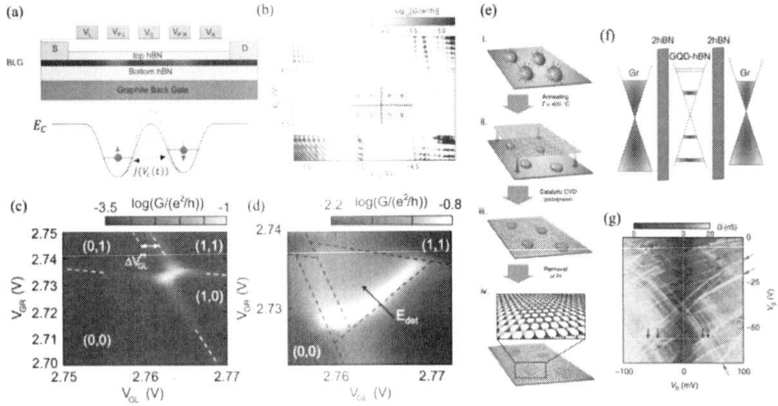

Figure 3. Graphene quantum dot devices for QIP. a) Schematic structure of a double graphene quantum dot (GQD) device and the corresponding band structure. b) Charge stability diagram of double quantum dot (DQD) device displaying ambipolar carrier confinement. c) Charge stability diagram at a bias of 0.1 mV demonstrating the single-electron regime in a graphene DQD device and d) a further zoomed in view of the charge stability diagram at a bias of 1 mV. The arrow shows the energy detuning along the axis of transition from then (0,1) to (1,0) state. e) Synthesis of in-plane GQD-*h*-BN heterostructures: i) Pt nanoparticles (NPs) are assembled using block copolymers, and the polymer is then removed by annealing; ii) monolayer *h*-BN is transferred onto the NPs; iii) catalytic conversion of *h*-BN to graphene occurs only in areas where there are NPs; iv) NPs are removed. f) Vertical single-electron tunnelling transistor energy band diagram. g) Conductance plot of low-density array of GQDs demonstrating single level transport. Reproduced with permission.[723]

Several researchers have primarily concentrated on gate-defined graphene quantum dots (GQDs), but an alternative bottom-up approach involves synthesizing graphene nanostructures. Theoretical work proposes the creation of Moiré potential-induced quantum dots in twisted bilayer graphene (tBLG) at low twist angles.[737] Controlled twist angle tBLG synthesis through CVD has been verified,[738] offering a pathway for realizing intrinsic GQDs experimentally. Geometric confinement is explored through in-plane *h*-BN-G heterostructures with clean edges formed by catalytic conversion of *h*-BN to graphene using patterned Pt layers.[739] Kim *et al.*[740] applied this method to synthesize lateral heterostructures of sub-15 nm GQDs embedded in an *h*-BN host layer, enabling the fabrication of vertical single-electron tunnelling transistors as seen in figure 3e. The resulting structure, presented as 30 nm *h*-BN/G/2L-*h*-BN/GQD-*h*-BN/2L-*h*-BN/G/20 nm $h$-BN on $SiO_2$ (figure 3f), displayed Coulomb diamonds as shown in Figure 3g, confirming single electron transport. Establishing and monitoring tunnel coupling in a comparable matrix structure could pave the way for scalable many-qubit systems, offering potential applications in quantum computation, simulation, and fundamental investigation on many-body systems in coupled GQDs.

The forthcoming objectives in GQD-based qubits involve measuring the coherence times for spin/valley states and achieving coherent qubit control. Using pulsed-gate spectroscopy, the minimum relaxation time for charge[741] and spin states[742] has been approximated to be in the range of hundreds of nanoseconds, although this estimate is probably constrained by the measurement method. Achieving single-shot readout is essential to precisely determine the lifetimes and coherence times of states in GQDs.[742]

### 9.3.1.2 TMD QDs

Other promising hosts for gate-defined QD spin-valley qubits are monolayer and bilayer TMDs and their heterostructures.[743] Current theoretical investigations have provided insights into gate operations on single spin qubits,[744] as well as one- and two-spin-valley[745] and valley[746] qubit systems in TMDs. The atomic structure of a prototypical TMD, 2H-$MoS_2$, is depicted in Figure 4a. TMDs exhibit a robust intrinsic SOC due to the d orbital electrons of the metal atoms, offering the potential for all-electrical qubit control via EDSR. Moreover, TMDs demonstrate Rashba SOC, which can be modulated by an electric field.[743] The strong SOC also leads to significant spin-valley coupling. Monolayer group-IV TMDs (MX2 where M = Mo, W and X = S, Se) possess a direct bandgap at the K and −K valleys and exhibit intrinsic broken inversion symmetry, resulting in valley-dependent optical selection rules for circularly polarized light, as illustrated schematically in Figure 4b.[747,748] This phenomenon enables selective optical initialization and read-out of spins, facilitated by the combination of SOC and valley-dependent optical selection rules.[749] This intrinsic spin-photon interface stands as one of the key advantages of TMD spin-valley qubits.

Gate-defined quantum dots have been successfully realized in monolayers of WS2,[750] $WSe_2$,[751,752] and $MoS_2$,[753,754,755] as well as in $MoS_2$ nanotubes.[756] Davari *et al.* presented a gate-defined quantum dot in $WSe_2$ encapsulated between layers of $h$-BN (refer to Figure 4c).[752] Their observations revealed a Coulomb blockade in the conductance plot of the device (Figure 4d). Moreover, they noted that the blockade peak exhibited a linear increase with temperature at zero bias, indicating single-level transport. However, despite indications of single-level transport, achieving confinement down to the single electron/hole limit in TMD-based gate-defined quantum dots remains an ongoing challenge. There exist potential strategies for confining carriers in TMDs beyond the scope of electrostatic gating. Quantum dots comprising TMDs, characterized by small radii that enable single carrier confinement, could be achieved through advancements in materials processing, coupled with tunnel barriers, optical channels, or alternative means. Recent advancements include the creation of sub-20 nm $MoS_2$ quantum dots via rapid thermal annealing[757] and the insight of heterostructures featuring $ReS_2$ quantum dots within a $MoS_2$ matrix.[758] Envisioning the confinement of single spins within bottom-up made TMD quantum dots holds promise for QIP. Additionally, akin to graphene, moiré potentials generated in twisted bilayers of TMDs or TMD heterostructures present another avenue for charge confinement within TMDs. An area of active investigation within TMD spin-valley devices involves the utilization of ferromagnetic and/or ferroelectric heterostructures to eliminate the valley degeneracy of the TMD layer, inducing a spontaneous valley polarization. Encouraging heterostructures have been pinpointed through ab initio methods, such as $WSe_2/CrI_3$[759] and $WSe_2/CrSnSe_3$.[760] Theoretical predictions have also identified "Ferrovalley" 2D materials-direct bandgap semiconductors that exhibit intrinsic valley polarization due to their ferromagnetic or ferroelectric properties, notably 2H-$VSe_2$.[761,762,763] This material, a ferromagnetic semiconductor, can be electrically manipulated by heterostructuring it with $Sc_2CO_2$.[763] While ab initio projections for 2H-$VSe_2$ are promising, the 1T phase remains the most thermodynamically stable. Therefore, synthesis of 2H phase $VSe_2$ monolayers must precede any experimental exploration of its potential for valleytronics or valley qubits.

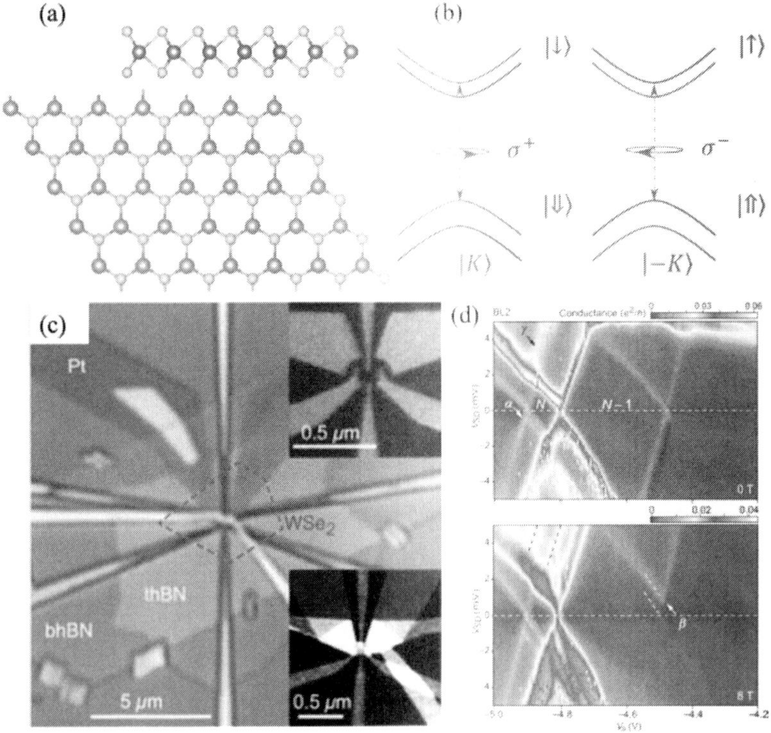


Figure 4. Toward spin-valley qubits in transition metal dichalcogenide quantum dots. a) Structure of 2H-$MoS_2$, a prototypical TMD. Yellow atoms are Sulfur, light purple atoms are molybdenum.[764] b) Valley-dependent optical selection rules in TMDs. c) Gate-defined quantum dot device structure in an *h*-BN-encapsulated $WSe_2$ flake and d) Coulomb blockade in the conductance plot. Reproduced with permission.[723]

Research into spin-valley qubits within TMD-based gate-defined quantum dots is still in its early stages. Demonstrating the confinement of single free electrons or holes in TMD quantum dots remains a pending task, and the coherence times of spin and valley states have yet to be determined. Blockade measurements on TMD quantum dots face challenges due to difficulties in establishing ohmic contacts to TMDs at low temperatures.[765] Therefore, identifying suitable ohmic contacts between 2D and 3D materials is a critical materials hurdle for TMD-based quantum dot devices. Optical readout presents a potential alternative. Also, synthesizing high-quality, large-area TMD monolayers poses a challenge. Although this is not currently a limiting factor, as exfoliated flakes have been used in individual devices, future scalability of TMD quantum devices will necessitate large-area, single-crystalline monolayers with minimal defects.

## 9.4 Quantum Algorithms and Applications

Classical algorithms lack the capability to accurately simulate the intricate dynamics of physical quantum systems,[766] especially those exhibiting multipartite entanglement and other highly non-classical behaviors. Consequently, quantum computers emerge as a promising platform for effectively simulating such non-classical behavior observed in physical systems. The introduction of quantum algorithms builds upon classical algorithms by exploiting the unique properties of quantum mechanics. Classical algorithms operate on bits, which can be either 0 or 1, whereas quantum algorithms operate on quantum bits or qubits, which can represent both 0 and 1 simultaneously due to superposition. This allows quantum algorithms to explore multiple computational paths simultaneously, leading to potentially exponential speedups for certain problems.

Quantum algorithms represent a paradigm shift in computing, offering the potential to solve certain problems exponentially faster than classical algorithms. They leverage the principles of quantum mechanics, such as superposition and entanglement, to perform computations in fundamentally different ways than classical computers. Quantum computers run eigensolvers[767] and factorization algorithms[768,769] to tackle problems that would otherwise be too sluggish or impractical to solve using classical computers. Quantum computers provide exponentially larger computational spaces per bit compared to classical systems, enhancing computational power and efficiency.

One of the most famous quantum algorithms is Shor's algorithm, which efficiently factors large numbers into their prime components.[770] This has significant implications for cryptography, as many encryption schemes rely on the difficulty of factoring large numbers. Shor's algorithm demonstrates how quantum computers can solve problems that are intractable for classical computers in a reasonable amount of time. Another notable quantum algorithm is Grover's algorithm, which provides a quadratic speedup for searching unsorted databases.[771] This has applications in areas such as optimization, where finding the optimal solution among a large number of possibilities is a common challenge. Moreover, quantum algorithms make use of entanglement, which allows qubits to become correlated in such a way that the state of one qubit depends on the state of another, even when they are physically separated.[772] This facilitates quantum computers[773] to accomplish parallel computations and achieve greater computational power than classical computers. Quantum logic gates, unlike classical gates, are unitary operators that act on qubits, enabling operations on single or multiple qubits for quantum computations. The demonstration of "quantum supremacy" in 2019 highlighted the significant progress in quantum information processing, showcasing the potential of quantum computers to outperform classical systems in specific tasks.[774] The goal is to develop fault-tolerant quantum systems that can correct errors efficiently and maintain coherence over extended periods for reliable computation and communication tasks.[775]

Despite their promise, quantum algorithms also face challenges. One major challenge is the requirement for error correction, as quantum systems are inherently susceptible to noise and decoherence, which can lead to errors in computation. Developing robust error correction techniques is crucial for realizing the full potential of quantum algorithms. However, quantum algorithms offer unparalleled advantages over classical algorithms by leveraging the principles of superposition, entanglement, and advanced gate operations. The transition from classical to quantum algorithms marks a significant advancement in computational capabilities, promising efficient simulations, factorization algorithms, and expanded computational spaces for solving complex problems that are beyond the reach of classical computing systems. While still in the early stages of development, quantum algorithms have the potential to revolutionize fields such as cryptography, optimization, and machine learning.

### 9.4.1 Applications of 2D Materials Quantum Algorithms and Computing

2D materials play a crucial role in advancing quantum algorithms and quantum computing, offering unique properties that enhance computational capabilities. Here is an overview:

1. Quantum Gates: 2D materials can serve as platforms for implementing quantum gates, which are fundamental building blocks of quantum circuits. Their atomic-scale thickness and electronic properties make them suitable for manipulating quantum states and performing logical operations.

2. Qubit Implementation: Qubits, the quantum analogs of classical bits, can be realized using 2D materials. These materials provide a

coherent quantum platform where qubits can be encoded and manipulated, enabling quantum information processing tasks.

3. Quantum Simulation: 2D materials can simulate quantum systems due to their ability to emulate different physical phenomena, such as electron-electron interactions, spin dynamics, and quantum phase transitions. Quantum simulators based on 2D materials offer insights into complex quantum systems that are challenging to study using classical computers.

4. Quantum Communication: 2D materials play a crucial role in quantum communication protocols. Their unique optical and electronic properties enable the generation, transmission, and manipulation of quantum information carriers, facilitating secure communication channels based on quantum principles.

5. Quantum Sensing: Quantum sensors made from 2D materials can achieve high sensitivity and resolution in detecting physical quantities such as magnetic fields, electric fields, and light. These sensors exploit quantum phenomena such as spin-dependent effects and quantum interference, offering superior performance compared to classical sensors.

6. Error Correction: 2D materials are explored for their potential in implementing error-correction schemes essential for fault-tolerant quantum computing. Their coherent quantum states and controllable properties are leveraged to develop robust error-correction techniques, mitigating the effects of noise and decoherence in quantum systems.

7. Quantum Algorithms: Researchers are developing quantum algorithms tailored for 2D material-based quantum computers. These algorithms exploit the unique properties of 2D materials to perform computational tasks efficiently, offering advantages over classical algorithms in certain problem domains.

8. Quantum Dot Platforms:

    - Quantum Dot Qubits: QDs within the range of 10-100 nm in diameter serve as a promising platform for quantum computing, enabling simulations of complex many-body systems.[776]

- Quantum Simulation: QD-based quantum computers excel at simulating challenging problems like Fermi-Hubbard models[777] and Nagaoka ferromagnetism,[778] which are computationally intractable on classical systems.

9. Advantages of 2D Materials:

- Intrinsic Quantum Properties: Nanomaterials with intrinsic quantum confinement, such as 2D materials, offer advantages over conventional materials for quantum information processing due to their unique electronic structures and physical properties.

- Surface Passivation: Nanomaterials provide passivated surfaces that enhance purity and reduce defects, crucial for minimizing sources of decoherence in quantum systems.

10. Quantum Computing Enhancements:

- Quantum Supremacy: The demonstration of "quantum supremacy" highlights the significant progress in quantum information processing, showcasing the potential of quantum computers to outperform classical systems in specific tasks.[774]

- Error Correction and Fault Tolerance: Developing fault-tolerant quantum systems is essential to correct errors efficiently and maintain coherence over extended periods for reliable computation and communication tasks.[774]

In conclusion, the integration of 2D materials like TMDs with quantum dot qubits offers a promising pathway for enhancing quantum algorithms and computing capabilities. These materials provide unique advantages such as intrinsic quantum properties, surface passivation, and the potential for fault-tolerant systems, paving the way for more efficient simulations, factorization algorithms, and expanded computational spaces in quantum information processing.

# CHAPTER 10

# EMERGING TOPICS

In recent years, the exploration of 2D materials has ignited a flurry of research activities aimed at uncovering their extraordinary properties and harnessing their potential for various applications. These materials, which consist of single or few atom-thick layers, exhibit remarkable electronic, optical, and mechanical characteristics that set them apart from their bulk counterparts. Graphene, the first discovered 2D material, opened the door to a vast array of possibilities, followed by other 2D materials like TMDs, *h*-BN, and beyond. The allure of 2D materials lies in their unique atomic-scale thickness, large surface-to-volume ratio, and tunable properties, making them highly desirable for diverse fields such as electronics, photonics, sensing, and energy conversion. This chapter delves into four emerging topics in the realm of 2D materials: Spintronics, Twistronics, Valleytronics, and Charge Density Wave (CDW) & Superconductivity each offering exciting prospects for advancing fundamental science and technological innovation.

Spintronics, an interdisciplinary field that utilizes the spin of electrons to store, process, and transmit information, has found promising avenues in 2D materials. Here, we report the manipulation and detection of electron spins in graphene, TMDs, and other 2D materials, along with the development of spin-based devices for next-generation electronics.

The concept of twistronics, which involves stacking layers of 2D materials at specific angles to engineer unique electronic properties, has emerged as a frontier in 2D material research. This section investigates the tunability of electronic band structures, moiré superlattices, and correlated phenomena in twisted bilayer graphene and other twisted van der Waals heterostructures.

Valleytronics, a promising paradigm for encoding and manipulating information using the valley degree of freedom in materials, has garnered significant interest in 2D semiconductors. Here, we delve into the exploitation of valley polarization, valley-dependent optical properties, and valleytronic devices in materials such as monolayer (1L) TMDs.

CDW and superconductivity phenomena in 2D materials have emerged as intriguing areas of study with profound implications for fundamental physics and technological applications. This section explores the discovery and manipulation of CDW and superconducting states in various 2D material systems, paving the way for novel quantum devices and high-performance electronics.

Overall, this chapter provides a comprehensive overview of the latest advancements and emerging topics in the dynamic field of 2D materials, highlighting their potential for revolutionizing electronics, spintronics, and quantum technologies.

## 10.1 Spintronics in 2D materials

Theoretical investigations into the electronic properties of mesoscopic condensed matter systems have traditionally focused on the charge characteristics of electrons, with less emphasis placed on their spin properties. However, an increasing body of experimental evidence suggests that electron spin presents unique opportunities[779,780,781] for discovering novel mechanisms in information processing and transmission. Particularly noteworthy are spin-related experiments conducted in quantum-confined nanostructures, revealing remarkably extended spin dephasing times reaching microseconds,[780,781,782] along with phase-coherent spin transport over distances exceeding 100 μm.[781] Beyond their intrinsic scientific intrigue, spin-related phenomena hold promise for future applications in two key domains: conventional[779] and quantum computer[783] hardware. In conventional computing, leveraging electron spin is anticipated to enhance the performance of quantum electronic devices, including spin-based transistors utilizing spin currents and injection, non-volatile memories, and the realization of single-spin configurations as the ultimate frontier in information storage. The burgeoning realm of quantum computing[784] and communication[785] necessitates a paradigm shift in hardware design. Electron spin emerges as a prime candidate for the qubit, the elemental unit of quantum information.[783] These spin qubits, when situated within quantum-confined structures like semiconductor quantum dots or atoms and molecules, fulfill all prerequisites for a scalable quantum computing architecture.[783] Furthermore, these spin qubits, linked to electrons with orbital degrees of freedom, can traverse conducting pathways between diverse subunits within a quantum network.[786] Notably, spin-entangled electrons, generated in interconnected quantum dots, serve as mobile Einstein-Podolsky-Rosen (EPR) pairs, furnishing the essential resources for quantum communication.[786]

Spintronics, an emerging computing paradigm, harnesses electron spin (±1/2) as a carrier for both information processing and storage. While its applications span diverse fields such as logic operations, memory, artificial intelligence (AI), and quantum computing (QC), all applications necessitate an effective method to manipulate electron spin while maintaining inherently high spin lifetimes, characterized by extensive spin diffusion lengths and robust spin injection efficiencies. As discussed in previous chapter, 2D-layered materials exhibit promising attributes compared to conventional 3D semiconductors in this regard. Their intrinsic higher strength of spin-orbit coupling (SOC) and superior electrostatics enable effective spin modulation. Additionally, the presence of van der Waals (vdW) gaps in their out-of-plane direction offers potential for achieving high spin-injection efficiencies.[787]

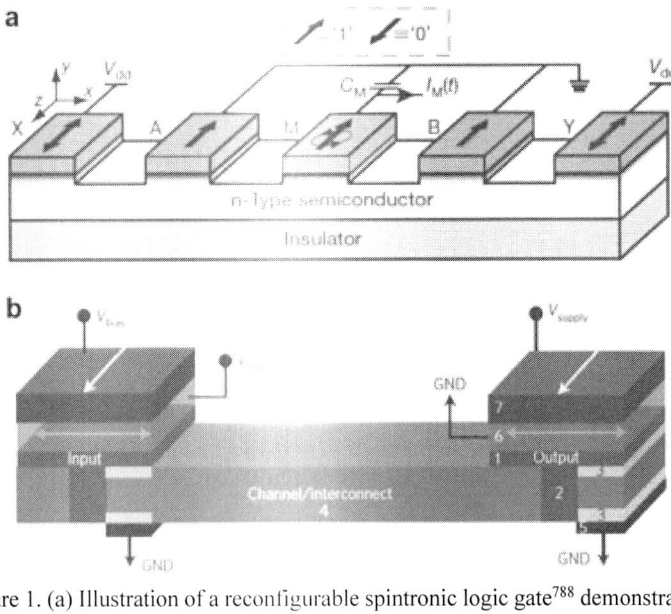


Figure 1. (a) Illustration of a reconfigurable spintronic logic gate[788] demonstrating a NAND operation between inputs X and Y. The magnetization alignments of A and B can be adjusted to configure different logic operations. (b) An all-spin logic (ASL) switch,[789] wherein nonlocal pure spin currents facilitate information transfer

between nanomagnets, functioning akin to digital spin capacitors. Reproduced with permission.[31]

### 10.1.1 Graphene based Spintronics

Amidst the array of concepts for post-CMOS logic switches,[790] there's a notable focus on spin-based, non-volatile digital logic[791,792] (see Figure 1a), which has garnered significant attention owing to its potential for reducing power consumption by employing magnetic elements in building logic units. In particular, the proposition of an all-spin logic (ASL) switch[789] stands out as an appealing prospect for future computing paradigms due to its capability to operate with exceptionally low power consumption.[793,794] This is attributed to the utilization of nonlocal spin currents in ASL,[795,796] enabling the selective routing of communication signals between input and output magnets without necessitating spin-to-charge conversion (refer to Fig. 1b). However, a substantial hurdle lies in developing an appropriate spin transport channel material that can achieve both high spin injection efficiency (from ferromagnets to the channel material) and prolonged spin propagation over considerable distances within the channel material. Graphene, characterized by its hexagonal 2D arrangement of $sp^2$-bonded carbon atoms, emerges as a promising candidate for this purpose due to its potential for (1) relatively extended spin diffusion length and lifetime (as predicted theoretically and under experimental scrutiny) and (2) facile integration with diverse material classes, including ferromagnets.

Graphene presents appealing properties for spintronic device applications.[797,798] With inherently low SOC, the spin-orbit splitting of the $\sigma$ band at the $\Gamma$ point is estimated to be as minimal as 8.8 meV.[799,800] This characteristic holds the potential for maintaining spin coherence over extended distances. Additionally, graphene boasts ultrahigh carrier mobility, reaching upwards of 200,000 $cm^2V^{-1}s^{-1}$ in suspended device configurations,[801] thereby facilitating enhanced diffusive spin transport. Progress in spin injection mechanisms into graphene, experimental elucidation of graphene's spin transport properties, and comprehension of spin relaxation mechanisms within graphene are poised to significantly contribute to the evolution of spin logic devices leveraging graphene as a spin transport channel.

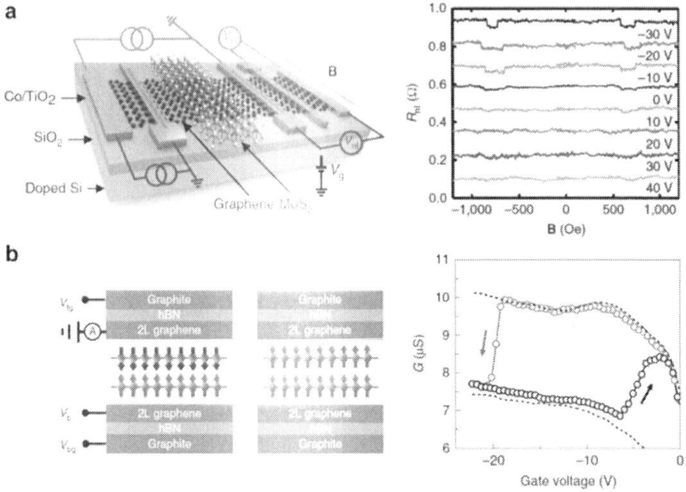


Figure 2. Spin Transistors Utilizing 2D Materials. (a) Schematic representation of the $MoS_2$/graphene heterostructure-based spin transistor device (left), alongside the recorded nonlocal resistances ($R_{nl}$) under varying gate voltage biases (Vg) applied (right).[802] (b) The spin tunnel FET (spin TFET) concept is proposed as an alternative way to implement spin transistor (left).[803] Reproduced with permission.[31]

### 10.1.2 Spintronics based on vdW Heterostructures

Yan *et al.* explored the vdW heterostructure composed of graphene/$MoS_2$ as an innovative spin transport channel.[802] Their investigation sought to leverage the distinct spin-related characteristics of both graphene and TMDs; graphene offers extended spin transport distances owing to its minimal SOC, [799,800] while TMDs exhibit robust SOC arising from the d-orbitals of heavy metals (resulting in a spin-orbit splitting of 148 meV in 1L $MoS_2$)[804], thereby holding potential for effective spin manipulation. In this research, graphene flakes were initially exfoliated onto a highly doped silicon substrate coated with a 300 nm-thick $SiO_2$ layer. Subsequently, a few-layer $MoS_2$ flake was transferred using an all-dry viscoelastic stamping method to establish a gate-tunable spin sink. Finally, multiple FM electrodes (Co) with $TiO_2$ serving as the tunnel barrier were patterned and deposited to construct the nonlocal spin valve (NLSV) configuration. As depicted in Figure 2a, the key finding demonstrated the ability to modulate the spin current (as determined by the nonlocal resistance measured in the NLSV) by adjusting the back-gate voltage ($V_g$) applied to the substrate. With a negative $V_g$ below −10 V, the low sheet conductance of $MoS_2$ facilitated efficient spin current flow within the adjacent graphene channel.

Conversely, for a positive $V_g$ exceeding 10 V, the increased sheet conductance of $MoS_2$ resulted in the absorption of a substantial portion of the spin current from graphene.

A recent endeavour aimed at realizing spin transistors utilizing 2D materials has resulted in the creation of a novel spin transistor variant known as the spin TFET.[803] In contrast to the conventional spin transistor concept proposed by Datta and Das, which hinges on the electrical manipulation of spin current within the high SOC channel material, the spin TFET is engineered by integrating emerging 2D magnetic crystals as the magnetic tunnel barrier (TB) (see Fig. 2b). Among the array of 2D magnetic materials uncovered recently,[805,806,807] the layered magnetic insulator, $CrI_3$, presents an intriguing thickness-dependent magnetic ordering phenomenon; 1L $CrI_3$ exhibits perpendicular ferromagnetism, whereas bilayer $CrI_3$ adopts an antiferromagnetic state comprising two ferromagnetic monolayers with antiparallel magnetic orientations. Notably, under a modest applied magnetic field (below 1 T), bilayer $CrI_3$ can undergo a spin-flip transition, transitioning to a ferromagnetic state. Furthermore, research has shown that the antiferromagnetic to ferromagnetic transition in bilayer $CrI_3$ can be induced by an electric field.[808] Therefore, in this study, a vertical junction consisting of graphene/$CrI_3$/graphene was fabricated to investigate the electrical switching behavior of $CrI_3$, interfaced with graphene top (source) and bottom (drain) contacts (as illustrated in Fig. 2b). Building upon previous demonstrations of the spin-filtering effect in $CrI_3$, which yields substantial tunnel magnetoresistance (TMR) signals in tunnel junctions,[809,810,811] this study represents a significant advancement in spin transistor research by unveiling an ambipolar transistor behavior with a remarkably high ON/OFF ratio approaching 400%. This achievement was realized by electrically manipulating (gate tuning) the magnetization configuration of the 2D magnetic insulator, rather than directly tempering the spin current.

## 10.2 Valleytronics in 2D Materials

Recent studies have revealed TMDs, characterized by the absence of a center of inversion (lack of crystallographic point symmetry)[812] but maintaining time-reversal symmetry (where energy bands at opposite momentum vectors have similar energies), such as 1L TMDs (Figure 3a), exhibit energy degenerate characteristics. Specifically, they possess valence and conduction band maxima and minima, respectively, at the K and K′ points in the Brillouin zone (as illustrated in Figure 3b). However, the electron spins occupying these minima and maxima are inequivalent. This intriguing property, known as the valley degree of freedom, offers a unique avenue for information storage and transmission, potentially surpassing conventional charge- and spin-based semiconductor technologies. Valleytronics, an emerging computational approach leveraging the spin–valley locking principle in systems without inversion symmetry, is notably enhanced in the VB of 2D-TMDs due to their substantial SOC, resulting in a valley-dependent optical selection rule. Here, right ($\sigma^+$)- and left ($\sigma^-$)-circularly polarized light induces electron–hole pair generation in the K′ and K valleys, respectively as depicted in Figure 3c, laying the foundation for valley-Hall measurements. Moreover, the unique valleytronic properties of 2D-TMDs hold promise for designing logic devices, offering avenues for low-energy computing.

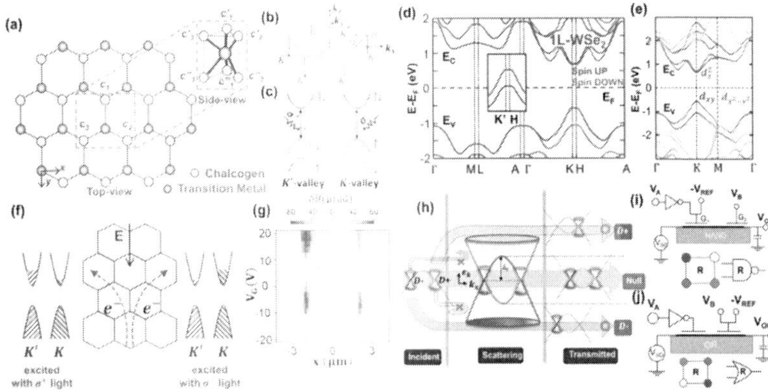


Figure 3. Valleytronic properties in 2D materials. (a) The atomic arrangement of the chalcogen (yellow) and the transition metals (red) in a 1L-2D-TMD showing the absence of the inversion symmetry in these materials, the key to obtaining valleytronic properties. (b) Brillouin zone (BZ) of a general hexagonal lattice showing symmetry points of Γ, M, K, K′ and H. (c) Degenerate spin-bands corresponding to K and K′ valleys in the BZ, and the valley dependent optical selection rules. (d) E–k band-structure of a conventional 1L-2D-TMD, WSe2, showing the SOC split spin-up (blue) and spin-down (red) bands. Inset figure shows the energy band splitting in the K′ point. (e) Orbital composition of the energy bands of a 1L-WSe2 flake showing the dominant contribution of the d-orbital of the transition metal in defining the CBM and VBM. (f) Illustration of the concept of valley-Hall effect. (g) Valley-Hall effect probed through Kerr measurements of a bilayer MoS2 sample. (h) Band diagram and working mechanism of a valley-filter. (i, j) Physical implementation of NAND (i) and OR (j) gates, designed by utilizing the three states of the basic valley filter. Reproduced with permission.[813] Copyright 2022, Wiley-VCH GmbH.

The spin-resolved band structure of a conventional 1L 2D-TMD $WSe_2$, illustrated in Figure 3d, serves as evidence for the phenomenon of spin-valley locking in TMDs with broken symmetry (as shown in Figure 3a). This locking mechanism is demonstrated by the distinct occupation of the spin-up and spin-down bands at the Valence Band maxima (VBM), situated at the K and K′ points of the BZ, respectively. The underlying physics of this spin–valley locking is influenced by the contribution of atomic orbitals to band definition in the BZ,[814] as depicted in Figure 3e.

The relationship between the valley and polarized light can be assessed through valley-Hall measurements, akin to Hall Effect measurements in charged systems and spin-Hall effect measurements in spin-polarized systems. However, unlike conventional Hall measurement systems where external magnetic fields are used as perturbations, the valley-Hall effect is observed upon the excitation of a semiconductor slab. This effect is evident in materials like monolayer 2D-TMDs, which exhibit both time-reversal symmetry and inversion asymmetry. Circularly polarized light is employed to excite the semiconductor slab, leading to a population imbalance of generated carriers in the conduction band of the K and K′ valleys as depicted in Figure 3f. These excited photocarriers encounter an effective magnetic field, which varies between the K and K′ valleys due to their distinct Berry curvatures resulting from opposite momentum. Consequently, this leads to the development of valley-contrasting Hall current.[815]

In samples featuring inversion symmetry, such as even-layered 2D-TMDs, the valley-Hall effect can still occur upon breaking their symmetry. This can be achieved either through the application of an external out-of-plane electric field[816] or through exchange interaction[817] with an adjacent foreign lattice. For instance, references[818,819] initially demonstrated the valley-Hall effect in a bilayer graphene (bilayer-Gr) and $MoS_2$ Hall-bar system, respectively as shown in Figure 3g. In the case of bilayer $MoS_2$, the inversion symmetry was broken by applying an out-of-plane electric field, which raised the potential of one layer relative to the other. Alternatively, references[820,821] utilized the exchange interaction of an *h*-BN substrate on graphene to disrupt the inversion symmetry in graphene.

An intriguing application of valleytronic devices is proposed in Ang *et al.*,[822] where valley polarization can be attained by manipulating valley-spin interactions in materials featuring merged Dirac cones (as depicted in Figure 3h). Examples of such materials include monolayer black phosphorus (BP), Weyl semimetals, graphene, and antimonene. By applying a gate voltage, the Fermi level can be shifted either into the conduction band or valence band, thus favouring the transmission of valley-polarized carriers. As a result, the device switches on and functions as a valley filter. When two such valley filters are connected in series, they form a valley valve, capable of generating a significant valley-pseudo magnetoresistance ratio exceeding 10,000%. Moreover, by adjusting the relative gate voltages applied to a series or parallel connection of these valley valves, their switching behavior can be modified to design fundamental NAND and OR gates, as illustrated in Figures 3i and 3j.

In summary, the field of valleytronics in 2D materials continues to advance rapidly, driven by ongoing research efforts to explore and harness the unique valley properties of these materials. With further developments in materials engineering, device fabrication, and characterization techniques, valleytronics holds great promise for revolutionizing future electronic technologies.

## 10.3 Twistronics in 2D Materials

Twistronics is a term coined to describe the study and manipulation of electronic properties in two-dimensional materials through the application of a twist or relative rotation between individual atomic layers.[823] It all began with the ground-breaking discovery in 2010 by Andre Geim and Konstantin Novoselov,[824] who observed that by stacking two layers of graphene and aligning them precisely, they could create a moiré pattern, a visually striking interference pattern that arises when two regular patterns are overlaid with a slight misalignment or twist. This field emerged from the discovery that stacking different layers of two-dimensional materials at a specific relative angle, known as the twist angle, can dramatically alter their electronic behavior, leading to novel physical phenomena and potential technological applications.

This seemingly simple twist in the graphene layers gave rise to a plethora of unexpected and fascinating phenomena. The twist-induced moiré pattern acts as a periodic potential landscape, effectively modulating the electronic structure of the graphene layers. As a result, the electronic properties of the material are profoundly altered compared to pristine graphene. This twist-induced modulation of electronic properties has since been extended to various other two-dimensional materials beyond graphene, such as TMDs, *h*-BN, and BP among others.

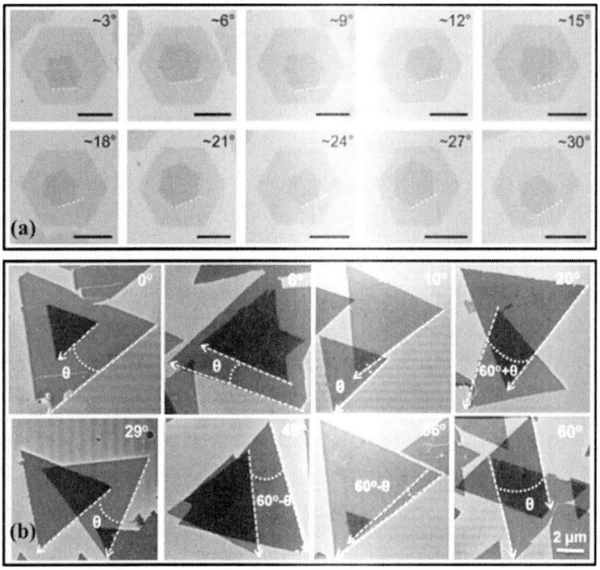


Figure 4. (a) Optical microscopy (OM) images showcasing as-grown twisted Bilayer Graphene with various twist angles, including approximately 3°, 6°, 9°, 12°, 15°, 18°, 21°, 24°, 27°, and 30°. The scale bars in the images are equivalent to 10 μm. (b) SEM images of $MoSe_2/WSe_2$ heterostructures with variety of twist angles in the range $0° \leq \theta \leq 60°$. The flakes with lower contrast are the $WSe_2$ layer (bottom layer) and with higher contrast are $MoSe_2$ layer (top layer). (a) Adapted with permission from ref.[825] Copyright 2024 American Chemical Society. (b) Adapted with permission from ref.[826]

CVD stands out as a promising method for creating moiré superlattices within twisted stacking configurations, owing to its capacity to facilitate the formation of clean interfaces and its scalability. Nevertheless, achieving precise control over twist angles presents a notable technical hurdle within the CVD approach. Typically, multilayer samples produced through CVD exhibit fixed twist angles of either 0° or 60°, underscoring the existing constraints in twist-angle manipulation. The optimization of catalytic substrates and meticulous management of physical conditions during synthesis are crucial in the fabrication of high-quality twisted bilayer graphene (t-BLG) using CVD. Furthermore, adjusting the growth parameters such as temperature, flow rate, and vacuum pressure is another critical aspect for enhancing the yield of small angle t-BLG (<30°). In Figure 4a, optical microscopy (OM) images showcase high-quality t-BLG with various twist angles ranging from approximately 3° to 30°, grown using the CVD method. Figure 4b illustrates the scanning electron microscopy (SEM) characterization of $MoSe_2/WSe_2$ heterostructures (HSs) featuring a range of twist angles between 0° and 60°. The flakes with lower contrast correspond to the WSe2 layer (bottom layer), while those with higher contrast represent the MoSe2 layer (top layer). In some regions, the top layer perfectly overlaps with the bottom layer, while in others, it exhibits partial overlap. The determination of the twist angle (θ) involves considering the clockwise orientation of the top MoSe2 triangle relative to the bottom WSe2 triangle, with an approximate error margin of 2°. Given the threefold symmetry, the maximum angle between the two triangles is 60°. When the angle exceeds 60° or falls below 0°, it is recorded as (60° + θ) or (60° - θ), respectively, where θ denotes the angle between the two layers. The error estimation is based on the disparities observed in the three twist angles determined from each edge (totalling three edges) in the two triangles.

In stacked $MoS_2$ bilayers featuring CVD-grown monolayers, van der Zande and colleagues uncovered twist angle-controlled adjustments in the indirect optical transition energy and second-harmonic generation (SHG), alongside slight variations in direct optical transition energy and Raman frequency,[827] as illustrated in Fig. 5(a). The displacement of indirect optical transition energies was ascribed to the angle-dependent interlayer separation, influenced by repulsion between sulfur (S) atoms [refer to Fig. 5(b)]. The findings suggested that the effective interlayer interaction was optimized in the aligned (θ = 0°) or anti-aligned (θ = 60°) configuration, while minimized at θ = 30°.

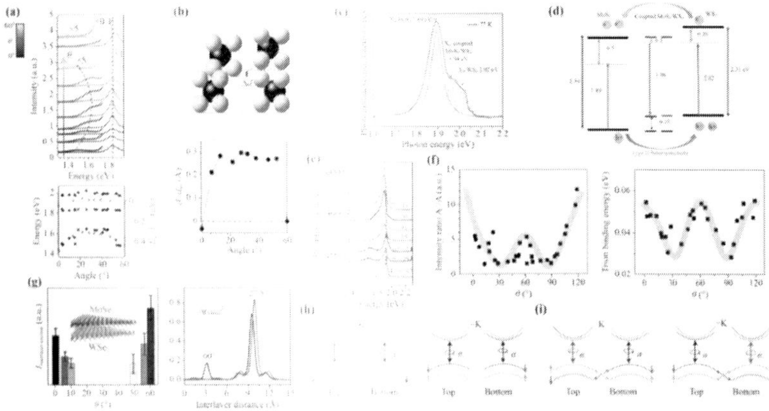

Figure 5. Experimental observations of stacking-angle modulated interlayer interactions are depicted in the figures as follows: (a) The upper panel illustrates photoluminescence (PL) spectroscopy data for both the bottom (as grown, gray line)

and top (transferred, green line) monolayers, along with twisted bilayers (t-BLs) of MoS2 at various twist angles (θ). The lower panel presents the measured (black) and simulated (magenta) peak energies plotted against θ for the indirect (circles), A (squares), and B (triangles) transitions. (b) The schematic illustrates the interlayer separation of t-BL MoS2, depending on the stacking patterns (upper panel) and the twisting angle θ (lower panel) relative to the 60° separation. (c) PL spectra are shown with Lorentzian fitting at 77 K. (d) The bandgap alignment of coupled as-grown MoS2/WS2 heterostructures is depicted, with solid black, green, and orange double-arrow lines representing the energy gap, optical gap, and exciton binding energy, respectively. (e) PL spectra of randomly twisted WS2 bilayers are compared with monolayers. (f) The twist angle dependence of the trion to exciton PL intensity ratio (left panel) and trion binding energy (right panel) are presented. (g) The intensity of the interlayer exciton peak versus the twist angle (left panel) and the plane-averaged partial charge density of each band edge state (valence band maximum and conduction band minimum at K) versus the interlayer distance for θ = 60° and 27.8° (right panel) of MoSe2/WSe2 heterostructures are shown. (h–i) Schematic representations illustrate valley and spin polarizations excited by circularly polarized light in bilayers with suppressed (h) interlayer hopping and (i) strong interlayer coupling. Adapted with permission from ref.[828]

Furthermore, Zheng *et al.*[829] documented modifications in bandgap structures and phonon vibrations in a CVD-grown WS2 bilayer. These bilayer WS2 structures, exhibiting random twist angles formed at elevated growth temperatures, demonstrated a significantly heightened photoluminescence (PL) intensity of peak A and an additional peak A1 compared to aligned stacking scenarios (θ = 0° or 60°), with the simultaneous absence of the indirect transition peak I (interlayer exciton) as seen in Fig. 5(e). These alterations were attributed to two mechanisms. Firstly, a larger interlayer distance weakened the coupling strength, as evidenced by the broadening and redshift of the out-of-plane A1g mode in Raman spectra. Secondly, symmetry breaking in the randomly twisted bilayers increased the likelihood of interlayer hopping, a phenomenon suppressed by spin-layer locking in the aligned scenario. Zhang *et al.*[830] also observed an additional peak in photoluminescence (PL) spectra at 77 K in the as-grown $MoS_2/WS_2$ heterostructure with type II band alignment, identified as the recombination of interlayer carriers [refer to Figs. 5(c) and (d)].

Huang *et al.*[831] focused their investigation on trions and excitons within stacked bilayers of CVD-grown MoS2, varying twist angles between 0° and 120°. In PL spectra, both the trion ($A^-$)-to-exciton (A) ratio and the trion binding energy exhibited periodic and oscillatory patterns in relation to the twist angle θ, as depicted in Fig. 5(f). These parameters peaked at aligned configurations (θ = 0° or 60°) and reached their lowest points at θ = 30° or

90°, reflecting the angle-dependent variations in interlayer coupling strength. By employing vdW-corrected density functional theory (DFT) calculations, Nayak *et al.*[832] discovered that the states at the valence band edges were influenced by the twist angle. In coherently stacked MoSe2/WSe2 configurations, these states displayed an elongated tail extending over another layer, whereas in incoherent configurations, they were localized within individual layers, as illustrated in Fig. 5(g). The lattice mismatch, regulated by the twist angle, would also have a substantial impact on the properties of devices relying on vdW heterostructures.

Circularly polarized PL spectra exhibited varying degrees of polarization (DOP) across different stacking bilayers, with folded bilayers showing DOP levels comparable to monolayers and significantly higher than those observed in natural 2H bilayers. The notable DOP observed in folded bilayers confirmed the suppression of interlayer hopping and the creation of a spin-valley configuration locked to the layer index, attributable to broken inversion symmetry. Conversely, in the centrosymmetric 2H stacking bilayer, the interlayer hopping amplitude of holes was estimated to be on par with the strength of spin-orbit coupling (SOC), resulting in the mixing of layer pseudospins and the dampening of valley- and spin-selective circular dichroism, as depicted in Figures 5(h) and (i).

## 10.4 Charge Density Waves and Superconductivity in 2D Materials

Recent advancements in synthesis techniques for 2D layered materials have opened up opportunities to explore material systems that were previously challenging to study.[833,834] TMDs, in particular, have garnered renewed interest due to their unique 2D confined structures.[835,836] TMDs exhibit a diverse range of intriguing properties, including semiconducting,[837,838,839] superconducting,[840] and charge density wave (CDW)[841,842] behaviors. A CDW state refers to a ground state characterized by a periodic modulation of charge density, coupled with a periodic distortion of the crystal lattice.[843,844,845] The transition to the CDW state is often attributed to Fermi surface nesting with a wave vector Q = 2kF and is associated with an energy gap at the Fermi energy in the quasi-1D counterpart, as illustrated in Figure 6.[846,847]

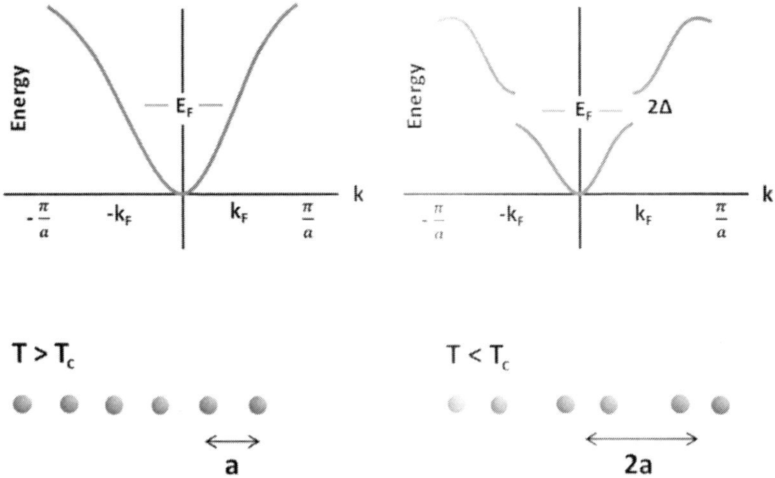


Figure 6. Visual representation depicting the transition from the normal state to the charge-density-wave state in one dimension. The displacement of atoms causes the formation of a "superlattice," leading to the opening of an electronic energy band gap. Adapted with permission from ref.[848]

A significant group of 2D CDW materials consists of TMDs, structured in the form of X-M-X, where M represents transition metals like V, Nb, Ta, or Ti, and X denotes chalcogen atoms such as S, Se, or Te. These materials feature chalcogen atoms arranged in two hexagonal planes, separated by a plane of metal atoms situated in the middle. The layers in X-M-X are weakly bonded together and may exhibit a different coordination of metal atoms within the layer and distinct stacking arrangements between layers. Metal atom coordination configurations in 2D CDW materials can take the form of octahedral (T) or trigonal prismatic (H) configurations. Various CDW phases, including commensurate CDW (C-CDW), nearly-commensurate CDW (NC-CDW), and incommensurate CDW (IC-CDW), can coexist at different temperatures. Despite over three decades of exhaustive experimental and theoretical investigations, the driving force behind the CDW transition remains a subject of debate. Intercalating with elements such as copper,[849] selenium,[850] or tellurium,[851] iron,[852] or applying pressure,[853] disrupts the CDW ordering, leading to the emergence of superconductivity with critical transition temperatures of several Kelvin. This phenomenon directly underscores the competition between CDW and superconductivity in 2D CDW materials. In atomically thin 2D CDW materials, phase transitions can induce significant alterations in electronic structure, paving the way for novel electronic devices. Recent years have seen successful demonstrations of electronic applications like oscillators[854] and non-volatile memory storage,[855,856] all based on phase transitions associated with charge density waves.

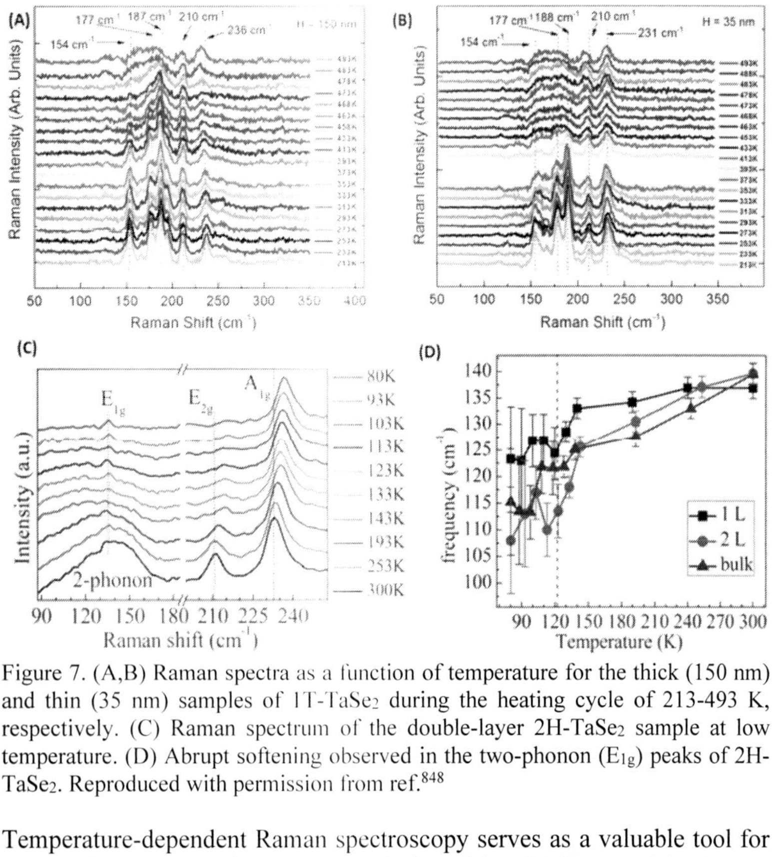

Figure 7. (A,B) Raman spectra as a function of temperature for the thick (150 nm) and thin (35 nm) samples of 1T-$TaSe_2$ during the heating cycle of 213-493 K, respectively. (C) Raman spectrum of the double-layer 2H-$TaSe_2$ sample at low temperature. (D) Abrupt softening observed in the two-phonon ($E_{1g}$) peaks of 2H-$TaSe_2$. Reproduced with permission from ref.[848]

Temperature-dependent Raman spectroscopy serves as a valuable tool for unravelling the vibrational characteristics of 2D CDW materials. Recent studies have delved into the temperature-dependent Raman spectra of various compounds, including 1T-VSe2,[857] 1T-TaS2,[858] 1T-TaSe2,[859] 2H-TaSe2,[860] 1T-TiSe2,[861] and 2H-NbSe2.[841] TMDs with metallic properties can undergo a Peierls transition,[862] leading to the formation of a superlattice atop the underlying crystal lattice at specific critical temperatures, thereby lowering the system's overall energy. Figure 7 illustrates the temperature-dependent Raman spectra of 1T-TaSe2 and 2H-TaSe2 samples within the temperature ranges of 213-493 K and 80-300 K, respectively. Bulk 1T-TaSe2 crystals transition into the C-CDW phase at 473 K and then into the

IC-CDW phase at 600 K.[859] Significant changes in the Raman spectra around 473 K for 1T-$TaSe_2$ are evident (Figure 7A). The merging of two Raman peaks at 177 and 187 $cm^{-1}$ into a single broad peak, accompanied by reduced intensity, marks the transition into the IC-CDW phase. Notably, at 473 K (IC-CDW phase), the spectra of 1T-$TaSe_2$ appear more continuous rather than consisting of narrow peaks. This behavior can be attributed to symmetry considerations. In the IC-CDW state, the lattice's translational equilibrium diminishes, and the phonon momentum components are no longer favourable quantum numbers when aligned parallel to the IC-CDW distortions. In the case of a thinner sample (35 nm) (Figure 7B), it is also observed that the two Raman peaks begin to merge, albeit at a lower transition temperature (413 K). These findings clearly demonstrate that the transition temperature between the C-CDW and IC-CDW phases decreases for thinner 1T-TaSe2 films.

In bulk 2H-TaSe2, a transition occurs from the normal metal phase to the IC-CDW phase at 123 K, followed by a transformation to the C-CDW phase at 90 K.[860] As depicted in Figure 7C, the intensities of the two-phonon peak and $E_{2g}$ mode of 2H-TaSe2 decrease as the temperature decreases. Conversely, the intensity of the two-phonon mode ($E_{1g}$) increases with decreasing temperatures. This phenomenon can be explained by the reduction of the background signal from the two-phonon process, leading to increased visibility of this peak, as well as strain-induced effects due to differences in thermal expansion coefficients between thinner films and the substrate.[863,864] The frequency changes of the two-phonon modes for monolayer, bilayers, and bulk samples are plotted as a function of temperature (Figure 7D), revealing frequency jumps at 120 K, indicating the presence of the IC-CDW state in the thinner crystals of 2H-$TaSe_2$.

Superconductivity in 2D materials represents a burgeoning field of research at the forefront of condensed matter physics and materials science. Historically, superconductivity was predominantly studied in bulk materials, but recent advancements have enabled the exploration of this phenomenon in atomically thin layers,[865] opening up new avenues for both fundamental research and technological applications. The discovery of superconductivity in 2D materials has sparked immense interest due to several compelling features. One of the most notable aspects is the potential for higher transition temperatures (Tc) compared to their bulk counterparts.[866] This heightened Tc is attributed to quantum confinement effects,[867] enhanced electron-electron interactions,[868] and reduced electron scattering in the 2D plane,[869] leading to more robust superconducting states.

Moreover, the 2D nature of these materials allows for precise control over their properties through methods such as electrostatic gating,[870] strain engineering,[871] and chemical doping.[872] This tunability offers opportunities to tailor the superconducting behavior for specific applications and to explore novel physical phenomena.

Different types of superconductivity have been observed in 2D materials, each with its own unique characteristics. Conventional Bardeen-Cooper-Schrieffer (BCS) superconductivity has been observed in ultrathin films of certain materials,[873,874] exhibiting properties consistent with classic superconductors. Additionally, high-Tc superconductivity has been reported[875] in specific 2D compounds,[876] such as Fe-based[877,878] and Cu-based superconductors,[879,880] which hold promise for practical applications due to their elevated transition temperatures.

Unconventional superconductivity has also been discovered in twist-angle bilayer graphene, where moiré patterns formed by stacking two graphene layers at a small twist angle induce flat electronic bands and strong electronic correlations,[881,882,883] leading to exotic superconducting states. Furthermore, there are efforts[884,885] to engineer superconductivity with enhanced critical magnetic fields (Bc) and to explore topological superconductivity in certain 2D materials, which could have profound implications for quantum computing[886,887] and topological quantum technologies.

In summary, superconductivity in 2D materials presents a rich and diverse landscape for scientific exploration and technological innovation. Continued research in this field holds the promise of uncovering new physics, advancing our understanding of superconductivity, and realizing novel applications in areas such as energy-efficient electronics, quantum computing, and beyond.

# CHAPTER 11

# CHALLENGES, FUTURE DIRECTIONS AND CONCLUSION

In this chapter, we report the key challenges in the field of 2D quantum materials and discuss potential future directions for overcoming these obstacles. By addressing these challenges and advancing the state-of-the-art in materials synthesis, characterization, and device fabrication, researchers and engineers can unlock new opportunities for innovation and drive technological advancements that benefit society as a whole. Through collaboration between academia, industry, and government agencies, we can overcome these challenges and pave the way for the widespread adoption and application of 2D quantum materials in various industries and sectors.

## 11.1 Scalability and Integration Challenges

2D quantum materials have emerged as promising candidates for various applications in electronics, optoelectronics, and quantum technologies. These materials, such as graphene, TMDs, and black phosphorus, exhibit unique electronic and optical properties due to their atomically thin nature. However, despite their immense potential, there are several challenges related to scalability and integration that need to be addressed for their successful implementation in practical devices. In the next section, we will delve into the scalability and integration challenges facing 2D quantum materials and discuss potential strategies to overcome them.

### 11.1.1 Scalability Challenges

One of the primary challenges in the scalability of 2D quantum materials is the reproducible synthesis of large-area, high-quality samples. While mechanical exfoliation can yield atomically thin flakes of materials like graphene, it is not scalable for industrial applications. CVD has emerged as a scalable technique for synthesizing large-area 2D materials, but achieving high-quality monolayers with uniform thickness and low defect density remains challenging. Moreover, the scalability of CVD depends on the availability of precursors and the optimization of growth parameters.

Another scalability challenge is the development of reliable methods for transferring and patterning 2D materials onto substrates. The transfer process can introduce defects and contamination, adversely affecting the material's electronic properties. Additionally, precise patterning of 2D materials is crucial for device fabrication, but conventional lithography techniques may not be suitable for atomically thin materials. Alternative patterning methods, such as electron beam lithography or nanoimprint lithography, need to be developed to achieve high-resolution patterning of 2D materials at scale.

### 11.1.2 Integration Challenges

Integration of 2D quantum materials into existing device architectures poses several challenges due to differences in material properties and interfaces. One major challenge is the integration of 2D materials with complementary metal-oxide-semiconductor (CMOS) technology, which forms the backbone of modern electronics. CMOS-compatible processing techniques are needed to incorporate 2D materials into existing silicon-based technology without compromising device performance or reliability.

Furthermore, achieving reliable electrical contacts between 2D materials and metal electrodes is critical for device operation. The contact resistance at the metal-2D material interface can significantly impact device performance, particularly in high-speed electronics and quantum devices. Developing low-resistance, ohmic contacts with minimal Schottky barriers is essential for realizing the full potential of 2D materials in practical devices.

Another integration challenge is the hetero-integration of different 2D materials to create complex device architectures with tailored functionalities. Heterostructures composed of stacked layers of different 2D materials can exhibit unique electronic, optical, and mechanical properties not present in individual materials. However, achieving precise alignment and intimate coupling between different layers while maintaining high-quality interfaces remains a significant challenge.

## 11.2 Strategies to Overcome the Challenges

Addressing scalability and integration challenges in 2D quantum materials requires a multidisciplinary approach involving materials science, device physics, and engineering. One promising strategy is the development of scalable synthesis techniques that can produce high-quality 2D materials with minimal defects and uniform thickness over large areas. Advances in CVD growth, as well as bottom-up synthesis methods, such as molecular beam epitaxy (MBE) and atomic layer deposition (ALD), hold promise for achieving scalable production of 2D materials.

Improving transfer and patterning techniques is essential for integrating 2D materials into device architectures. Novel transfer methods, such as dry transfer techniques and stamping methods, can minimize contamination and defects during the transfer process. Furthermore, the development of advanced patterning techniques, such as directed self-assembly and block copolymer lithography, can enable high-resolution patterning of 2D materials with precise control over feature size and spacing.

To address integration challenges, researchers are exploring novel device architectures and materials combinations that are compatible with existing CMOS technology. Hybrid integration approaches, such as van der Waals heterostructures and hybrid materials, can facilitate the seamless integration of 2D materials into CMOS-compatible devices. Moreover, advances in contact engineering and interface engineering are crucial for reducing contact resistance and improving device performance.

In conclusion, scalability and integration challenges pose significant hurdles in the realization of practical devices based on 2D quantum materials. However, with continued research and development efforts, innovative solutions are being developed to overcome these challenges. By leveraging advances in materials synthesis, device fabrication, and integration techniques, researchers can unlock the full potential of 2D quantum materials for next-generation electronics, photonics, and quantum technologies.

## 11.3 Environmental and Toxicity Considerations

2D materials hold immense potential for various applications, ranging from electronics and optoelectronics to energy storage and biomedicine. However, as with any emerging technology, the environmental and toxicity considerations associated with the synthesis, use, and disposal of 2D quantum materials need to be carefully evaluated. We explore the environmental and toxicity challenges facing 2D quantum materials in the next part and discuss potential strategies to address them.

### 11.3.1 Environmental Impact of Synthesis Methods

The synthesis of 2D quantum materials often involves energy-intensive processes and the use of hazardous chemicals, raising concerns about their environmental footprint. For example, CVD, a common technique for synthesizing large-area graphene and TMDs, requires high temperatures and precursor gases, which can contribute to greenhouse gas emissions and air pollution. Similarly, liquid-phase exfoliation methods may involve the use of organic solvents, some of which are toxic and harmful to the environment if not properly managed.

Moreover, the production of precursor materials for 2D quantum materials, such as graphite for graphene or metal oxides for TMDs, may involve mining and extraction processes that can cause environmental degradation and habitat destruction. Additionally, the disposal of waste materials generated during the synthesis process poses challenges for waste management and environmental sustainability.

### 11.3.2 Toxicity Considerations

While 2D quantum materials themselves are often considered biocompatible and inert, the functionalization and modification of these materials for specific applications can introduce toxicity concerns. For example, the functionalization of graphene with chemical groups or nanoparticles may alter its biocompatibility and introduce potential health risks. Similarly, the use of TMDs in biomedical applications, such as drug delivery and biosensing, raises questions about their long-term effects on human health and the environment.

Furthermore, the release of 2D quantum materials into the environment, either during manufacturing processes or through the disposal of end-of-life products, may pose risks to ecosystems and wildlife. Studies have shown that certain nanoparticles, including graphene and TMDs, can accumulate in soil and water systems, potentially affecting microbial communities and disrupting ecological balances.

### 11.3.3 Strategies to Address Environmental and Toxicity Challenges

To mitigate the environmental impact of synthesis methods, researchers are exploring greener approaches to produce 2D quantum materials. Green synthesis methods, such as microwave-assisted exfoliation and electrochemical synthesis, offer alternative routes that reduce energy consumption and minimize the use of hazardous chemicals. Moreover, the recycling and reuse of precursor materials and waste products can minimize environmental impact and promote sustainability in the production process.

In terms of toxicity considerations, researchers are investigating ways to modify 2D quantum materials to enhance their biocompatibility and reduce potential health risks. Surface functionalization with biocompatible polymers or biomolecules can improve the stability and compatibility of these materials for biomedical applications. Furthermore, rigorous toxicity assessments and risk analyses are essential to evaluate the safety of 2D quantum materials and ensure their responsible use in various applications.

Additionally, the development of nanomaterials regulations and guidelines can help govern the responsible synthesis, handling, and disposal of 2D quantum materials. Collaboration between researchers, industry stakeholders, and regulatory agencies is essential to establish standards and protocols for the safe and sustainable use of these materials.

In a nutshell we can suggest that environmental and toxicity considerations are critical aspects that must be addressed in the development and utilization of 2D quantum materials. By adopting greener synthesis methods, minimizing waste generation, and ensuring the biocompatibility and safety of these materials, researchers can mitigate their environmental impact and promote sustainable development. Collaboration across disciplines and sectors is key to addressing these challenges and realizing the full potential of 2D quantum materials for transformative applications in science, technology, and society.

## 11.4 Future Prospects and Potential Breakthroughs

In the following section, we explore the future prospects and potential breakthroughs in the field of 2D quantum materials, focusing on key challenges and opportunities for further advancement.

### 11.4.1 Electronics and Optoelectronics

One of the most promising applications of 2D quantum materials is in the field of electronics and optoelectronics. Graphene, with its high carrier mobility and transparency, has the potential to revolutionize the semiconductor industry by enabling faster and more energy-efficient electronic devices. Additionally, TMDs offer unique properties for optoelectronic applications, such as photodetection and light emission, due to their direct bandgap and strong light-matter interactions.

Future breakthroughs in this area may include the development of novel electronic and optoelectronic devices based on 2D quantum materials, such as flexible and transparent displays, ultrafast transistors, and efficient solar cells. Furthermore, the integration of 2D materials with other functional materials, such as organic semiconductors and quantum dots, could lead to the development of hybrid devices with enhanced performance and functionality.

### 11.4.2 Energy Storage and Conversion

Another promising area for 2D quantum materials is in energy storage and conversion technologies. Graphene, with its high surface area and mechanical strength, has shown potential for applications in supercapacitors and batteries, offering high energy density and fast charging capabilities. Additionally, TMDs have been investigated for their potential use in catalysis and hydrogen evolution reactions, owing to their unique electronic and chemical properties.

Future breakthroughs in this field may involve the development of advanced energy storage devices based on 2D quantum materials, such as flexible and lightweight batteries and supercapacitors with improved energy density and cycling stability. Moreover, the integration of 2D materials with other nanomaterials, such as carbon nanotubes and metal nanoparticles, could lead to the development of hybrid materials with enhanced performance for energy conversion applications.

### 11.4.3 Biomedical Applications

2D quantum materials also hold promise for biomedical applications, including drug delivery, bio-sensing, and tissue engineering. Graphene, with its biocompatibility and large surface area, has shown potential for applications in drug delivery systems and biosensors, enabling targeted drug delivery and sensitive detection of biomolecules. Similarly, TMDs have been explored for their potential use in biomedical imaging and cancer therapy, owing to their strong light-matter interactions and tunable optical properties.

Future breakthroughs in this area may involve the development of novel biomedical devices and therapeutics based on 2D quantum materials, such as implantable sensors for monitoring physiological parameters and drug delivery platforms for targeted cancer therapy. Moreover, the functionalization and modification of 2D materials with biomolecules and targeting ligands could further enhance their biocompatibility and specificity for biomedical applications.

## 11.5 Challenges and Opportunities

Despite the promising prospects of 2D quantum materials, several challenges remain to be addressed for their widespread adoption and commercialization. These challenges include scalability and reproducibility of synthesis methods, environmental and toxicity considerations, and integration with existing technologies. Additionally, fundamental research is needed to fully understand the underlying physics and chemistry of 2D materials and explore new applications and functionalities.

However, with continued investment in research and development, collaboration between academia and industry, and advancements in materials characterization and fabrication techniques, the field of 2D quantum materials is poised for significant growth and innovation in the coming years. Potential breakthroughs in this field have the potential to revolutionize various industries and drive technological advancements that benefit society as a whole.

## 11.6 Conclusion

The book "Two-Dimensional Quantum Materials" offers a comprehensive exploration of the multifaceted world of 2D materials, spanning from their fundamental properties to cutting-edge applications and emerging research areas. Through a series of meticulously curated chapters, the book provides a thorough understanding of the synthesis, characterization, and utilization of 2D materials, while also addressing the challenges and future directions in this rapidly evolving field.

- Introduction to Quantum Materials:

The introductory chapter provides a foundational overview of quantum materials, elucidating their classification based on dimensional confinement and tracing the historical journey that led to the discovery of graphene. It highlights the significance and broad applications of 2D materials across various disciplines.

- Crystal Structure and Properties:

This chapter delves into the atomic structure, bonding mechanisms, and fundamental properties of 2D materials, shedding light on their electronic band structures and diverse optical, electrical, and mechanical characteristics.

- Synthesis and Fabrication Techniques:

Exploring the methods of synthesis and fabrication, this chapter discusses both top-down and bottom-up approaches, offering insights into the techniques used to produce high-quality 2D materials.

- Characterization Methods:

The characterization methods chapter explores advanced techniques such as Raman spectroscopy, PL spectroscopy, and Transmission Electron Microscopy, essential for understanding the structural and functional properties of 2D materials.

- Quantum Phenomena:

Unravelling the intriguing quantum phenomena exhibited by 2D materials, this chapter discusses phenomena like Quantum Conductance, Quantum Hall Effect, and the presence of Dirac and Weyl fermions, shedding light on the quantum nature of these materials.

- Applications in Nano-electronics:

Highlighting the practical applications of 2D materials in nano-electronics, this chapter explores their role in devices such as Field-Effect Transistors (FETs), Thin Film Transistors (TFTs), and Flexible Electronics, showcasing their potential for revolutionizing electronic technologies.

➢ Optoelectronic Devices:

This chapter delves into the optoelectronic applications of 2D materials, including photodetectors, solar cells, and light-emitting diodes (LEDs), emphasizing their importance in the field of photonics and optoelectronics.

➢ Energy Storage and Conversion:

Exploring the role of 2D materials in energy storage and conversion technologies, this chapter discusses their applications in supercapacitors, batteries, hydrogen production, and fuel cells, highlighting their potential for addressing global energy challenges.

➢ Quantum Computing:

Discussing the suitability of 2D materials for quantum computing applications, this chapter explores their role in single quantum emission, quantum entanglement, and the development of qubits and quantum algorithms, paving the way for next-generation computing technologies.

➢ Emerging Topics:

The chapter on emerging topics explores cutting-edge research areas such as spintronics, twistronics, valleytronics, and the emergence of charge density wave (CDW) and superconductivity in 2D materials, showcasing the exciting avenues for future exploration and innovation.

➢ Challenges and Future Directions:

In the final chapter, the book addresses the challenges and future directions in the field of 2D quantum materials, discussing scalability and integration challenges, environmental considerations, and potential breakthroughs, providing a roadmap for advancing the field in the years to come.

***In conclusion, "Two-Dimensional Quantum Materials" serves as an indispensable resource for researchers, students, and professionals interested in delving into the fascinating world of 2D materials, offering a comprehensive overview of their synthesis, properties, applications, and future prospects.***

# REFERENCES


1 Furness, James W., Ruiqi Zhang, Jamin Kidd, and Jianwei Sun. "Emergence of competing electronic states from non-integer nuclear charges." Communications Physics 6, no. 1 (2023): 246.

2 Liu, Jinglu, Yiran Ding, Mengqi Zeng, and Lei Fu. "Chemical insights into two-dimensional quantum materials." Matter 5, no. 7 (2022): 2168-2189.

3 De Broglie, M. "La relation hν= ε dans les phénomenes photoélectriques." Atomes et electrons (Paris: Gauthier, 1923) (1923): 80-100.

4 McNaught, Alan D. Compendium of chemical terminology. Vol. 1669. Oxford: Blackwell Science, 1997.

5 Brus, Louis E. "Electron–electron and electron-hole interactions in small semiconductor crystallites: The size dependence of the lowest excited electronic state." The Journal of chemical physics 80, no. 9 (1984): 4403-4409.

6 Kailasa, Suresh Kumar, Kuang-Hung Cheng, and Hui-Fen Wu. "Semiconductor nanomaterials-based fluorescence spectroscopic and matrix-assisted laser desorption/ionization (MALDI) mass spectrometric approaches to proteome analysis." Materials 6, no. 12 (2013): 5763-5795.

7 Wallace, Philip Richard. "The band theory of graphite." Physical review 71, no. 9 (1947): 622.

8 Brodie, Benjamin Collins. "XIII. On the atomic weight of graphite." Philosophical transactions of the Royal Society of London 149 (1859): 249-259.

9 Staudenmaier, L. "Process for the preparation of graphitic acid." Ber. Dtsch. Chem. Ges 31 (1898): 1481-1487.

10 Hummers Jr, William S., and Richard E. Offeman. "Preparation of graphitic oxide." Journal of the american chemical society 80, no. 6 (1958): 1339-1339.

11 Boehm, Hanns-Peter, A. Clauss, G. O. Fischer, and U. Hofmann. "Dünnste kohlenstoff-folien." Zeitschrift Für Naturforschung B 17, no. 3 (1962): 150-153.

12 Avouris, Phaedon. "Graphene: electronic and photonic properties and devices." Nano letters 10, no. 11 (2010): 4285-4294.

13 Neto, AH Castro, Francisco Guinea, Nuno MR Peres, Kostya S. Novoselov, and Andre K. Geim. "The electronic properties of graphene." Reviews of modern physics 81, no. 1 (2009): 109.

14 Karu, Alexander E., and Michael Beer. "Pyrolytic formation of highly crystalline graphite films." Journal of Applied Physics 37, no. 5 (1966): 2179-2181.

15 Lang, B. "A LEED study of the deposition of carbon on platinum crystal surfaces." Surface Science 53, no. 1 (1975): 317-329.

16 Eizenberg, M., and J. M. Blakely. "Carbon monolayer phase condensation on Ni (111)." Surface Science 82, no. 1 (1979): 228-236.

[17] Land, T. A., Th Michely, R. J. Behm, J. C. Hemminger, and G. Comsa. "STM investigation of single layer graphite structures produced on Pt (111) by hydrocarbon decomposition." Surface science 264, no. 3 (1992): 261-270.

[18] Van Bommel, A. J., J. E. Crombeen, and A. Van Tooren. "LEED and Auger electron observations of the SiC (0001) surface." Surface Science 48, no. 2 (1975): 463-472.

[19] Forbeaux, I., J-M. Themlin, and J-M. Debever. "Heteroepitaxial graphite on 6 H− SiC (0001): Interface formation through conduction-band electronic structure." Physical Review B 58, no. 24 (1998): 16396.

[20] Lu, Xuekun, Minfeng Yu, Hui Huang, and Rodney S. Ruoff. "Tailoring graphite with the goal of achieving single sheets." Nanotechnology 10, no. 3 (1999): 269.

[21] Novoselov, Kostya S., Andre K. Geim, Sergei V. Morozov, De-eng Jiang, Yanshui Zhang, Sergey V. Dubonos, Irina V. Grigorieva, and Alexandr A. Firsov. "Electric field effect in atomically thin carbon films." science 306, no. 5696 (2004): 666-669.

[22] Booth, Tim J., Peter Blake, Rahul R. Nair, Da Jiang, Ernie W. Hill, Ursel Bangert, Andrew Bleloch et al. "Macroscopic graphene membranes and their extraordinary stiffness." Nano letters 8, no. 8 (2008): 2442-2446.

[23] Lee, Changgu, Xiaoding Wei, Jeffrey W. Kysar, and James Hone. "Measurement of the elastic properties and intrinsic strength of monolayer graphene." science 321, no. 5887 (2008): 385-388.

[24] Geng, Dechao, Bin Wu, Yunlong Guo, Liping Huang, Yunzhou Xue, Jianyi Chen, Gui Yu, Lang Jiang, Wenping Hu, and Yunqi Liu. "Uniform hexagonal graphene flakes and films grown on liquid copper surface." Proceedings of the National Academy of Sciences 109, no. 21 (2012): 7992-7996.

[25] Alaskar, Yazeed, Shamsul Arafin, Darshana Wickramaratne, Mark A. Zurbuchen, Liang He, Jeff McKay, Qiyin Lin, Mark S. Goorsky, Roger K. Lake, and Kang L. Wang. "Towards van der Waals epitaxial growth of GaAs on Si using a graphene buffer layer." Advanced Functional Materials 24, no. 42 (2014): 6629-6638.

[26] Chhowalla, Manish, Hyeon Suk Shin, Goki Eda, Lain-Jong Li, Kian Ping Loh, and Hua Zhang. "The chemistry of two-dimensional layered transition metal dichalcogenide nanosheets." Nature chemistry 5, no. 4 (2013): 263-275.

[27] Mak, Kin Fai, Changgu Lee, James Hone, Jie Shan, and Tony F. Heinz. "Atomically thin MoS2: a new direct-gap semiconductor." Physical review letters 105, no. 13 (2010): 136805.

[28] Li, Na, Qinqin Wang, Cheng Shen, Zheng Wei, Hua Yu, Jing Zhao, Xiaobo Lu et al. "Large-scale flexible and transparent electronics based on monolayer molybdenum disulfide field-effect transistors." Nature Electronics 3, no. 11 (2020): 711-717.

[29] Rawat, Brajesh, M. M. Vinaya, and Roy Paily. "Transition metal dichalcogenide-based field-effect transistors for analog/mixed-signal applications." IEEE Transactions on Electron Devices 66, no. 5 (2019): 2424-2430.

30 Sortino, Luca, P. G. Zotev, S. Mignuzzi, J. Cambiasso, D. Schmidt, A. Genco, M. Aßmann et al. "Enhanced light-matter interaction in an atomically thin semiconductor coupled with dielectric nano-antennas." Nature communications 10, no. 1 (2019): 5119.

31 Ahn, Ethan C. "2D materials for spintronic devices." npj 2D Materials and Applications 4, no. 1 (2020): 17.

32 Kuc, Agnieszka, Nourdine Zibouche, and Thomas Heine. "Influence of quantum confinement on the electronic structure of the transition metal sulfide T S 2." Physical review B 83, no. 24 (2011): 245213.

33 Li, Likai, Yijun Yu, Guo Jun Ye, Qingqin Ge, Xuedong Ou, Hua Wu, Donglai Feng, Xian Hui Chen, and Yuanbo Zhang. "Black phosphorus field-effect transistors." Nature nanotechnology 9, no. 5 (2014): 372-377.

34 Chen, Xiaolong, Xiaobo Lu, Bingchen Deng, Ofer Sinai, Yuchuan Shao, Cheng Li, Shaofan Yuan et al. "Widely tunable black phosphorus mid-infrared photodetector." Nature communications 8, no. 1 (2017): 1672.

35 Valt, Matteo, Maria Caporali, Barbara Fabbri, Andrea Gaiardo, Soufiane Krik, Erica Iacob, Lia Vanzetti et al. "Air stable nickel-decorated black phosphorus and its room-temperature chemiresistive gas sensor capabilities." ACS Applied Materials & Interfaces 13, no. 37 (2021): 44711-44722.

36 Ye, Fan, Qingchang Liu, Baoxing Xu, Philip X-L. Feng, and Xian Zhang. "Ultra-High Interfacial Thermal Conductance via Double hBN Encapsulation for Efficient Thermal Management of 2D Electronics." Small 19, no. 12 (2023): 2205726.

37 Lee, Gun-Hee, Tran-Viet Cuong, Dong-Kyu Yeo, Hyunjin Cho, Beo-Deul Ryu, Eun-Mi Kim, Tae-Sik Nam, Eun-Kyung Suh, Tae-Hoon Seo, and Chang-Hee Hong. "Hexagonal Boron Nitride Passivation Layer for Improving the Performance and Reliability of InGaN/GaN Light-Emitting Diodes." Applied Sciences 11, no. 19 (2021): 9321.

38 Senyk, Szymon, Arkadiusz Chodkiewicz, Krzysztof Gocman, Barbara Szczęśniak, and Tadeusz Kałdoński. "Hexagonal nano and micro boron nitride: Properties and lubrication applications." Materials 15, no. 3 (2022): 955.

39 Malko, Daniel, Christian Neiss, Francesc Vines, and Andreas Görling. "Competition for graphene: graphynes with direction-dependent dirac cones." Physical review letters 108, no. 8 (2012): 086804.

40 Elias, Christine, Pierre Valvin, Thomas Pelini, A. Summerfield, C. J. Mellor, T. S. Cheng, L. Eaves et al. "Direct band-gap crossover in epitaxial monolayer boron nitride." Nature communications 10, no. 1 (2019): 2639.

41 Kang, Mingu, Beomyoung Kim, Sae Hee Ryu, Sung Won Jung, Jimin Kim, Luca Moreschini, Chris Jozwiak, Eli Rotenberg, Aaron Bostwick, and Keun Su Kim. "Universal mechanism of band-gap engineering in transition-metal dichalcogenides." Nano letters 17, no. 3 (2017): 1610-1615.

42 Kim, Jimin, Seung Su Baik, Sae Hee Ryu, Yeongsup Sohn, Soohyung Park, Byeong-Gyu Park, Jonathan Denlinger, Yeonjin Yi, Hyoung Joon Choi, and Keun Su Kim. "Observation of tunable band gap and anisotropic Dirac semimetal state in black phosphorus." Science 349, no. 6249 (2015): 723-726.

43 Lee, Jae Yoon, Jun-Hwan Shin, Gwan-Hyoung Lee, and Chul-Ho Lee. "Two-dimensional semiconductor optoelectronics based on van der Waals heterostructures." Nanomaterials 6, no. 11 (2016): 193.

44 Rahmani, Komeil, Saeed Mohammadi, and Mohammad Danaie. "Analytical investigation on the electro-optical characteristics of white graphene." Journal of Computational Electronics 20, no. 5 (2021): 1860-1867.

45 Xu, Yong, Binghai Yan, Hai-Jun Zhang, Jing Wang, Gang Xu, Peizhe Tang, Wenhui Duan, and Shou-Cheng Zhang. "Large-gap quantum spin Hall insulators in Tin films." Physical review letters 111, no. 13 (2013): 136804.

46 Zhao, Chen-Xiao, and Jin-Feng Jia. "Stanene: A good platform for topological insulator and topological superconductor." Frontiers of Physics 15 (2020): 1-15.

47 Xing, Dan-Xu, Ceng-Ceng Ren, Shu-Feng Zhang, Yong Feng, Xin-Lian Chen, Chang-Wen Zhang, and Pei-Ji Wang. "Tunable electronic and magnetic properties in stanene by 3d transition metal atoms absorption." Superlattices and Microstructures 103 (2017): 139-144.

48 Liu, Chao, Peng-Fei Liu, Xin-Hai Tu, Ling Shang, Ping Zhang, Wei Ren, and Bao-Tian Wang. "First-principles prediction of superconductivity in high-buckled two-dimensional tin." ACS Applied Electronic Materials 4, no. 4 (2022): 2062-2069.

49 Xu, Ming, and Min Wei. "Layered double hydroxide-based catalysts: recent advances in preparation, structure, and applications." Advanced Functional Materials 28, no. 47 (2018): 1802943.

50 Shirin, VK Ameena, Renu Sankar, Asha P. Johnson, H. V. Gangadharappa, and K. Pramod. "Advanced drug delivery applications of layered double hydroxide." Journal of controlled release 330 (2021): 398-426.

51 Gao, Yanshan, Jingwen Wu, Qiang Wang, Charles A. Wilkie, and Dermot O'Hare. "Flame retardant polymer/layered double hydroxide nanocomposites." Journal of Materials Chemistry A 2, no. 29 (2014): 10996-11016.

52 Liu, Zhaoping, Renzhi Ma, Minoru Osada, Nobuo Iyi, Yasuo Ebina, Kazunori Takada, and Takayoshi Sasaki. "Synthesis, anion exchange, and delamination of Co− Al layered double hydroxide: assembly of the exfoliated nanosheet/polyanion composite films and magneto-optical studies." Journal of the American Chemical Society 128, no. 14 (2006): 4872-4880.

53 Sajid, Muhammad, Shehzada Muhammad Sajid Jillani, Nadeem Baig, and Khalid Alhooshani. "Layered double hydroxide-modified membranes for water treatment: Recent advances and prospects." Chemosphere 287 (2022): 132140.

54 Naguib, Michael, Murat Kurtoglu, Volker Presser, Jun Lu, Junjie Niu, Min Heon, Lars Hultman, Yury Gogotsi, and Michel W. Barsoum. "Two-dimensional nanocrystals produced by exfoliation of Ti3AlC2." In MXenes, pp. 15-29. Jenny Stanford Publishing, 2011.

55 Seh, Zhi Wei, Kurt D. Fredrickson, Babak Anasori, Jakob Kibsgaard, Alaina L. Strickler, Maria R. Lukatskaya, Yury Gogotsi, Thomas F. Jaramillo, and Aleksandra Vojvodic. "Two-dimensional molybdenum carbide (MXene) as an efficient electrocatalyst for hydrogen evolution." ACS Energy Letters 1, no. 3 (2016): 589-594.

56 Bai, Saishuai, Meiqing Yang, Jizhou Jiang, Xiaomiao He, Jing Zou, Zhiguo Xiong, Guodong Liao, and Song Liu. "Recent advances of MXenes as electrocatalysts for hydrogen evolution reaction." npj 2D Materials and Applications 5, no. 1 (2021): 78.

57 Pei, Yangyang, Xiaoli Zhang, Zengyu Hui, Jinyuan Zhou, Xiao Huang, Gengzhi Sun, and Wei Huang. "$Ti_3C_2T_X$ MXene for sensing applications: recent progress, design principles, and future perspectives." ACS nano 15, no. 3 (2021): 3996-4017.

58 Han, Meikang, Christopher Eugene Shuck, Roman Rakhmanov, David Parchment, Babak Anasori, Chong Min Koo, Gary Friedman, and Yury Gogotsi. "Beyond $Ti_3C_2T_X$: MXenes for electromagnetic interference shielding." ACS nano 14, no. 4 (2020): 5008-5016.

59 Xie, Xiuqiang, Chi Chen, Nan Zhang, Zi-Rong Tang, Jianjun Jiang, and Yi-Jun Xu. "Microstructure and surface control of MXene films for water purification." Nature Sustainability 2, no. 9 (2019): 856-862.

60 Ronchi, Rodrigo Mantovani, Jeverson Teodoro Arantes, and Sydney Ferreira Santos. "Synthesis, structure, properties and applications of MXenes: Current status and perspectives." Ceramics International 45, no. 15 (2019): 18167-18188.

61 Feng, Baojie, Zijing Ding, Sheng Meng, Yugui Yao, Xiaoyue He, Peng Cheng, Lan Chen, and Kehui Wu. "Evidence of silicene in honeycomb structures of silicon on Ag (111)." Nano letters 12, no. 7 (2012): 3507-3511.

62 Kharadi, Mubashir A., Gul Faroz A. Malik, Farooq A. Khanday, Khurshed A. Shah, Sparsh Mittal, and Brajesh Kumar Kaushik. "Silicene: From material to device applications." ECS Journal of Solid State Science and Technology 9, no. 11 (2020): 115031.

63 Dávila, M. E., Lede Xian, Seymur Cahangirov, Angel Rubio, and Guy Le Lay. "Germanene: a novel two-dimensional germanium allotrope akin to graphene and silicene." New Journal of Physics 16, no. 9 (2014): 095002.

64 Zheng, Jun, Yang Xiang, Chunlei Li, Ruiyang Yuan, Feng Chi, and Yong Guo. "Multichannel depletion-type field-effect transistor based on ferromagnetic germanene." Physical Review Applied 16, no. 2 (2021): 024046.

65 Chowdhury, Suman, Arka Bandyopadhyay, Namrata Dhar, and Debnarayan Jana. "Optical and magnetic properties of free-standing silicene, germanene and T-graphene system." Physical Sciences Reviews 2, no. 5 (2017): 20165102.

66 Trivedi, Shyam, Anurag Srivastava, and Rajnish Kurchania. "Silicene and germanene: a first principle study of electronic structure and effect of hydrogenation-passivation." Journal of Computational and Theoretical Nanoscience 11, no. 3 (2014): 781-788.

67 Wittemeier, Nils, Pablo Ordejón, and Zeila Zanolli. "Tuning the topological band gap of bismuthene with silicon-based substrates." Journal of Physics: Materials 5, no. 3 (2022): 035002.

68 Reis, F., G. Li, L. Dudy, M. Bauernfeind, S. Glass, W. Hanke, R. Thomale, J. Schäfer, and R. Claessen. "Bismuthene on a SiC substrate: A candidate for a

high-temperature quantum spin Hall material." Science 357, no. 6348 (2017): 287-290.

69 Li, Yaqi, Jingwei Zhang, Xun Xu, Weichang Hao, Jincheng Zhuang, and Yi Du. "Advances in bismuth-based topological quantum materials by scanning tunneling microscopy." Materials Futures 1, no. 3 (2022): 032202.

70 Chuang, Hsun-Jen, Bhim Chamlagain, Michael Koehler, Meeghage Madusanka Perera, Jiaqiang Yan, David Mandrus, David Tomanek, and Zhixian Zhou. "Low-resistance 2D/2D ohmic contacts: a universal approach to high-performance WSe2, MoS2, and MoSe2 transistors." Nano letters 16, no. 3 (2016): 1896-1902.

71 Tosun, Mahmut, Steven Chuang, Hui Fang, Angada B. Sachid, Mark Hettick, Yongjing Lin, Yuping Zeng, and Ali Javey. "High-gain inverters based on WSe2 complementary field-effect transistors." ACS nano 8, no. 5 (2014): 4948-4953.

72 Han, Tae-Hee, Hobeom Kim, Sung-Joo Kwon, and Tae-Woo Lee. "Graphene-based flexible electronic devices." Materials Science and Engineering: R: Reports 118 (2017): 1-43.

73 Nazir, Ghazanfar, Adeela Rehman, and Soo-Jin Park. "Energy-efficient tunneling field-effect transistors for low-power device applications: challenges and opportunities." ACS applied materials & interfaces 12, no. 42 (2020): 47127-47163.

74 Wang, Gang, Miao Zhang, Da Chen, Qinglei Guo, Xuefei Feng, Tianchao Niu, Xiaosong Liu et al. "Seamless lateral graphene p–n junctions formed by selective in situ doping for high-performance photodetectors." Nature Communications 9, no. 1 (2018): 5168.

75 Ji, Xuan, Zongqi Bai, Fang Luo, Mengjian Zhu, Chucai Guo, Zhihong Zhu, and Shiqiao Qin. "High-performance photodetectors based on MoTe2–MoS2 van der Waals heterostructures." ACS omega 7, no. 12 (2022): 10049-10055.

76 Kwak, Do-Hyun, Parthiban Ramasamy, Yang-Soo Lee, Min-Hye Jeong, and Jong-Soo Lee. "High-performance hybrid InP QDs/black phosphorus photodetector." ACS applied materials & interfaces 11, no. 32 (2019): 29041-29046.

77 Long, Mingsheng, Peng Wang, Hehai Fang, and Weida Hu. "Progress, challenges, and opportunities for 2D material based photodetectors." Advanced Functional Materials 29, no. 19 (2019): 1803807.

78 Yin, Jianbo, Zhenjun Tan, Hao Hong, Jinxiong Wu, Hongtao Yuan, Yujing Liu, Cheng Chen et al. "Ultrafast and highly sensitive infrared photodetectors based on two-dimensional oxyselenide crystals." Nature communications 9, no. 1 (2018): 3311.

79 Kwon, Hyeokjae, Sourav Garg, Ji Hoon Park, Yeonsu Jeong, Sanghyuck Yu, Seongsin M. Kim, Patrick Kung, and Seongil Im. "Monolayer MoS2 field-effect transistors patterned by photolithography for active matrix pixels in organic light-emitting diodes." npj 2D Materials and Applications 3, no. 1 (2019): 9.

80 Wang, Xiaomu, and Fengnian Xia. "Stacked 2D materials shed light." Nature materials 14, no. 3 (2015): 264-265.

81 Ma, Qijie, Guanghui Ren, Kai Xu, and Jian Zhen Ou. "Tunable optical properties of 2D materials and their applications." Advanced Optical Materials 9, no. 2 (2021): 2001313.

82 Wang, Qixing, Jun Guo, Zijing Ding, Dianyu Qi, Jizhou Jiang, Zhuo Wang, Wei Chen, Yuanjiang Xiang, Wenjing Zhang, and Andrew TS Wee. "Fabry–Perot cavity-enhanced optical absorption in ultrasensitive tunable photodiodes based on hybrid 2D materials." Nano Letters 17, no. 12 (2017): 7593-7598.

83 Verduci, Rosaria, Valentino Romano, Giuseppe Brunetti, Narges Yaghoobi Nia, Aldo Di Carlo, Giovanna D'Angelo, and Caterina Ciminelli. "Solar energy in space applications: review and technology perspectives." Advanced Energy Materials 12, no. 29 (2022): 2200125.

84 Suragtkhuu, Selengesuren, Suvdanchimeg Sunderiya, Purevlkham Myagmarsereejid, Solongo Purevdorj, Abdulaziz SR Bati, Bujinlkham Bold, Yu Lin Zhong, Sarangerel Davaasambuu, and Munkhbayar Batmunkh. "Graphene-like monoelemental 2D materials for perovskite solar cells." Advanced Energy Materials 13, no. 12 (2023): 2204074.

85 Wu, Zhong-Shuai, Guangmin Zhou, Li-Chang Yin, Wencai Ren, Feng Li, and Hui-Ming Cheng. "Graphene/metal oxide composite electrode materials for energy storage." Nano Energy 1, no. 1 (2012): 107-131.

86 Stoller, Meryl D., Sungjin Park, Yanwu Zhu, Jinho An, and Rodney S. Ruoff. "Graphene-based ultracapacitors." Nano letters 8, no. 10 (2008): 3498-3502.

87 Kumar, Madhu Raj, Sangeeta Singh, Heba Mohamed Fahmy, Neeraj K. Jaiswal, Seckin Akin, Ahmed Esmail Shalan, Senentxu Lanceros-Mendez, and Manuel Salado. "Next generation 2D materials for anodes in battery applications." Journal of Power Sources 556 (2023): 232256.

88 Tomy, Merin, Athira Ambika Rajappan, Vimuna VM, and Xavier Thankappan Suryabai. "Emergence of novel 2D materials for high-performance supercapacitor electrode applications: a brief review." Energy & Fuels 35, no. 24 (2021): 19881-19900.

89 Divya, P., and S. Ramaprabhu. "Platinum–graphene hybrid nanostructure as anode and cathode electrocatalysts in proton exchange membrane fuel cells." Journal of Materials Chemistry A 2, no. 14 (2014): 4912-4918.

90 Khan, Karim, Ayesha Khan Tareen, Muhammad Aslam, Yupeng Zhang, Renheng Wang, Zhengbiao Ouyang, Zhongyi Gou, and Han Zhang. "Recent advances in two-dimensional materials and their nanocomposites in sustainable energy conversion applications." Nanoscale 11, no. 45 (2019): 21622-21678.

91 Quan, Quan, and Johnny C. Ho. "Recent advances in the construction of 2D heterostructures for electrocatalytic water splitting." Advanced Energy and Sustainability Research 3, no. 11 (2022): 2200059.

92 Chakraborty, Chitraleema, Nick Vamivakas, and Dirk Englund. "Advances in quantum light emission from 2D materials." Nanophotonics 8, no. 11 (2019): 2017-2032.

93 Gupta, Sunny, Wenjing Wu, Shengxi Huang, and Boris I. Yakobson. "Single-photon emission from two-dimensional materials, to a brighter future." The Journal of Physical Chemistry Letters 14, no. 13 (2023): 3274-3284.

[94] Lee, Kan-Heng, Srivatsan Chakram, Shi En Kim, Fauzia Mujid, Ariana Ray, Hui Gao, Chibeom Park et al. "Two-dimensional material tunnel barrier for Josephson junctions and superconducting qubits." Nano letters 19, no. 11 (2019): 8287-8293.

[95] Chen, Huawei, Xiaoyong Xue, Chunsen Liu, Jinbei Fang, Zhen Wang, Jianlu Wang, David Wei Zhang, Weida Hu, and Peng Zhou. "Logic gates based on neuristors made from two-dimensional materials." Nature Electronics 4, no. 6 (2021): 399-404.

[96] Jing, Fang-Ming, Zhuo-Zhi Zhang, Guo-Quan Qin, Gang Luo, Gang Cao, Hai-Ou Li, Xiang-Xiang Song, and Guo-Ping Guo. "Gate-Controlled Quantum Dots Based on 2D Materials." Advanced Quantum Technologies 5, no. 6 (2022): 2100162.

[97] Kennes, Dante M., Martin Claassen, Lede Xian, Antoine Georges, Andrew J. Millis, James Hone, Cory R. Dean, D. N. Basov, Abhay N. Pasupathy, and Angel Rubio. "Moiré heterostructures as a condensed-matter quantum simulator." Nature Physics 17, no. 2 (2021): 155-163.

[98] Sajjan, Manas, Shree Hari Sureshbabu, and Sabre Kais. "Quantum machine-learning for eigenstate filtration in two-dimensional materials." Journal of the American Chemical Society 143, no. 44 (2021): 18426-18445.

[99] Xiang, Heng, Yu-Chieh Chien, Yufei Shi, and Kah-Wee Ang. "Application of 2D Materials in Hardware Security for Internet-of-Things: Progress and Perspective." Small Structures 3, no. 8 (2022): 2200060.

[100] Wen, Chao, Xuehua Li, Tommaso Zanotti, Francesco Maria Puglisi, Yuanyuan Shi, Fernan Saiz, Aleandro Antidormi et al. "Advanced Data Encryption using 2D Materials." Advanced Materials 33, no. 27 (2021): 2100185.

[101] Lee, Changgu, Xiaoding Wei, Jeffrey W. Kysar, and James Hone. "Measurement of the elastic properties and intrinsic strength of monolayer graphene." science 321, no. 5887 (2008): 385-388.

[102] Yang, Gao, Lihua Li, Wing Bun Lee, and Man Cheung Ng. "Structure of graphene and its disorders: a review." Science and technology of advanced materials 19, no. 1 (2018): 613-648.

[103] Jhang, Sung Ho, M. F. Craciun, Silvia Schmidmeier, S. Tokumitsu, Saverio Russo, M. Yamamoto, Yurii Skourski et al. "Stacking-order dependent transport properties of trilayer graphene." Physical Review B 84, no. 16 (2011): 161408.

[104] Koshino, Mikito. "Interlayer screening effect in graphene multilayers with A B A and A B C stacking." Physical Review B 81, no. 12 (2010): 125304.

[105] Lui, Chun Hung, Zhiqiang Li, Zheyuan Chen, Paul V. Klimov, Louis E. Brus, and Tony F. Heinz. "Imaging stacking order in few-layer graphene." Nano letters 11, no. 1 (2011): 164-169.

[106] Dreyer, Daniel R., Sungjin Park, Christopher W. Bielawski, and Rodney S. Ruoff. "The chemistry of graphene oxide." Chemical Society Reviews 39, no. 1 (2010): 228-240.

[107] Latil, Sylvain, and Luc Henrard. "Charge carriers in few-layer graphene films." Physical review letters 97, no. 3 (2006): 036803.

108 Menezes, Marcos G., Rodrigo B. Capaz, and Steven G. Louie. "Ab initio quasiparticle band structure of ABA and ABC-stacked graphene trilayers." Physical Review B 89, no. 3 (2014): 035431.

109 https://phys.org/news/2017-01-scientists-bilayer-graphene.html (Credit: http://mipt.ru/en/)

110 Cassabois, Guillaume, Pierre Valvin, and Bernard Gil. "Hexagonal boron nitride is an indirect bandgap semiconductor." Nature photonics 10, no. 4 (2016): 262-266.

111 Wang, Jingang, Fengcai Ma, and Mengtao Sun. "Graphene, hexagonal boron nitride, and their heterostructures: properties and applications." RSC advances 7, no. 27 (2017): 16801-16822.

112 Topsakal, Mehmet, E. Aktürk, and S. J. P. R. B. Ciraci. "First-principles study of two-and one-dimensional honeycomb structures of boron nitride." Physical Review B 79, no. 11 (2009): 115442.

113 Bhattacharya, A., S. Bhattacharya, and G. P. Das. "Band gap engineering by functionalization of BN sheet." Physical Review B 85, no. 3 (2012): 035415.

114 Roy, Soumyabrata, Xiang Zhang, Anand B. Puthirath, Ashokkumar Meiyazhagan, Sohini Bhattacharyya, Muhammad M. Rahman, Ganguli Babu et al. "Structure, properties and applications of two-dimensional hexagonal boron nitride." Advanced Materials 33, no. 44 (2021): 2101589.

115 Salavati, Mohammad, Arvin Mojahedin, and Ali Hossein Nezhad Shirazi. "Mechanical responses of pristine and defective hexagonal boron-nitride nanosheets: A molecular dynamics investigation." Frontiers of Structural and Civil Engineering 14 (2020): 623-631.

Majety, S., X. K. Cao, R. Dahal, B. N. Pantha, J. Li, J. Y. Lin, and H. X. Jiang. "Semiconducting hexagonal boron nitride for deep ultraviolet photonics." In Quantum Sensing and Nanophotonic Devices IX, vol. 8268, pp. 607-614. SPIE, 2012.

117 Wilson, Jl A., F. J. Di Salvo, and S. Mahajan. "Charge-density waves and superlattices in the metallic layered transition metal dichalcogenides." Advances in Physics 24, no. 2 (1975): 117-201.

118 Wang, Haotian, Zhiyi Lu, Desheng Kong, Jie Sun, Thomas M. Hymel, and Yi Cui. "Electrochemical tuning of MoS2 nanoparticles on three-dimensional substrate for efficient hydrogen evolution." ACS nano 8, no. 5 (2014): 4940-4947.

119 Eda, Goki, Hisato Yamaguchi, Damien Voiry, Takeshi Fujita, Mingwei Chen, and Manish Chhowalla. "Photoluminescence from chemically exfoliated MoS2." Nano letters 11, no. 12 (2011): 5111-5116.

120 Sandoval, S. Jiménez, D. Yang, R. F. Frindt, and J. C. Irwin. "Raman study and lattice dynamics of single molecular layers of MoS 2." Physical Review B 44, no. 8 (1991): 3955.

121 Chhowalla, Manish, Hyeon Suk Shin, Goki Eda, Lain-Jong Li, Kian Ping Loh, and Hua Zhang. "The chemistry of two-dimensional layered transition metal dichalcogenide nanosheets." Nature chemistry 5, no. 4 (2013): 263-275.

[122] Wilson, Jl A., and A. D. Yoffe. "The transition metal dichalcogenides discussion and interpretation of the observed optical, electrical and structural properties." Advances in Physics 18, no. 73 (1969): 193-335.

[123] Coogan, Áine, and Yurii K. Gun'ko. "Solution-based “bottom-up” synthesis of group VI transition metal dichalcogenides and their applications." Materials Advances 2, no. 1 (2021): 146-164.

[124] He, Jiangang, Kerstin Hummer, and Cesare Franchini. "Stacking effects on the electronic and optical properties of bilayer transition metal dichalcogenides MoS2, MoSe2, WS2, and WSe2." Physical Review B 89, no. 7 (2014): 075409.

[125] Neto, AH Castro, Francisco Guinea, Nuno MR Peres, Kostya S. Novoselov, and Andre K. Geim. "The electronic properties of graphene." Reviews of modern physics 81, no. 1 (2009): 109.

[126] Wallace, Philip Richard. "The band theory of graphite." Physical review 71, no. 9 (1947): 622.

[127] Zhan, Da, Jiaxu Yan, Linfei Lai, Zhenhua Ni, Lei Liu, and Zexiang Shen. "Engineering the electronic structure of graphene." Advanced Materials 24, no. 30 (2012): 4055-4069.

[128] Zhang, Yuanbo, Yan-Wen Tan, Horst L. Stormer, and Philip Kim. "Experimental observation of the quantum Hall effect and Berry's phase in graphene." nature 438, no. 7065 (2005): 201-204.

[129] Schedin, Fredrik, Andrei Konstantinovich Geim, Sergei Vladimirovich Morozov, Ew W. Hill, Peter Blake, Mi I. Katsnelson, and Kostya Sergeevich Novoselov. "Detection of individual gas molecules adsorbed on graphene." Nature materials 6, no. 9 (2007): 652-655.

[130] Mak, Kin Fai, Changgu Lee, James Hone, Jie Shan, and Tony F. Heinz. "Atomically thin MoS 2: a new direct-gap semiconductor." Physical review letters 105, no. 13 (2010): 136805.

[131] Eda, Goki, Hisato Yamaguchi, Damien Voiry, Takeshi Fujita, Mingwei Chen, and Manish Chhowalla. "Photoluminescence from chemically exfoliated MoS2." Nano letters 11, no. 12 (2011): 5111-5116.

[132] Tongay, Sefaattin, Jian Zhou, Can Ataca, Kelvin Lo, Tyler S. Matthews, Jingbo Li, Jeffrey C. Grossman, and Junqiao Wu. "Thermally driven crossover from indirect toward direct bandgap in 2D semiconductors: MoSe2 versus MoS2." Nano letters 12, no. 11 (2012): 5576-5580.

[133] Zhao, Weijie, Zohreh Ghorannevis, Leiqiang Chu, Minglin Toh, Christian Kloc, Ping-Heng Tan, and Goki Eda. "Evolution of electronic structure in atomically thin sheets of WS2 and WSe2." ACS nano 7, no. 1 (2013): 791-797.

[134] Splendiani, Andrea, Liang Sun, Yuanbo Zhang, Tianshu Li, Jonghwan Kim, Chi-Yung Chim, Giulia Galli, and Feng Wang. "Emerging photoluminescence in monolayer MoS2." Nano letters 10, no. 4 (2010): 1271-1275.

[135] Xiao, Di, Gui-Bin Liu, Wanxiang Feng, Xiaodong Xu, and Wang Yao. "Coupled spin and valley physics in monolayers of MoS 2 and other group-VI dichalcogenides." Physical review letters 108, no. 19 (2012): 196802.

[136] Cao, Ting, Gang Wang, Wenpeng Han, Huiqi Ye, Chuanrui Zhu, Junren Shi, Qian Niu et al. "Valley-selective circular dichroism of monolayer molybdenum disulphide." Nature communications 3, no. 1 (2012): 887.

[137] Zeng, Hualing, Junfeng Dai, Wang Yao, Di Xiao, and Xiaodong Cui. "Valley polarization in MoS2 monolayers by optical pumping." Nature nanotechnology 7, no. 8 (2012): 490-493.

[138] Mak, Kin Fai, Keliang He, Jie Shan, and Tony F. Heinz. "Control of valley polarization in monolayer MoS2 by optical helicity." Nature nanotechnology 7, no. 8 (2012): 494-498.

[139] Lee, Jae Yoon, Jun-Hwan Shin, Gwan-Hyoung Lee, and Chul-Ho Lee. "Two-dimensional semiconductor optoelectronics based on van der Waals heterostructures." Nanomaterials 6, no. 11 (2016): 193.

[140] Osada, Minoru, and Takayoshi Sasaki. "Two-dimensional dielectric nanosheets: novel nanoelectronics from nanocrystal building blocks." Advanced Materials 24, no. 2 (2012): 210-228.

[141] Kang, Jun, Sefaattin Tongay, Jian Zhou, Jingbo Li, and Junqiao Wu. "Band offsets and heterostructures of two-dimensional semiconductors." Applied Physics Letters 102, no. 1 (2013).

[142] Radisavljevic, Branimir, Aleksandra Radenovic, Jacopo Brivio, Valentina Giacometti, and Andras Kis. "Single-layer MoS2 transistors." Nature nanotechnology 6, no. 3 (2011): 147-150.

[143] Sup Choi, Min, Gwan-Hyoung Lee, Young-Jun Yu, Dae-Yeong Lee, Seung Hwan Lee, Philip Kim, James Hone, and Won Jong Yoo. "Controlled charge trapping by molybdenum disulphide and graphene in ultrathin heterostructured memory devices." Nature communications 4, no. 1 (2013): 1624.

[144] Mak, Kin Fai, and Jie Shan. "Photonics and optoelectronics of 2D semiconductor transition metal dichalcogenides." Nature Photonics 10, no. 4 (2016): 216-226.

[145] Sun, Zhipei, Amos Martinez, and Feng Wang. "Optical modulators with 2D layered materials." Nature Photonics 10, no. 4 (2016): 227-238.

[146] Mak, Kin Fai, Changgu Lee, James Hone, Jie Shan, and Tony F. Heinz. "Atomically thin MoS 2: a new direct-gap semiconductor." Physical review letters 105, no. 13 (2010): 136805.

[147] Zeng, Hualing, Junfeng Dai, Wang Yao, Di Xiao, and Xiaodong Cui. "Valley polarization in MoS2 monolayers by optical pumping." Nature nanotechnology 7, no. 8 (2012): 490-493.

[148] Britnell, L.; Ribeiro, R.; Eckmann, A.; Jalil, R.; Belle, B.; Mishchenko, A.; Kim, Y.-J.; Gorbachev, R.; Georgiou, T.; Morozov, S. Strong light-matter interactions in heterostructures of atomically thin films. Science 2013, 340, 1311–1314.

[149] Ramasubramaniam, Ashwin. "Large excitonic effects in monolayers of molybdenum and tungsten dichalcogenides." Physical Review B 86, no. 11 (2012): 115409.

[150] Ugeda, Miguel M., Aaron J. Bradley, Su-Fei Shi, Felipe H. Da Jornada, Yi Zhang, Diana Y. Qiu, Wei Ruan et al. "Giant bandgap renormalization and

excitonic effects in a monolayer transition metal dichalcogenide semiconductor." Nature materials 13, no. 12 (2014): 1091-1095.
151 Li, Likai, Yijun Yu, Guo Jun Ye, Qingqin Ge, Xuedong Ou, Hua Wu, Donglai Feng, Xian Hui Chen, and Yuanbo Zhang. "Black phosphorus field-effect transistors." Nature nanotechnology 9, no. 5 (2014): 372-377.
152 Xia, Fengnian, Han Wang, and Yichen Jia. "Rediscovering black phosphorus as an anisotropic layered material for optoelectronics and electronics." Nature communications 5, no. 1 (2014): 4458.
153 Xu, Mingsheng, Tao Liang, Minmin Shi, and Hongzheng Chen. "Graphene-like two-dimensional materials." Chemical reviews 113, no. 5 (2013): 3766-3798.
154 Dean, Cory R., Andrea F. Young, Inanc Meric, Chris Lee, Lei Wang, Sebastian Sorgenfrei, Kenji Watanabe et al. "Boron nitride substrates for high-quality graphene electronics." Nature nanotechnology 5, no. 10 (2010): 722-726.
155 Geim, Andre K., and Irina V. Grigorieva. "Van der Waals heterostructures." Nature 499, no. 7459 (2013): 419-425.
156 Li, Henan, Yumeng Shi, Ming-Hui Chiu, and Lain-Jong Li. "Emerging energy applications of two-dimensional layered transition metal dichalcogenides." Nano Energy 18 (2015): 293-305.
157 Xia, Fengnian, Han Wang, Di Xiao, Madan Dubey, and Ashwin Ramasubramaniam. "Two-dimensional material nanophotonics." Nature photonics 8, no. 12 (2014): 899-907.
158 Rasmussen, Filip A., and Kristian S. Thygesen. "Computational 2D materials database: electronic structure of transition-metal dichalcogenides and oxides." The Journal of Physical Chemistry C 119, no. 23 (2015): 13169-13183.
159 Alam, Shahinoor, Mohammad Asaduzzaman Chowdhury, Abdus Shahid, Rubel Alam, and Abdur Rahim. "Synthesis of emerging two-dimensional (2D) materials–Advances, challenges and prospects." FlatChem 30 (2021): 100305.
160 Moss, David. "Fabrication methods for integrating 2D materials." arXiv preprint arXiv:2201.00964 (2022).
161 Shanmugam, Vigneshwaran, Rhoda Afriyie Mensah, Karthik Babu, Sidique Gawusu, Avishek Chanda, Yongming Tu, Rasoul Esmaeely Neisiany, Michael Försth, Gabriel Sas, and Oisik Das. "A review of the synthesis, properties, and applications of 2D materials." Particle & Particle Systems Characterization 39, no. 6 (2022): 2200031.
162 Naikoo, Gowhar A., Fareeha Arshad, Muniba Almas, Israr U. Hassan, Mona Z. Pedram, Alaa AA Aljabali, Vijay Mishra et al. "2D materials, synthesis, characterization and toxicity: A critical review." Chemico-Biological Interactions 365 (2022): 110081.
163 Novoselov, K. S., and AH Castro Neto. "Two-dimensional crystals-based heterostructures: materials with tailored properties." Physica Scripta 2012, no. T146 (2012): 014006.
164 Tiwari, Sourabh, Anushka Purabgola, and Balasubramanian Kandasubramanian. "Functionalised graphene as flexible electrodes for polymer photovoltaics." Journal of Alloys and Compounds 825 (2020): 153954.

165 Zeng, Zhiyuan, Zongyou Yin, Xiao Huang, Hai Li, Qiyuan He, Gang Lu, Freddy Boey, and Hua Zhang. "Single-layer semiconducting nanosheets: high-yield preparation and device fabrication." Angewandte Chemie International Edition 47, no. 50 (2011): 11093-11097.

166 Rehman, Adeela, and Soo-Jin Park. "State of the art two-dimensional materials-based photodetectors: Prospects, challenges and future outlook." Journal of Industrial and Engineering Chemistry 89 (2020): 28-46.

167 Shanmugam, Vigneshwaran, Rhoda Afriyie Mensah, Karthik Babu, Sidique Gawusu, Avishek Chanda, Yongming Tu, Rasoul Esmaeely Neisiany, Michael Försth, Gabriel Sas, and Oisik Das. "A review of the synthesis, properties, and applications of 2D materials." Particle & Particle Systems Characterization 39, no. 6 (2022): 2200031.

168 Nagaraju, Chandu, Chandu VV Muralee Gopi, Jin-Woo Ahn, and Hee-Je Kim. "Hydrothermal synthesis of MoS2 and WS2 nanoparticles for high-performance supercapacitor applications." New Journal of Chemistry 42, no. 15 (2018): 12357-12360.

169 Cai, Zhengyang, Bilu Liu, Xiaolong Zou, and Hui-Ming Cheng. "Chemical vapor deposition growth and applications of two-dimensional materials and their heterostructures." Chemical reviews 118, no. 13 (2018): 6091-6133.

170 Geng, Dechao, Bin Wu, Yunlong Guo, Liping Huang, Yunzhou Xue, Jianyi Chen, Gui Yu, Lang Jiang, Wenping Hu, and Yunqi Liu. "Uniform hexagonal graphene flakes and films grown on liquid copper surface." Proceedings of the National Academy of Sciences 109, no. 21 (2012): 7992-7996.

171 Yeh, Chao-Hui, Yung-Chang Lin, Pramoda K. Nayak, Chun-Chieh Lu, Zheng Liu, Kazu Suenaga, and Po-Wen Chiu. "Probing interlayer coupling in twisted single-crystal bilayer graphene by Raman spectroscopy." Journal of Raman Spectroscopy 45, no. 10 (2014): 912-917.

172 Nayak, Pramoda K., Yevhen Horbatenko, Seongjoon Ahn, Gwangwoo Kim, Jae-Ung Lee, Kyung Yeol Ma, A-Rang Jang et al. "Probing evolution of twist-angle-dependent interlayer excitons in MoSe2/WSe2 van der Waals heterostructures." ACS nano 11, no. 4 (2017): 4041-4050.

173 Alexeev, Evgeny M., Alessandro Catanzaro, Oleksandr V. Skrypka, Pramoda K. Nayak, Seongjoon Ahn, Sangyeon Pak, Juwon Lee et al. "Imaging of interlayer coupling in van der Waals heterostructures using a bright-field optical microscope." Nano letters 17, no. 9 (2017): 5342-5349.

174 Zuo, Zheng, Zhongguang Xu, Renjing Zheng, Alireza Khanaki, Jian-Guo Zheng, and Jianlin Liu. "In-situ epitaxial growth of graphene/h-BN van der Waals heterostructures by molecular beam epitaxy." Scientific reports 5, no. 1 (2015): 14760.

175 Dabrowski, J., Gunther Lippert, J. Avila, Jens Baringhaus, I. Colambo, Yu S. Dedkov, Felix Herziger et al. "Understanding the growth mechanism of graphene on Ge/Si (001) surfaces." Scientific reports 6, no. 1 (2016): 31639.

176 Cheng, Peng, Wen Zhang, Lei Zhang, Jian Gou, Ping Kwan Johnny Wong, and Lan Chen. "Molecular beam epitaxy fabrication of two-dimensional materials." In 2D semiconductor materials and devices, pp. 103-134. Elsevier, 2020.

[177] Seredyński, Bartłomiej, Zuzanna Ogorzałek, Wiktoria Zajkowska, Rafał Bożek, Mateusz Tokarczyk, Jan Suffczynski, Sławomir Kret, Janusz Sadowski, Marta Gryglas-Borysiewicz, and Wojciech Pacuski. "Molecular beam epitaxy of a 2D material nearly lattice matched to a 3D substrate: NiTe2 on GaAs." Crystal Growth & Design 21, no. 10 (2021): 5773-5779.

[178] Ermolaev, Georgy A., Marwa A. El-Sayed, Dmitry I. Yakubovsky, Kirill V. Voronin, Roman I. Romanov, Mikhail K. Tatmyshevskiy, Natalia V. Doroshina et al. "Optical constants and structural properties of epitaxial MoS2 monolayers." Nanomaterials 11, no. 6 (2021): 1411.

[179] Ermolaev, Georgy A., Yury V. Stebunov, Andrey A. Vyshnevyy, Dmitry E. Tatarkin, Dmitry I. Yakubovsky, Sergey M. Novikov, Denis G. Baranov et al. "Broadband optical properties of monolayer and bulk MoS2." npj 2D Materials and Applications 4, no. 1 (2020): 21.

[180] Lee, Changgu, Hugen Yan, Louis E. Brus, Tony F. Heinz, James Hone, and Sunmin Ryu. "Anomalous lattice vibrations of single-and few-layer MoS2." ACS nano 4, no. 5 (2010): 2695-2700.

[181] Li, Hai, Jumiati Wu, Xiao Huang, Gang Lu, Jian Yang, Xin Lu, Qihua Xiong, and Hua Zhang. "Rapid and reliable thickness identification of two-dimensional nanosheets using optical microscopy." ACS nano 7, no. 11 (2013): 10344-10353.

[182] Ohtake, Akihiro, and Yoshiki Sakuma. "Two-dimensional WSe2/MoSe2 heterostructures grown by molecular-beam epitaxy." The Journal of Physical Chemistry C 125, no. 20 (2021): 11257-11261.

[183] Onomitsu, Koji, Aleksandra Krajewska, Ryan AE Neufeld, Fumihiko Maeda, Kazuhide Kumakura, and Hideki Yamamoto. "Epitaxial growth of monolayer MoSe2 on GaAs." Applied Physics Express 9, no. 11 (2016): 115501.

[184] Zhang, Fang, Wei Li, and Xianqi Dai. "Modulation of electronic structures of MoSe2/WSe2 van der Waals heterostructure by external electric field." Solid State Communications 266 (2017): 11-15.

[185] Wang, K., G. Tai, K. H. Wong, Shu Ping Lau, and W. Guo. "Ni induced few-layer graphene growth at low temperature by pulsed laser deposition." Aip Advances 1, no. 2 (2011).

[186] Qian, Min, Yun Shen Zhou, Yang Gao, Jong Bok Park, Tao Feng, Su Mei Huang, Zhuo Sun, Lan Jiang, and Yong Feng Lu. "Formation of graphene sheets through laser exfoliation of highly ordered pyrolytic graphite." Applied Physics Letters 98, no. 17 (2011).

[187] Serrao, Claudy R., Anthony M. Diamond, Shang-Lin Hsu, Long You, Sushant Gadgil, James Clarkson, Carlo Carraro, Roya Maboudian, Chenming Hu, and Sayeef Salahuddin. "Highly crystalline MoS2 thin films grown by pulsed laser deposition." Applied Physics Letters 106, no. 5 (2015).

[188] Loh, Tamie AJ, Daniel HC Chua, and Andrew TS Wee. "One-step synthesis of few-layer WS2 by pulsed laser deposition." Scientific reports 5, no. 1 (2015): 18116.

[189] Yao, J. D., Z. Q. Zheng, and G. W. Yang. "Production of large-area 2D materials for high-performance photodetectors by pulsed-laser deposition." Progress in Materials Science 106 (2019): 100573.

[190] Yang, Zhibin, and Jianhua Hao. "Progress in pulsed laser deposited two-dimensional layered materials for device applications." Journal of Materials Chemistry C 4, no. 38 (2016): 8859-8878.

[191] Harris, Sumner B., Arpan Biswas, Seok Joon Yun, Christopher M. Rouleau, Alexander A. Puretzky, Rama K. Vasudevan, David B. Geohegan, and Kai Xiao. "Autonomous synthesis of thin film materials with pulsed laser deposition enabled by in situ spectroscopy and automation." arXiv preprint arXiv:2308.08700 (2023).

[192] Niketa, A. K., Md Aasif Ikbal, Susmitha Kothapalli, and Shishir Kumar. "An automated chemical vapor deposition setup for 2D materials." HardwareX 9 (2021): e00165.

[193] Ferrari, Andrea C., Jannik C. Meyer, Vittorio Scardaci, Cinzia Casiraghi, Michele Lazzeri, Francesco Mauri, Stefano Piscanec et al. "Raman spectrum of graphene and graphene layers." Physical review letters 97, no. 18 (2006): 187401.

[194] Niu, Yue, Sergio Gonzalez-Abad, Riccardo Frisenda, Philipp Marauhn, Matthias Drüppel, Patricia Gant, Robert Schmidt et al. "Thickness-dependent differential reflectance spectra of monolayer and few-layer MoS2, MoSe2, WS2 and WSe2." Nanomaterials 8, no. 9 (2018): 725.

[195] Najmaei, Sina, Zheng Liu, Wu Zhou, Xiaolong Zou, Gang Shi, Sidong Lei, Boris I. Yakobson, Juan-Carlos Idrobo, Pulickel M. Ajayan, and Jun Lou. "Vapour phase growth and grain boundary structure of molybdenum disulphide atomic layers." Nature materials 12, no. 8 (2013): 754-759.

[196] Van Der Zande, Arend M., Pinshane Y. Huang, Daniel A. Chenet, Timothy C. Berkelbach, YuMeng You, Gwan-Hyoung Lee, Tony F. Heinz, David R. Reichman, David A. Muller, and James C. Hone. "Grains and grain boundaries in highly crystalline monolayer molybdenum disulphide." Nature materials 12, no. 6 (2013): 554-561.

[197] Terrones, H., E. Del Corro, S. Feng, J. M. Poumirol, D. Rhodes, D. Smirnov, N. R. Pradhan et al. "New first order Raman-active modes in few layered transition metal dichalcogenides." Scientific reports 4, no. 1 (2014): 4215.

[198] Gutiérrez, Humberto R., Nestor Perea-López, Ana Laura Elías, Ayse Berkdemir, Bei Wang, Ruitao Lv, Florentino López-Urías, Vincent H. Crespi, Humberto Terrones, and Mauricio Terrones. "Extraordinary room-temperature photoluminescence in triangular WS2 monolayers." Nano letters 13, no. 8 (2013): 3447-3454.

[199] Berkdemir, Ayse, Humberto R. Gutiérrez, Andrés R. Botello-Méndez, Néstor Perea-López, Ana Laura Elías, Chen-Ing Chia, Bei Wang et al. "Identification of individual and few layers of WS2 using Raman Spectroscopy." Scientific reports 3, no. 1 (2013): 1755.

[200] Tonndorf, Philipp, Robert Schmidt, Philipp Böttger, Xiao Zhang, Janna Börner, Andreas Liebig, Manfred Albrecht et al. "Photoluminescence emission and

Raman response of monolayer MoS 2, MoSe 2, and WSe 2." Optics express 21, no. 4 (2013): 4908-4916.

201 Lu, Xin, M. Iqbal Bakti Utama, Junhao Lin, Xue Gong, Jun Zhang, Yanyuan Zhao, Sokrates T. Pantelides et al. "Large-area synthesis of monolayer and few-layer MoSe2 films on SiO2 substrates." Nano letters 14, no. 5 (2014): 2419-2425.

202 Chiu, Ming-Hui, Ming-Yang Li, Wengjing Zhang, Wei-Ting Hsu, Wen-Hao Chang, Mauricio Terrones, Humberto Terrones, and Lain-Jong Li. "Spectroscopic signatures for interlayer coupling in MoS2–WSe2 van der Waals stacking." ACS nano 8, no. 9 (2014): 9649-9656.

203 Luo, Xin, Yanyuan Zhao, Jun Zhang, Minglin Toh, Christian Kloc, Qihua Xiong, and Su Ying Quek. "Effects of lower symmetry and dimensionality on Raman spectra in two-dimensional WSe 2." Physical Review B 88, no. 19 (2013): 195313.

204 Puretzky, Alexander A., Liangbo Liang, Xufan Li, Kai Xiao, Bobby G. Sumpter, Vincent Meunier, and David B. Geohegan. "Twisted MoSe2 bilayers with variable local stacking and interlayer coupling revealed by low-frequency raman spectroscopy." ACS nano 10, no. 2 (2016): 2736-2744.

205 Puretzky, Alexander A., Liangbo Liang, Xufan Li, Kai Xiao, Kai Wang, Masoud Mahjouri-Samani, Leonardo Basile et al. "Low-frequency Raman fingerprints of two-dimensional metal dichalcogenide layer stacking configurations." ACS nano 9, no. 6 (2015): 6333-6342.

206 Gong, Yongji, Junhao Lin, Xingli Wang, Gang Shi, Sidong Lei, Zhong Lin, Xiaolong Zou et al. "Vertical and in-plane heterostructures from WS 2/MoS 2 monolayers." Nature materials 13, no. 12 (2014): 1135-1142.

207 Nayak, Pramoda K., Yevhen Horbatenko, Seongjoon Ahn, Gwangwoo Kim, Jae-Ung Lee, Kyung Yeol Ma, A-Rang Jang et al. "Probing evolution of twist-angle-dependent interlayer excitons in MoSe2/WSe2 van der Waals heterostructures." ACS nano 11, no. 4 (2017): 4041-4050.

208 Tongay, Sefaattin, Joonki Suh, Can Ataca, Wen Fan, Alexander Luce, Jeong Seuk Kang, Jonathan Liu et al. "Defects activated photoluminescence in two-dimensional semiconductors: interplay between bound, charged and free excitons." Scientific reports 3, no. 1 (2013): 2657.

209 Mak, Kin Fai, Changgu Lee, James Hone, Jie Shan, and Tony F. Heinz. "Atomically thin MoS 2: a new direct-gap semiconductor." Physical review letters 105, no. 13 (2010): 136805.

210 Tongay, Sefaattin, Jian Zhou, Can Ataca, Kelvin Lo, Tyler S. Matthews, Jingbo Li, Jeffrey C. Grossman, and Junqiao Wu. "Thermally driven crossover from indirect toward direct bandgap in 2D semiconductors: MoSe2 versus MoS2." Nano letters 12, no. 11 (2012): 5576-5580.

211 Zhao, Weijie, Zohreh Ghorannevis, Leiqiang Chu, Minglin Toh, Christian Kloc, Ping-Heng Tan, and Goki Eda. "Evolution of electronic structure in atomically thin sheets of WS2 and WSe2." ACS nano 7, no. 1 (2013): 791-797.

212 Nayak, Pramoda K., Chao-Hui Yeh, Yu-Chen Chen, and Po-Wen Chiu. "Layer-dependent optical conductivity in atomic thin WS2 by reflection contrast

spectroscopy." ACS Applied Materials & Interfaces 6, no. 18 (2014): 16020-16026.

213 Frey, G. L., Reshef Tenne, M. J. Matthews, M. S. Dresselhaus, and G. Dresselhaus. "Optical properties of MS2 (M= Mo, W) inorganic fullerenelike and nanotube material optical absorption and resonance Raman measurements." Journal of materials research 13, no. 9 (1998): 2412-2417.

214 Ballif, C., M. Regula, P. E. Schmid, M. Remškar, R. Sanjines, and F. Lévy. "Preparation and characterization of highly oriented, photoconducting WS 2 thin films." Applied Physics A 62 (1996): 543-546.

215 Wang, Qing Hua, Kourosh Kalantar-Zadeh, Andras Kis, Jonathan N. Coleman, and Michael S. Strano. "Electronics and optoelectronics of two-dimensional transition metal dichalcogenides." Nature nanotechnology 7, no. 11 (2012): 699-712.

216 Zeng, Hualing, Gui-Bin Liu, Junfeng Dai, Yajun Yan, Bairen Zhu, Ruicong He, Lu Xie et al. "Optical signature of symmetry variations and spin-valley coupling in atomically thin tungsten dichalcogenides." Scientific reports 3, no. 1 (2013): 1608.

217 Cheiwchanchamnangij, Tawinan, and Walter RL Lambrecht. "Quasiparticle band structure calculation of monolayer, bilayer, and bulk MoS 2." Physical Review B 85, no. 20 (2012): 205302.

218 Liu, Kaihui, Liming Zhang, Ting Cao, Chenhao Jin, Diana Qiu, Qin Zhou, Alex Zettl, Peidong Yang, Steve G. Louie, and Feng Wang. "Evolution of interlayer coupling in twisted molybdenum disulfide bilayers." Nature communications 5, no. 1 (2014): 4966.

219 Amani, Matin, Robert A. Burke, Xiang Ji, Peida Zhao, Der-Hsien Lien, Peyman Taheri, Geun Ho Ahn et al. "High luminescence efficiency in MoS2 grown by chemical vapor deposition." ACS nano 10, no. 7 (2016): 6535-6541.

220 Amani, Matin, Der-Hsien Lien, Daisuke Kiriya, Jun Xiao, Angelica Azcatl, Jiyoung Noh, Surabhi R. Madhvapathy et al. "Near-unity photoluminescence quantum yield in MoS2." Science 350, no. 6264 (2015): 1065-1068.

221 Zhao, Weijie, Zohreh Ghorannevis, Leiqiang Chu, Minglin Toh, Christian Kloc, Ping-Heng Tan, and Goki Eda. "Evolution of electronic structure in atomically thin sheets of WS2 and WSe2." ACS nano 7, no. 1 (2013): 791-797.

222 Fang, Hui, Corsin Battaglia, Carlo Carraro, Slavomir Nemsak, Burak Ozdol, Jeong Seuk Kang, Hans A. Bechtel et al. "Strong interlayer coupling in van der Waals heterostructures built from single-layer chalcogenides." Proceedings of the National Academy of Sciences 111, no. 17 (2014): 6198-6202..

223 Rivera, Pasqual, John R. Schaibley, Aaron M. Jones, Jason S. Ross, Sanfeng Wu, Grant Aivazian, Philip Klement et al. "Observation of long-lived interlayer excitons in monolayer MoSe2–WSe2 heterostructures." Nature communications 6, no. 1 (2015): 6242.

224 Van Der Zande, Arend M., Jens Kunstmann, Alexey Chernikov, Daniel A. Chenet, YuMeng You, XiaoXiao Zhang, Pinshane Y. Huang et al. "Tailoring the electronic structure in bilayer molybdenum disulfide via interlayer twist." Nano letters 14, no. 7 (2014): 3869-3875.

[225] Liu, Kaihui, Liming Zhang, Ting Cao, Chenhao Jin, Diana Qiu, Qin Zhou, Alex Zettl, Peidong Yang, Steve G. Louie, and Feng Wang. "Evolution of interlayer coupling in twisted molybdenum disulfide bilayers." Nature communications 5, no. 1 (2014): 4966.

[226] Ahn, Sung Joon, Pilkyung Moon, Tae-Hoon Kim, Hyun-Woo Kim, Ha-Chul Shin, Eun Hye Kim, Hyun Woo Cha et al. "Dirac electrons in a dodecagonal graphene quasicrystal." Science 361, no. 6404 (2018): 782-786.

[227] Sutter, Peter, Rina Ibragimova, Hannu-Pekka Komsa, Bruce A. Parkinson, and Eli Sutter. "Self-organized twist-heterostructures via aligned van der Waals epitaxy and solid-state transformations." Nature Communications 10, no. 1 (2019): 5528.

[228] Kim, Kwanpyo, Zonghoon Lee, William Regan, C. Kisielowski, M. F. Crommie, and Alex Zettl. "Grain boundary mapping in polycrystalline graphene." ACS nano 5, no. 3 (2011): 2142-2146.

[229] Kim, Kwanpyo, Sinisa Coh, Liang Z. Tan, William Regan, Jong Min Yuk, Eric Chatterjee, M. F. Crommie, Marvin L. Cohen, Steven G. Louie, and Alex Zettl. "Raman spectroscopy study of rotated double-layer graphene: misorientation-angle dependence of electronic structure." Physical review letters 108, no. 24 (2012): 246103.

[230] Sun, Luzhao, Zihao Wang, Yuechen Wang, Liang Zhao, Yanglizhi Li, Buhang Chen, Shenghong Huang et al. "Hetero-site nucleation for growing twisted bilayer graphene with a wide range of twist angles." Nature communications 12, no. 1 (2021): 2391.

[231] Matte, HSS Ramakrishna, A. Gomathi, Arun K. Manna, Dattatray J. Late, Ranjan Datta, Swapan K. Pati, and C. N. R. Rao. "MoS 2 and WS 2 analogues of graphene." Angewandte Chemie International Edition 24, no. 49 (2010): 4059-4062.

[232] Choudhary, Nitin, Juhong Park, Jun Yeon Hwang, Hee-Suk Chung, Kenneth H. Dumas, Saiful I. Khondaker, Wonbong Choi, and Yeonwoong Jung. "Centimeter scale patterned growth of vertically stacked few layer only 2D MoS2/WS2 van der Waals heterostructure." Scientific reports 6, no. 1 (2016): 25456.

[233] Wang, Kai, Bing Huang, Mengkun Tian, Frank Ceballos, Ming-Wei Lin, Masoud Mahjouri-Samani, Abdelaziz Boulesbaa et al. "Interlayer coupling in twisted WSe2/WS2 bilayer heterostructures revealed by optical spectroscopy." ACS nano 10, no. 7 (2016): 6612-6622.

[234] Tao, Nongjian. "Electrochemical fabrication of metallic quantum wires." (2005): 720.

[235] Van Wees, B. J., H. Van Houten, C. W. J. Beenakker, J. Gr Williamson, L. P. Kouwenhoven, D. Van der Marel, and C. T. Foxon. "Quantized conductance of point contacts in a two-dimensional electron gas." Physical Review Letters 60, no. 9 (1988): 848.

[236] Li, C. Z., H. X. He, A. Bogozi, J. S. Bunch, and N. J. Tao. "Molecular detection based on conductance quantization of nanowires." Applied Physics Letters 76, no. 10 (2000): 1333-1335.

[237] Yoon, Young-Gui, Mario SC Mazzoni, and Steven G. Louie. "Quantum conductance of carbon nanotube peapods." Applied physics letters 83, no. 25 (2003): 5217-5219.

[238] Ott, F., S. Barberan, J. G. Lunney, J. M. D. Coey, P. Berthet, A. M. de Leon-Guevara, and A. Revcolevschi. "Quantized conductance in a contact between metallic oxide crystals." Physical Review B 58, no. 8 (1998): 4656.

[239] Taylor, Simon. "Quantum Hall effect." (2015) URN: urn:nbn:se:uu:diva-255261.

[240] v Klitzing, K., and G. Ebert. "Application of the quantum Hall effect in metrology." Metrologia 21, no. 1 (1985): 11.

[241] Sarma, Sankar Das, and Aron Pinczuk, eds. Perspectives in quantum hall effects: Novel quantum liquids in low-dimensional semiconductor structures. John Wiley & Sons, 2008.

[242] Zhang, Yuanbo, Yan-Wen Tan, Horst L. Stormer, and Philip Kim. "Experimental observation of the quantum Hall effect and Berry's phase in graphene." nature 438, no. 7065 (2005): 201-204.

[243] Stern, Ady. "Anyons and the quantum Hall effect—A pedagogical review." Annals of Physics 323, no. 1 (2008): 204-249.

[244] Novoselov, Kostya S., Andre K. Geim, Sergei Vladimirovich Morozov, Dingde Jiang, Michail I. Katsnelson, Irina V. Grigorieva, SVb Dubonos, and and AA Firsov. "Two-dimensional gas of massless Dirac fermions in graphene." nature 438, no. 7065 (2005): 197-200.

[245] Gusynin, V. P., and S. G. Sharapov. "Unconventional integer quantum Hall effect in graphene." Physical review letters 95, no. 14 (2005): 146801.

[246] Checkelsky, Joseph G., Lu Li, and N. P. Ong. "Zero-energy state in graphene in a high magnetic field." Physical review letters 100, no. 20 (2008): 206801.

[247] Zhang, Y., Z. Jiang, J. P. Small, M. S. Purewal, Y-W. Tan, M. Fazlollahi, J. D. Chudow, J. A. Jaszczak, H. L. Stormer, and P. Kim. "Landau-level splitting in graphene in high magnetic fields." Physical review letters 96, no. 13 (2006): 136806.

[248] Comtet, Alain, T. Jolicoeur, S. Ouvry, and F. David, eds. Aspects topologiques de la physique en basse dimension. Topological Aspects of Low Dimensional Systems. Vol. 69. Springer Science & Business Media, 2000.

[249] Novoselov, Konstantin S., Z-F. Jiang, Y-S. Zhang, S. V. Morozov, Horst L. Stormer, U. Zeitler, J. C. Maan, G. S. Boebinger, Philip Kim, and Andre K. Geim. "Room-temperature quantum Hall effect in graphene." science 315, no. 5817 (2007): 1379-1379.

[250] Bolotin, Kirill I., Fereshte Ghahari, Michael D. Shulman, Horst L. Stormer, and Philip Kim. "Observation of the fractional quantum Hall effect in graphene." Nature 462, no. 7270 (2009): 196-199.

[251] Li, J. I. A., C. Tan, S. Chen, Y. Zeng, T. Taniguchi, K. Watanabe, J. Hone, and C. R. Dean. "Even-denominator fractional quantum Hall states in bilayer graphene." Science 358, no. 6363 (2017): 648-652.

[252] Diankov, Georgi, Chi-Te Liang, François Amet, Patrick Gallagher, Menyoung Lee, Andrew J. Bestwick, Kevin Tharratt et al. "Robust fractional quantum Hall

effect in the N= 2 Landau level in bilayer graphene." Nature Communications 7, no. 1 (2016): 13908.

253 Maher, Patrick, Lei Wang, Yuanda Gao, Carlos Forsythe, Takashi Taniguchi, Kenji Watanabe, Dmitry Abanin et al. "Tunable fractional quantum Hall phases in bilayer graphene." Science 345, no. 6192 (2014): 61-64.

254 Sanchez-Yamagishi, Javier D., Jason Y. Luo, Andrea F. Young, Benjamin M. Hunt, Kenji Watanabe, Takashi Taniguchi, Raymond C. Ashoori, and Pablo Jarillo-Herrero. "Helical edge states and fractional quantum Hall effect in a graphene electron–hole bilayer." Nature nanotechnology 12, no. 2 (2017): 118-122.

255 Jariwala, Deep, Vinod K. Sangwan, Lincoln J. Lauhon, Tobin J. Marks, and Mark C. Hersam. "Emerging device applications for semiconducting two-dimensional transition metal dichalcogenides." ACS nano 8, no. 2 (2014): 1102-1120.

256 Feng, Wei, Wei Zheng, Wenwu Cao, and PingAn Hu. "Back gated multilayer InSe transistors with enhanced carrier mobilities via the suppression of carrier scattering from a dielectric interface." Advanced materials 26, no. 38 (2014): 6587-6593.

257 Datta, Supriyo. Electronic transport in mesoscopic systems. Cambridge university press, 1997.

258 Ma, Nan, and Debdeep Jena. "Charge scattering and mobility in atomically thin semiconductors." Physical Review X 4, no. 1 (2014): 011043.

259 Fang, Hui, Steven Chuang, Ting Chia Chang, Kuniharu Takei, Toshitake Takahashi, and Ali Javey. "High-performance single layered WSe2 p-FETs with chemically doped contacts." Nano letters 12, no. 7 (2012): 3788-3792.

260 Xiang, Du, Cheng Han, Jing Wu, Shu Zhong, Yiyang Liu, Jiadan Lin, Xue-Ao Zhang et al. "Surface transfer doping induced effective modulation on ambipolar characteristics of few-layer black phosphorus." Nature communications 6, no. 1 (2015): 6485.

261 Ye, Ziliang, Lutz Waldecker, Eric Yue Ma, Daniel Rhodes, Abhinandan Antony, Bumho Kim, Xiao-Xiao Zhang et al. "Efficient generation of neutral and charged biexcitons in encapsulated WSe2 monolayers." Nature communications 9, no. 1 (2018): 3718.

262 Xu, Shuigang, Zefei Wu, Huanhuan Lu, Yu Han, Gen Long, Xiaolong Chen, Tianyi Han et al. "Universal low-temperature Ohmic contacts for quantum transport in transition metal dichalcogenides." 2D Materials 3, no. 2 (2016): 021007.

263 Cui, Xu, Gwan-Hyoung Lee, Young Duck Kim, Ghidewon Arefe, Pinshane Y. Huang, Chul-Ho Lee, Daniel A. Chenet et al. "Multi-terminal transport measurements of MoS2 using a van der Waals heterostructure device platform." Nature nanotechnology 10, no. 6 (2015): 534-540.

264 Li, Likai, Fangyuan Yang, Guo Jun Ye, Zuocheng Zhang, Zengwei Zhu, Wenkai Lou, Xiaoying Zhou et al. "Quantum Hall effect in black phosphorus two-dimensional electron system." Nature nanotechnology 11, no. 7 (2016): 593-597.

265 Wu, Zefei, Shuigang Xu, Huanhuan Lu, Armin Khamoshi, Gui-Bin Liu, Tianyi Han, Yingying Wu et al. "Even–odd layer-dependent magnetotransport of high-mobility Q-valley electrons in transition metal disulfides." Nature communications 7, no. 1 (2016): 12955.
266 Yi, Ya, Zhanxu Chen, Xue-Feng Yu, Zhang-Kai Zhou, and Jia Li. "Recent advances in quantum effects of 2D materials." Advanced Quantum Technologies 2, no. 5-6 (2019): 1800111.
267 Long, Gen, Denis Maryenko, Junying Shen, Shuigang Xu, Jianqiang Hou, Zefei Wu, Wing Ki Wong et al. "Achieving ultrahigh carrier mobility in two-dimensional hole gas of black phosphorus." Nano Letters 16, no. 12 (2016): 7768-7773.
268 Chen, Xiaolong, Yingying Wu, Zefei Wu, Yu Han, Shuigang Xu, Lin Wang, Weiguang Ye et al. "High-quality sandwiched black phosphorus heterostructure and its quantum oscillations." Nature communications 6, no. 1 (2015): 7315.
269 Long, Gen, Denis Maryenko, Sergio Pezzini, Shuigang Xu, Zefei Wu, Tianyi Han, Jiangxiazi Lin et al. "Ambipolar quantum transport in few-layer black phosphorus." Physical Review B 96, no. 15 (2017): 155448.
270 Xu, Shuigang, Junying Shen, Gen Long, Zefei Wu, Zhi-qiang Bao, Cheng-Cheng Liu, Xiao Xiao et al. "Odd-integer quantum Hall states and giant spin susceptibility in p-type few-layer WSe 2." Physical Review Letters 118, no. 6 (2017): 067702.
271 Chen, Q. H., J. M. Lu, Lei Liang, Oleksandr Zheliuk, A. Ali, Ping Sheng, and J. T. Ye. "Inducing and manipulating heteroelectronic states in a single MoS 2 thin flake." Physical Review Letters 119, no. 14 (2017): 147002.
272 Pisoni, Riccardo, Yongjin Lee, Hiske Overweg, Marius Eich, Pauline Simonet, Kenji Watanabe, Takashi Taniguchi, Roman Gorbachev, Thomas Ihn, and Klaus Ensslin. "Gate-defined one-dimensional channel and broken symmetry states in MoS2 van der Waals heterostructures." Nano letters 17, no. 8 (2017): 5008-5011.
273 Pal, Palash B. "Dirac, majorana, and weyl fermions." American Journal of Physics 79, no. 5 (2011): 485-498.
274 Butler, Sheneve Z., Shawna M. Hollen, Linyou Cao, Yi Cui, Jay A. Gupta, Humberto R. Gutiérrez, Tony F. Heinz et al. "Progress, challenges, and opportunities in two-dimensional materials beyond graphene." ACS nano 7, no. 4 (2013): 2898-2926.
275 Xu, Mingsheng, Tao Liang, Minmin Shi, and Hongzheng Chen. "Graphene-like two-dimensional materials." Chemical reviews 113, no. 5 (2013): 3766-3798.
276 Wallace, Philip Richard. "The band theory of graphite." Physical review 71, no. 9 (1947): 622.
277 Cahangirov, Seymur, Mehmet Topsakal, Ethem Aktürk, Hasan Şahin, and Salim Ciraci. "Two-and one-dimensional honeycomb structures of silicon and germanium." Physical review letters 102, no. 23 (2009): 236804.
278 Malko, Daniel, Christian Neiss, Francesc Vines, and Andreas Görling. "Competition for graphene: graphynes with direction-dependent dirac cones." Physical review letters 108, no. 8 (2012): 086804.

[279] Zhou, Xiang-Feng, Xiao Dong, Artem R. Oganov, Qiang Zhu, Yongjun Tian, and Hui-Tian Wang. "Semimetallic two-dimensional boron allotrope with massless Dirac fermions." Physical Review Letters 112, no. 8 (2014): 085502.

[280] Wang, Z. F., Zheng Liu, and Feng Liu. "Organic topological insulators in organometallic lattices." Nature communications 4, no. 1 (2013): 1471.

[281] Gomes, Kenjiro K., Warren Mar, Wonhee Ko, Francisco Guinea, and Hari C. Manoharan. "Designer Dirac fermions and topological phases in molecular graphene." Nature 483, no. 7389 (2012): 306-310.

[282] Ullal, Chaitanya K., Jian Shi, and Ravishankar Sundararaman. "Electron mobility in graphene without invoking the Dirac equation." American Journal of Physics 87, no. 4 (2019): 291-295.

[283] Du, Xu, Ivan Skachko, Fabian Duerr, Adina Luican, and Eva Y. Andrei. "Fractional quantum Hall effect and insulating phase of Dirac electrons in graphene." Nature 462, no. 7270 (2009): 192-195.

[284] Wang, Jinying, Shibin Deng, Zhongfan Liu, and Zhirong Liu. "The rare two-dimensional materials with Dirac cones." National Science Review 2, no. 1 (2015): 22-39.

[285] Young, Steve M., and Charles L. Kane. "Dirac semimetals in two dimensions." Physical review letters 115, no. 12 (2015): 126803.

[286] You, Jing-Yang, Cong Chen, Zhen Zhang, Xian-Lei Sheng, Shengyuan A. Yang, and Gang Su. "Two-dimensional Weyl half-semimetal and tunable quantum anomalous Hall effect." Physical Review B 100, no. 6 (2019): 064408.

[287] Hunt, Benjamin, Javier D. Sanchez-Yamagishi, Andrea F. Young, Matthew Yankowitz, Brian J. LeRoy, Kenji Watanabe, Takashi Taniguchi et al. "Massive Dirac fermions and Hofstadter butterfly in a van der Waals heterostructure." Science 340, no. 6139 (2013): 1427-1430.

[288] Palmieri, Tania, Edoardo Baldini, Alexander Steinhoff, Ana Akrap, Márton Kollár, Endre Horváth, László Forró, Frank Jahnke, and Majed Chergui. "Mahan excitons in room-temperature methylammonium lead bromide perovskites." Nature communications 11, no. 1 (2020): 850.

[289] Song, Yekai, Chunjing Jia, Hongyu Xiong, Binbin Wang, Zhicheng Jiang, Kui Huang, Jinwoong Hwang et al. "Signatures of the exciton gas phase and its condensation in monolayer 1T-ZrTe2." Nature communications 14, no. 1 (2023): 1116.

[290] Molas, Maciej R. "Excitons and Phonons in Two-Dimensional Materials: From Fundamental to Applications." Nanomaterials 13, no. 23 (2023): 3047.

[291] Xiao, Jun, Mervin Zhao, Yuan Wang, and Xiang Zhang. "Excitons in atomically thin 2D semiconductors and their applications." Nanophotonics 6, no. 6 (2017): 1309-1328.

[292] Mueller, Thomas, and Ermin Malic. "Exciton physics and device application of two-dimensional transition metal dichalcogenide semiconductors." npj 2D Materials and Applications 2, no. 1 (2018): 29.

[293] Xie, Yong, Gulsum Ersu, Thomas Pucher, Sruthi Kuriakose, Wenliang Zhang, Abdullah M. Al-Enizi, Hamad AH Albrithen et al. "Making exciton physics easy and affordable." European Journal of Physics 44, no. 5 (2023): 055501.

294 Xiao, Jun, Mervin Zhao, Yuan Wang, and Xiang Zhang. "Excitons in atomically thin 2D semiconductors and their applications." Nanophotonics 6, no. 6 (2017): 1309-1328.

295 Wang, Gang, Alexey Chernikov, Mikhail M. Glazov, Tony F. Heinz, Xavier Marie, Thierry Amand, and Bernhard Urbaszek. "Colloquium: Excitons in atomically thin transition metal dichalcogenides." Reviews of Modern Physics 90, no. 2 (2018): 021001.

296 Mueller, Thomas, and Ermin Malic. "Exciton physics and device application of two-dimensional transition metal dichalcogenide semiconductors." npj 2D Materials and Applications 2, no. 1 (2018): 29.

297 Jariwala, Deep, Vinod K. Sangwan, Lincoln J. Lauhon, Tobin J. Marks, and Mark C. Hersam. "Emerging device applications for semiconducting two-dimensional transition metal dichalcogenides." ACS nano 8, no. 2 (2014): 1102-1120.

298 Pu, Jiang, and Taishi Takenobu. "Monolayer transition metal dichalcogenides as light sources." Advanced materials 30, no. 33 (2018): 1707627.

299 Chakraborty, Chitraleema, Laura Kinnischtzke, Kenneth M. Goodfellow, Ryan Beams, and A. Nick Vamivakas. "Voltage-controlled quantum light from an atomically thin semiconductor." Nature nanotechnology 10, no. 6 (2015): 507-511.

300 Mak, Kin Fai, Keliang He, Changgu Lee, Gwan Hyoung Lee, James Hone, Tony F. Heinz, and Jie Shan. "Tightly bound trions in monolayer MoS2." Nature materials 12, no. 3 (2013): 207-211.

301 You, Yumeng, Xiao-Xiao Zhang, Timothy C. Berkelbach, Mark S. Hybertsen, David R. Reichman, and Tony F. Heinz. "Observation of biexcitons in monolayer WSe 2." Nature Physics 11, no. 6 (2015): 477-481.

302 Chen, Shao-Yu, Thomas Goldstein, Takashi Taniguchi, Kenji Watanabe, and Jun Yan. "Coulomb-bound four-and five-particle intervalley states in an atomically-thin semiconductor." Nature communications 9, no. 1 (2018): 3717.

303 Li, Zhipeng, Tianmeng Wang, Zhengguang Lu, Chenhao Jin, Yanwen Chen, Yuze Meng, Zhen Lian et al. "Revealing the biexciton and trion-exciton complexes in BN encapsulated WSe2." Nature communications 9, no. 1 (2018): 3719.

304 Kaplan, D., Y. Gong, K. Mills, V. Swaminathan, P. M. Ajayan, S. Shirodkar, and E. Kaxiras. "Excitation intensity dependence of photoluminescence from monolayers of MoS2 and WS2/MoS2 heterostructures." 2D Materials 3, no. 1 (2016): 015005.

305 Huang, Jiani, Thang B. Hoang, and Maiken H. Mikkelsen. "Probing the origin of excitonic states in monolayer WSe 2." Scientific reports 6, no. 1 (2016): 22414.

306 Kato, Toshiaki, and Toshiro Kaneko. "Transport dynamics of neutral excitons and trions in monolayer WS2." ACS nano 10, no. 10 (2016): 9687-9694.

307 Lee, Yongjun, Seok Joon Yun, Youngbum Kim, Min Su Kim, Gang Hee Han, A. K. Sood, and Jeongyong Kim. "Near-field spectral mapping of individual

exciton complexes of monolayer WS 2 correlated with local defects and charge population." Nanoscale 9, no. 6 (2017): 2272-2278.

308 Sie, Edbert J., Alex J. Frenzel, Yi-Hsien Lee, Jing Kong, and Nuh Gedik. "Intervalley biexcitons and many-body effects in monolayer MoS 2." Physical Review B 92, no. 12 (2015): 125417.

309 Plechinger, Gerd, Philipp Nagler, Ashish Arora, Robert Schmidt, Alexey Chernikov, Andrés Granados Del Águila, Peter CM Christianen, Rudolf Bratschitsch, Christian Schüller, and Tobias Korn. "Trion fine structure and coupled spin–valley dynamics in monolayer tungsten disulfide." Nature communications 7, no. 1 (2016): 12715.

310 Hao, Kai, Judith F. Specht, Philipp Nagler, Lixiang Xu, Kha Tran, Akshay Singh, Chandriker Kavir Dass et al. "Neutral and charged inter-valley biexcitons in monolayer MoSe2." Nature communications 8, no. 1 (2017): 15552.

311 Arora, Ankit, Pramoda K. Nayak, Tejendra Dixit, Kolla Lakshmi Ganapathi, Ananth Krishnan, and Mamidanna Sri Ramachandra Rao. "Stacking angle dependent multiple excitonic resonances in bilayer tungsten diselenide." Nanophotonics 9, no. 12 (2020): 3881-3887.

312 Wang, Yanlong, Chunxiao Cong, Jingzhi Shang, Mustafa Eginligil, Yuqi Jin, Gang Li, Yu Chen, Namphung Peimyoo, and Ting Yu. "Unveiling exceptionally robust valley contrast in AA-and AB-stacked bilayer WS 2." Nanoscale Horizons 4, no. 2 (2019): 396-403.

313 Das, Sarthak, Garima Gupta, and Kausik Majumdar. "Layer degree of freedom for excitons in transition metal dichalcogenides." Physical Review B 99, no. 16 (2019): 165411.

314 Jin, Chenhao, Eric Yue Ma, Ouri Karni, Emma C. Regan, Feng Wang, and Tony F. Heinz. "Ultrafast dynamics in van der Waals heterostructures." Nature nanotechnology 13, no. 11 (2018): 994-1003.

315 Jones, Aaron M., Hongyi Yu, Jason S. Ross, Philip Klement, Nirmal J. Ghimire, Jiaqiang Yan, David G. Mandrus, Wang Yao, and Xiaodong Xu. "Spin–layer locking effects in optical orientation of exciton spin in bilayer WSe2." Nature Physics 10, no. 2 (2014): 130-134.

316 Huang, Shengxi, Xi Ling, Liangbo Liang, Jing Kong, Humberto Terrones, Vincent Meunier, and Mildred S. Dresselhaus. "Probing the interlayer coupling of twisted bilayer MoS2 using photoluminescence spectroscopy." Nano letters 14, no. 10 (2014): 5500-5508.

317 Van Der Zande, Arend M., Jens Kunstmann, Alexey Chernikov, Daniel A. Chenet, YuMeng You, XiaoXiao Zhang, Pinshane Y. Huang et al. "Tailoring the electronic structure in bilayer molybdenum disulfide via interlayer twist." Nano letters 14, no. 7 (2014): 3869-3875.

318 Zheng, Shoujun, Linfeng Sun, Xiaohao Zhou, Fucai Liu, Zheng Liu, Zexiang Shen, and Hong Jin Fan. "Coupling and interlayer exciton in twist-stacked WS2 bilayers." Advanced Optical Materials 3, no. 11 (2015): 1600-1605.

319 Unuchek, Dmitrii, Alberto Ciarrocchi, Ahmet Avsar, Kenji Watanabe, Takashi Taniguchi, and Andras Kis. "Room-temperature electrical control of exciton flux in a van der Waals heterostructure." Nature 560, no. 7718 (2018): 340-344.

[320] Ciarrocchi, Alberto, Dmitrii Unuchek, Ahmet Avsar, Kenji Watanabe, Takashi Taniguchi, and Andras Kis. "Polarization switching and electrical control of interlayer excitons in two-dimensional van der Waals heterostructures." Nature photonics 13, no. 2 (2019): 131-136.

[321] Tahir, Muhammad, P. M. Krstajić, and Panagiotis Vasilopoulos. "Electrically controlled dc and ac transport in bilayer WSe 2." Physical Review B 98, no. 7 (2018): 075429.

[322] Qian, Qingkai, Jiacheng Lei, Jin Wei, Zhaofu Zhang, Gaofei Tang, Kailun Zhong, Zheyang Zheng, and Kevin J. Chen. "2D materials as semiconducting gate for field-effect transistors with inherent over-voltage protection and boosted ON-current." npj 2D Materials and Applications 3, no. 1 (2019): 24.

[323] Kanungo, Sayan, Gufran Ahmad, Parikshit Sahatiya, Arnab Mukhopadhyay, and Sanatan Chattopadhyay. "2D materials-based nanoscale tunneling field effect transistors: current developments and future prospects." npj 2D Materials and Applications 6, no. 1 (2022): 83.

[324] Kajale, Shivam Nitin, Shubham Yadav, Yubin Cai, Baju Joy, and Deblina Sarkar. "2D material based field effect transistors and nanoelectromechanical systems for sensing applications." IScience 24, no. 12 (2021).

[325] Roy, Tania, Mahmut Tosun, Jeong Seuk Kang, Angada B. Sachid, Sujay B. Desai, Mark Hettick, Chenming C. Hu, and Ali Javey. "Field-effect transistors built from all two-dimensional material components." ACS nano 8, no. 6 (2014): 6259-6264.

[326] Krsihna, B. Vamsi, S. Ravi, and M. Durga Prakash. "Recent developments in graphene based field effect transistors." Materials Today: Proceedings 45 (2021): 1524-1528.

[327] Krishnan, Siva Kumar, Nandini Nataraj, M. Meyyappan, and Umapada Pal. "Graphene-based field-effect transistors in biosensing and neural interfacing applications: recent advances and prospects." Analytical Chemistry 95, no. 5 (2023): 2590-2622.

[328] Wang, Shiyu, Zakir Hossain, Yan Zhao, Tao Han, Shiyu Wang, Zakir Hossain, Yan Zhao, and Tao Han. "Graphene Field-Effect Transistor Biosensor." Graphene Field-Effect Transistor Biosensors (2021): 45-67.

[329] Yu, Zhihao, Zhun-Yong Ong, Songlin Li, Jian-Bin Xu, Gang Zhang, Yong-Wei Zhang, Yi Shi, and Xinran Wang. "Analyzing the carrier mobility in transition-metal dichalcogenide MoS2 field-effect transistors." Advanced Functional Materials 27, no. 19 (2017): 1604093.

[330] Voiry, Damien, Jieun Yang, and Manish Chhowalla. "Recent strategies for improving the catalytic activity of 2D TMD nanosheets toward the hydrogen evolution reaction." Advanced materials 28, no. 29 (2016): 6197-6206.

[331] Susarla, Sandhya, Alex Kutana, Jordan A. Hachtel, Vidya Kochat, Amey Apte, Robert Vajtai, Juan Carlos Idrobo, Boris I. Yakobson, Chandra Sekhar Tiwary, and Pulickel M. Ajayan. "Quaternary 2D transition metal dichalcogenides (TMDs) with tunable bandgap." Advanced Materials 29, no. 35 (2017): 1702457.

[332] Daus, Alwin, Sam Vaziri, Victoria Chen, Çağıl Köroğlu, Ryan W. Grady, Connor S. Bailey, Hye Ryoung Lee, Kirstin Schauble, Kevin Brenner, and Eric Pop. "High-performance flexible nanoscale transistors based on transition metal dichalcogenides." Nature Electronics 4, no. 7 (2021): 495-501.

[333] Zhang, Qing, Chunsen Liu, and Peng Zhou. "2D materials readiness for the transistor performance breakthrough." Iscience (2023).

[334] Liao, Liping, Evgeniya Kovalska, Jakub Regner, Qunliang Song, and Zdeněk Sofer. "Two-Dimensional Van Der Waals Thin Film and Device." Small 20, no. 4 (2024): 2303638.

[335] Lee, Seoung-Ki, Ho Young Jang, Sukjae Jang, Euiyoung Choi, Byung Hee Hong, Jaichan Lee, Sungho Park, and Jong-Hyun Ahn. "All graphene-based thin film transistors on flexible plastic substrates." Nano letters 12, no. 7 (2012): 3472-3476.

[336] Li, X. A., R. Y. Ren, M. W. Ren, Y. W. Ma, J. P. Yang, and G. Bao. "Integration of graphene into thin film transistors." Materials Technology 28, no. 4 (2013): 175-180.

[337] Liu, Yuan. "Graphene as tunable contact for high performance thin film transistor." PhD diss., UCLA, 2015.

[338] Zhu, Zhongcheng, Imran Murtaza, Hong Meng, and Wei Huang. "Thin film transistors based on two dimensional graphene and graphene/semiconductor heterojunctions." RSC advances 7, no. 28 (2017): 17387-17397.

[339] Liu, Liting, Yuan Liu, and Xiangfeng Duan. "Graphene-based vertical thin film transistors." Science China Information Sciences 63 (2020): 1-12.

[340] Kim, Maria, David MA Mackenzie, Wonjae Kim, Kirill Isakov, and Harri Lipsanen. "All-parylene flexible wafer-scale graphene thin film transistor." Applied Surface Science 551 (2021): 149410.

[341] Zhao, Jing, Wei Chen, Jianling Meng, Hua Yu, Mengzhou Liao, Jianqi Zhu, Rong Yang, Dongxia Shi, and Guangyu Zhang. "Integrated flexible and high-quality thin film transistors based on monolayer MoS2." Advanced Electronic Materials 2, no. 3 (2016): 1500379.

[342] Mandyam, Srinivas V., Hyong M. Kim, and Marija Drndić. "Large area few-layer TMD film growths and their applications." Journal of Physics: Materials 3, no. 2 (2020): 024008.

[343] Sridevi, R., and J. Charles Pravin. "Two-Dimensional Transition Metal Dichalcogenide (TMD) Materials in Field-Effect Transistor (FET) Devices for Low Power Applications." In Semiconductor Devices and Technologies for Future Ultra Low Power Electronics, pp. 253-288. CRC Press, 2021.

[344] Sarkar, Debalina. "2D Materials for Field-Effect Transistor–Based Biosensors." In Fundamentals and Sensing Applications of 2D Materials, pp. 329-377. Woodhead Publishing, 2019.

[345] Choi, Woong, and Sunkook Kim. "Thin-Film Transistors Based on Transition Metal Dichalcogenides." Nanomaterials, Polymers, and Devices: Materials Functionalization and Device Fabrication (2015): 539-562.

[346] Rai, Suyash, Himanshu Mishra, Vijay K. Singh, Tejendra K. Gupta, and Anchal Srivastava. "2D materials for flexible electronics." In 2D Materials for Electronics, Sensors and Devices, pp. 169-206. Elsevier, 2023.

[347] Katiyar, Ajit Kumar, Anh Tuan Hoang, Duo Xu, Juyeong Hong, Beom Jin Kim, Seunghyeon Ji, and Jong-Hyun Ahn. "2D materials in flexible electronics: recent advances and future prospectives." Chemical Reviews 124, no. 2 (2023): 318-419.

[348] Han, Tae-Hee, Hobeom Kim, Sung-Joo Kwon, and Tae-Woo Lee. "Graphene-based flexible electronic devices." Materials Science and Engineering: R: Reports 118 (2017): 1-43.

[349] Gao, Wei, Hiroki Ota, Daisuke Kiriya, Kuniharu Takei, and Ali Javey. "Flexible electronics toward wearable sensing." Accounts of chemical research 52, no. 3 (2019): 523-533.

[350] Kim, Maria, David MA Mackenzie, Wonjae Kim, Kirill Isakov, and Harri Lipsanen. "All-parylene flexible wafer-scale graphene thin film transistor." Applied Surface Science 551 (2021): 149410.

[351] Jiang, Dongting, Zhiyuan Liu, Zhe Xiao, Zhengfang Qian, Yiling Sun, Zhiyuan Zeng, and Renheng Wang. "Flexible electronics based on 2D transition metal dichalcogenides." Journal of Materials Chemistry A 10, no. 1 (2022): 89-121.

[352] Akinwande, Deji, Nicholas Petrone, and James Hone. "Two-dimensional flexible nanoelectronics." Nature communications 5, no. 1 (2014): 5678.

[353] Hoang, Anh Tuan, Luhing Hu, Ajit Kumar Katiyar, and Jong-Hyun Ahn. "Two-dimensional layered materials and heterostructures for flexible electronics." Matter 5, no. 12 (2022): 4116-4132.

[354] Glavin, Nicholas R., Christopher Muratore, and Michael Snure. "Toward 2D materials for flexible electronics: opportunities and outlook." Oxford Open Materials Science 1, no. 1 (2021): itaa002.

[355] Kim, Beom Joon, Houk Jang, Seoung-Ki Lee, Byung Hee Hong, Jong-Hyun Ahn, and Jeong Ho Cho. "High-performance flexible graphene field effect transistors with ion gel gate dielectrics." Nano letters 10, no. 9 (2010): 3464-3466.

[356] Lee, Jongho, Tae-Jun Ha, Huifeng Li, Kristen N. Parrish, Milo Holt, Ananth Dodabalapur, Rodney S. Ruoff, and Deji Akinwande. "25 GHz embedded-gate graphene transistors with high-K dielectrics on extremely flexible plastic sheets." ACS nano 7, no. 9 (2013): 7744-7750.

[357] Chang, Hsiao-Yu, Shixuan Yang, Jongho Lee, Li Tao, Wan-Sik Hwang, Debdeep Jena, Nanshu Lu, and Deji Akinwande. "High-performance, highly bendable MoS2 transistors with high-k dielectrics for flexible low-power systems." ACS nano 7, no. 6 (2013): 5446-5452.

[358] Song, Won Geun, Hyuk-Jun Kwon, Jozeph Park, Junyeob Yeo, Minjeong Kim, Suntak Park, Sungryul Yun et al. "High-performance flexible multilayer MoS2 transistors on solution-based polyimide substrates." Advanced Functional Materials 26, no. 15 (2016): 2426-2434.

[359] Zhu, Weinan, Maruthi N. Yogeesh, Shixuan Yang, Sandra H. Aldave, Joon-Seok Kim, Sushant Sonde, Li Tao, Nanshu Lu, and Deji Akinwande. "Flexible

black phosphorus ambipolar transistors, circuits and AM demodulator." Nano letters 15, no. 3 (2015): 1883-1890.

360 Petrone, Nicholas, Tarun Chari, Inanc Meric, Lei Wang, Kenneth L. Shepard, and James Hone. "Flexible graphene field-effect transistors encapsulated in hexagonal boron nitride." ACS nano 9, no. 9 (2015): 8953-8959.

361 Wu, Liangmei, Jinan Shi, Zhang Zhou, Jiahao Yan, Aiwei Wang, Ce Bian, Jiajun Ma et al. "InSe/hBN/graphite heterostructure for high-performance 2D electronics and flexible electronics." Nano Research 13 (2020): 1127-1132.

362 Schwierz, Frank. "Graphene transistors." Nature nanotechnology 5, no. 7 (2010): 487-496.

363 Yu, Huihui, Zhihong Cao, Zheng Zhang, Xiankun Zhang, and Yue Zhang. "Flexible electronics and optoelectronics of 2D van der Waals materials." International Journal of Minerals, Metallurgy and Materials 29, no. 4 (2022): 671-690.

364 Lee, Seoung-Ki, Beom Joon Kim, Houk Jang, Sung Cheol Yoon, Changjin Lee, Byung Hee Hong, John A. Rogers, Jeong Ho Cho, and Jong-Hyun Ahn. "Stretchable graphene transistors with printed dielectrics and gate electrodes." Nano letters 11, no. 11 (2011): 4642-4646.

365 Fiori, Gianluca, Francesco Bonaccorso, Giuseppe Iannaccone, Tomás Palacios, Daniel Neumaier, Alan Seabaugh, Sanjay K. Banerjee, and Luigi Colombo. "Electronics based on two-dimensional materials." Nature nanotechnology 9, no. 10 (2014): 768-779.

366 Hosseini, Manouchehr, Mohammad Elahi, Mahdi Pourfath, and David Esseni. "Strain-Induced Modulation of Electron Mobility in Single-Layer Transition Metal Dichalcogenides MX2 (M = Mo, W; X = S, Se)." IEEE Transactions on Electron Devices 62, no. 10 (2015): 3192-3198.

367 Xiao, Jiankun, Zhuo Kang, Baishan Liu, Xiankun Zhang, Junli Du, Kuanglei Chen, Huihui Yu, Qingliang Liao, Zheng Zhang, and Yue Zhang. "Record-high saturation current in end-bond contacted monolayer MoS2 transistors." Nano Research 15, no. 1 (2022): 475-481.

368 Pu, Jiang, Yohei Yomogida, Keng-Ku Liu, Lain-Jong Li, Yoshihiro Iwasa, and Taishi Takenobu. "Highly flexible MoS2 thin-film transistors with ion gel dielectrics." Nano letters 12, no. 8 (2012): 4013-4017.

369 Rhyee, Jong-Soo, Junyeon Kwon, Piyush Dak, Jin Hee Kim, Seung Min Kim, Jozeph Park, Young Ki Hong et al. "High-mobility transistors based on large-area and highly crystalline CVD-grown MoSe2 films on insulating substrates." Advanced Materials 28, no. 12 (2016): 2316-2321.

370 Novoselov, K. S卌, Artem Mishchenko, Alexandra Carvalho, and A. H. Castro Neto. "2D materials and van der Waals heterostructures." Science 353, no. 6298 (2016): aac9439.

371 Liu, Chunsen, Huawei Chen, Shuiyuan Wang, Qi Liu, Yu-Gang Jiang, David Wei Zhang, Ming Liu, and Peng Zhou. "Two-dimensional materials for next-generation computing technologies." Nature Nanotechnology 15, no. 7 (2020): 545-557.

[372] Zhang, Xiwei, Jiahua Shao, Chenxi Yan, Ruijie Qin, Zhihui Lu, Huijuan Geng, Tongshui Xu, and Lin Ju. "A review on optoelectronic device applications of 2D transition metal carbides and nitrides." Materials & Design 200 (2021): 109452.

[373] Cheng, Zhao, Rui Cao, Kangkang Wei, Yuhan Yao, Xinyu Liu, Jianlong Kang, Jianji Dong, Zhe Shi, Han Zhang, and Xinliang Zhang. "2D materials enabled next-generation integrated optoelectronics: from fabrication to applications." Advanced Science 8, no. 11 (2021): 2003834.

[374] Tan, Teng, Xiantao Jiang, Cong Wang, Baicheng Yao, and Han Zhang. "2D material optoelectronics for information functional device applications: status and challenges." Advanced Science 7, no. 11 (2020): 2000058.

[375] Hills, Gage, Christian Lau, Andrew Wright, Samuel Fuller, Mindy D. Bishop, Tathagata Srimani, Pritpal Kanhaiya et al. "Modern microprocessor built from complementary carbon nanotube transistors." Nature 572, no. 7771 (2019): 595-602.

[376] Wang, Mengzhu, Yingying Xiao, Ye Li, Lu Han, Zhicheng Sun, Liang He, Ruping Liu, and Kuan Hu. "Recent progress on graphene flexible photodetectors." Materials 15, no. 14 (2022): 4820.

[377] Schall, Daniel, Emiliano Pallecchi, Guillaume Ducournau, Vanessa Avramovic, Martin Otto, and Daniel Neumaier. "Record high bandwidth integrated graphene photodetectors for communication beyond 180 Gb/s." In Optical Fiber Communication Conference, pp. M2I-4. Optica Publishing Group, 2018.

[378] Guo, Xitao, Wenhui Wang, Haiyan Nan, Yuanfang Yu, Jie Jiang, Weiwei Zhao, Jinhuan Li et al. "High-performance graphene photodetector using interfacial gating." Optica 3, no. 10 (2016): 1066-1070.

[379] Schroder, Dieter K. "Surface voltage and surface photovoltage: history, theory and applications." Measurement Science and Technology 12, no. 3 (2001): R16.

[380] Norton, Paul. "HgCdTe infrared detectors." Optoelectronics review 3 (2002): 159-174.

[381] Withers, Freddie, Thomas Hardisty Bointon, Monica Felicia Craciun, and Saverio Russo. "All-graphene photodetectors." ACS nano 7, no. 6 (2013): 5052-5057.

[382] Mueller, Thomas, and Ermin Malic. "Exciton physics and device application of two-dimensional transition metal dichalcogenide semiconductors." npj 2D Materials and Applications 2, no. 1 (2018): 29.

[383] Gong, Chuanhui, Yuxi Zhang, Wei Chen, Junwei Chu, Tianyu Lei, Junru Pu, Liping Dai et al. "Electronic and optoelectronic applications based on 2D novel anisotropic transition metal dichalcogenides." Advanced Science 4, no. 12 (2017): 1700231.

[384] Octon, Tobias J., K. Nagareddy, S. R. Russo, M. C. Craciun, and C. David Wright. "Fast high-responsivity few-layer MoTe2 photodetectors." (2016).

[385] Zhang, Enze, Yibo Jin, Xiang Yuan, Weiyi Wang, Cheng Zhang, Lei Tang, Shanshan Liu, Peng Zhou, Weida Hu, and Faxian Xiu. "ReS2-based field-effect transistors and photodetectors." Advanced Functional Materials 25, no. 26 (2015): 4076-4082.

[386] Hafeez, Muhammad, Lin Gan, Huiqiao Li, Ying Ma, and Tianyou Zhai. "Large-area bilayer ReS2 film/multilayer ReS2 flakes synthesized by chemical vapor deposition for high performance photodetectors." Advanced Functional Materials 26, no. 25 (2016): 4551-4560.

[387] Liu, Erfu, Mingsheng Long, Junwen Zeng, Wei Luo, Yaojia Wang, Yiming Pan, Wei Zhou et al. "High responsivity phototransistors based on few-layer ReS2 for weak signal detection." Advanced Functional Materials 26, no. 12 (2016): 1938-1944.

[388] Zhang, Wenxu, Zhishuo Huang, Wanli Zhang, and Yanrong Li. "Two-dimensional semiconductors with possible high room temperature mobility." Nano Research 7 (2014): 1731-1737.

[389] Yin, Lei, Xueying Zhan, Kai Xu, Feng Wang, Zhenxing Wang, Yun Huang, Qisheng Wang, Chao Jiang, and Jun He. "Ultrahigh sensitive MoTe2 phototransistors driven by carrier tunneling." Applied Physics Letters 108, no. 4 (2016).

[390] Aftab, Sikandar, and Hosameldin Helmy Hegazy. "Emerging Trends in 2D TMDs Photodetectors and Piezo-Phototronic Devices." Small 19, no. 18 (2023): 2205778.

[391] Hoang, Anh Tuan, Luhing Hu, Ajit Kumar Katiyar, and Jong-Hyun Ahn. "Two-dimensional layered materials and heterostructures for flexible electronics." Matter 5, no. 12 (2022): 4116-4132.

[392] Sarkar, Debalina. "2D Materials for Field-Effect Transistor–Based Biosensors." In Fundamentals and Sensing Applications of 2D Materials, pp. 329-377. Woodhead Publishing, 2019.

[393] Fonash, Stephen. Solar cell device physics. Elsevier, 2012.

[394] Battaglia, Corsin, Andres Cuevas, and Stefaan De Wolf. "High-efficiency crystalline silicon solar cells: status and perspectives." Energy & Environmental Science 9, no. 5 (2016): 1552-1576.

[395] Das, Sonali, Deepak Pandey, Jayan Thomas, and Tania Roy. "The role of graphene and other 2D materials in solar photovoltaics." Advanced Materials 31, no. 1 (2019): 1802722.

[396] Rai, Suyash, Himanshu Mishra, Vijay K. Singh, Tejendra K. Gupta, and Anchal Srivastava. "2D materials for flexible electronics." In 2D Materials for Electronics, Sensors and Devices, pp. 169-206. Elsevier, 2023.

[397] Cheng, Jinbing, Chunlan Wang, Xuming Zou, and Lei Liao. "Recent advances in optoelectronic devices based on 2D materials and their heterostructures." Advanced Optical Materials 7, no. 1 (2019): 1800441.

[398] Novoselov, Kostya S., Andre K. Geim, Sergei V. Morozov, De-eng Jiang, Yanshui Zhang, Sergey V. Dubonos, Irina V. Grigorieva, and Alexandr A. Firsov. "Electric field effect in atomically thin carbon films." science 306, no. 5696 (2004): 666-669.

[399] Li, Xinming, Hongwei Zhu, Kunlin Wang, Anyuan Cao, Jinquan Wei, Chunyan Li, Yi Jia, Zhen Li, Xiao Li, and Dehai Wu. "Graphene-on-silicon Schottky junction solar cells." Advanced materials 25, no. 22 (2010): 2743-2748.

400 Ahn, Jaehyun, Harry Chou, and Sanjay K. Banerjee. "Graphene-Al2O3-silicon heterojunction solar cells on flexible silicon substrates." Journal of Applied Physics 121, no. 16 (2017).

401 Li, Xinming, and Hongwei Zhu. "The graphene–semiconductor Schottky junction." Physics Today 69, no. 9 (2016): 46-51.

402 Bernardi, Marco, Maurizia Palummo, and Jeffrey C. Grossman. "Extraordinary sunlight absorption and one nanometer thick photovoltaics using two-dimensional monolayer materials." Nano letters 13, no. 8 (2013): 3664-3670.

403 Mak, Kin Fai, Changgu Lee, James Hone, Jie Shan, and Tony F. Heinz. "Atomically thin MoS 2: a new direct-gap semiconductor." Physical review letters 105, no. 13 (2010): 136805.

404 Tongay, Sefaattin, Jian Zhou, Can Ataca, Kelvin Lo, Tyler S. Matthews, Jingbo Li, Jeffrey C. Grossman, and Junqiao Wu. "Thermally driven crossover from indirect toward direct bandgap in 2D semiconductors: MoSe2 versus MoS2." Nano letters 12, no. 11 (2012): 5576-5580.

405 Gutiérrez, Humberto R., Nestor Perea-López, Ana Laura Elías, Ayse Berkdemir, Bei Wang, Ruitao Lv, Florentino López-Urías, Vincent H. Crespi, Humberto Terrones, and Mauricio Terrones. "Extraordinary room-temperature photoluminescence in triangular WS2 monolayers." Nano letters 13, no. 8 (2013): 3447-3454.

406 Britnell, Liam, Ricardo Mendes Ribeiro, Axel Eckmann, Rashid Jalil, Branson D. Belle, Artem Mishchenko, Y-J. Kim et al. "Strong light-matter interactions in heterostructures of atomically thin films." Science 340, no. 6138 (2013): 1311-1314.

407 Tsai, Meng-Lin, Sheng-Han Su, Jan-Kai Chang, Dung-Sheng Tsai, Chang-Hsiao Chen, Chih-I. Wu, Lain-Jong Li, Lih-Juann Chen, and Jr-Hau He. "Monolayer MoS2 heterojunction solar cells." ACS nano 8, no. 8 (2014): 8317-8322.

408 Rehman, Atteq ur, Muhammad Farooq Khan, Muhammad Arslan Shehzad, Sajjad Hussain, Muhammad Fahad Bhopal, Sang Hee Lee, Jonghwa Eom, Yongho Seo, Jongwan Jung, and Soo Hong Lee. "n-MoS2/p-Si solar cells with Al2O3 passivation for enhanced photogeneration." ACS applied materials & interfaces 8, no. 43 (2016): 29383-29390.

409 Shanmugam, Mariyappan, Chris A. Durcan, and Bin Yu. "Layered semiconductor molybdenum disulfide nanomembrane based Schottky-barrier solar cells." Nanoscale 4, no. 23 (2012): 7399-7405.

410 Wi, Sungjin, Hyunsoo Kim, Mikai Chen, Hongsuk Nam, L. Jay Guo, Edgar Meyhofer, and Xiaogan Liang. "Enhancement of photovoltaic response in multilayer MoS2 induced by plasma doping." ACS nano 8, no. 5 (2014): 5270-5281.

411 Jang, Min Seok, Victor W. Brar, Michelle C. Sherrott, Josue J. Lopez, Laura Kim, Seyoon Kim, Mansoo Choi, and Harry A. Atwater. "Tunable large resonant absorption in a midinfrared graphene Salisbury screen." Physical Review B 90, no. 16 (2014): 165409.

[412] Jeong, Hye Yun, Un Jeong Kim, Hyun Kim, Gang Hee Han, Hyangsook Lee, Min Su Kim, Youngjo Jin et al. "Optical gain in MoS2 via coupling with nanostructured substrate: Fabry–Perot interference and plasmonic excitation." ACS nano 10, no. 9 (2016): 8192-8198.

[413] Lee, Chul-Ho, Gwan-Hyoung Lee, Arend M. Van Der Zande, Wenchao Chen, Yilei Li, Minyong Han, Xu Cui et al. "Atomically thin p–n junctions with van der Waals heterointerfaces." Nature nanotechnology 9, no. 9 (2014): 676-681.

[414] Furchi, Marco M., Andreas Pospischil, Florian Libisch, Joachim Burgdorfer, and Thomas Mueller. "Photovoltaic effect in an electrically tunable van der Waals heterojunction." Nano letters 14, no. 8 (2014): 4785-4791.

[415] Roy, Tania, Mahmut Tosun, Xi Cao, Hui Fang, Der-Hsien Lien, Peida Zhao, Yu-Ze Chen, Yu-Lun Chueh, Jing Guo, and Ali Javey. "Dual-gated MoS2/WSe2 van der Waals tunnel diodes and transistors." ACS nano 9, no. 2 (2015): 2071-2079.

[416] Fang, Hui, Corsin Battaglia, Carlo Carraro, Slavomir Nemsak, Burak Ozdol, Jeong Seuk Kang, Hans A. Bechtel et al. "Strong interlayer coupling in van der Waals heterostructures built from single-layer chalcogenides." Proceedings of the National Academy of Sciences 111, no. 17 (2014): 6198-6202.

[417] Furchi, Marco M., Armin A. Zechmeister, Florian Hoeller, Stefan Wachter, Andreas Pospischil, and Thomas Mueller. "Photovoltaics in Van der Waals heterostructures." IEEE Journal of Selected Topics in Quantum Electronics 23, no. 1 (2016): 106-116.

[418] Novoselov, K. S., and AH Castro Neto. "Two-dimensional crystals-based heterostructures: materials with tailored properties." Physica Scripta 2012, no. T146 (2012): 014006.

[419] Novoselov, K. S., Artem Mishchenko, Alexandra Carvalho, and A. H. Castro Neto. "2D materials and van der Waals heterostructures." Science 353, no. 6298 (2016): aac9439.

[420] Xu, Zhongguang, Alireza Khanaki, Hao Tian, Renjing Zheng, Mohammad Suja, Jian-Guo Zheng, and Jianlin Liu. "Direct growth of hexagonal boron nitride/graphene heterostructures on cobalt foil substrates by plasma-assisted molecular beam epitaxy." Applied Physics Letters 109, no. 4 (2016).

[421] Xu, Zhongguang, Renjing Zheng, Alireza Khanaki, Zheng Zuo, and Jianlin Liu. "Direct growth of graphene on in situ epitaxial hexagonal boron nitride flakes by plasma-assisted molecular beam epitaxy." Applied Physics Letters 107, no. 21 (2015).

[422] Zhang, Chaohua, Shuli Zhao, Chuanhong Jin, Ai Leen Koh, Yu Zhou, Weigao Xu, Qiucheng Li, Qihua Xiong, Hailin Peng, and Zhongfan Liu. "Direct growth of large-area graphene and boron nitride heterostructures by a co-segregation method." Nature communications 6, no. 1 (2015): 6519.

[423] Pierucci, Debora, Hugo Henck, Carl H. Naylor, Haikel Sediri, Emmanuel Lhuillier, Adrian Balan, Julien E. Rault et al. "Large area molybdenum disulphide-epitaxial graphene vertical Van der Waals heterostructures." Scientific reports 6, no. 1 (2016): 26656.

[424] Yu, Lili, Yi-Hsien Lee, Xi Ling, Elton JG Santos, Yong Cheol Shin, Yuxuan Lin, Madan Dubey et al. "Graphene/MoS2 hybrid technology for large-scale two-dimensional electronics." Nano letters 14, no. 6 (2014): 3055-3063.

[425] Yan, Aiming, Jairo Velasco Jr, Salman Kahn, Kenji Watanabe, Takashi Taniguchi, Feng Wang, Michael F. Crommie, and Alex Zettl. "Direct growth of single-and few-layer MoS2 on h-BN with preferred relative rotation angles." Nano letters 15, no. 10 (2015): 6324-6331.

[426] Cattelan, Mattia, Brian Markman, Giacomo Lucchini, Pranab Kumar Das, Ivana Vobornik, Joshua Alexander Robinson, Stefano Agnoli, and Gaetano Granozzi. "New strategy for the growth of complex heterostructures based on different 2D materials." Chemistry of Materials 27, no. 11 (2015): 4105-4113.

[427] Gong, Yongji, Junhao Lin, Xingli Wang, Gang Shi, Sidong Lei, Zhong Lin, Xiaolong Zou et al. "Vertical and in-plane heterostructures from WS 2/MoS 2 monolayers." Nature materials 13, no. 12 (2014): 1135-1142.

[428] Kim, Young Duck, Hakseong Kim, Yujin Cho, Ji Hoon Ryoo, Cheol-Hwan Park, Pilkwang Kim, Yong Seung Kim et al. "Bright visible light emission from graphene." Nature nanotechnology 10, no. 8 (2015): 676-681.

[429] Gong, Yongji, Sidong Lei, Gonglan Ye, Bo Li, Yongmin He, Kunttal Keyshar, Xiang Zhang et al. "Two-step growth of two-dimensional WSe2/MoSe2 heterostructures." Nano letters 15, no. 9 (2015): 6135-6141.

[430] Tsai, Meng-Lin, Ming-Yang Li, José Ramón Durán Retamal, Kai-Tak Lam, Yung-Chang Lin, Kazu Suenaga, Lih-Juann Chen, Gengchiau Liang, Lain-Jong Li, and Jr-Hau He. "Single atomically sharp lateral monolayer p-n heterojunction solar cells with extraordinarily high power conversion efficiency." Advanced materials 29, no. 32 (2017): 1701168.

[431] Rivera, Pasqual, John R. Schaibley, Aaron M. Jones, Jason S. Ross, Sanfeng Wu, Grant Aivazian, Philip Klement et al. "Observation of long-lived interlayer excitons in monolayer MoSe2–WSe2 heterostructures." Nature communications 6, no. 1 (2015): 6242.

[432] Ye, Jian, Xueliang Li, Jianjun Zhao, Xuelan Mei, and Qian Li. "A facile way to fabricate high-performance solution-processed n-MoS 2/p-MoS 2 bilayer photodetectors." Nanoscale research letters 10 (2015): 1-7.

[433] Lee, Chul-Ho, Gwan-Hyoung Lee, Arend M. Van Der Zande, Wenchao Chen, Yilei Li, Minyong Han, Xu Cui et al. "Atomically thin p–n junctions with van der Waals heterointerfaces." Nature nanotechnology 9, no. 9 (2014): 676-681.

[434] Wang, Xiaomu, and Fengnian Xia. "Stacked 2D materials shed light." Nature materials 14, no. 3 (2015): 264-265.

[435] Radisavljevic, Branimir, Aleksandra Radenovic, Jacopo Brivio, Valentina Giacometti, and Andras Kis. "Single-layer MoS2 transistors." Nature nanotechnology 6, no. 3 (2011): 147-150.

[436] Wu, Wenzhuo, Lei Wang, Yilei Li, Fan Zhang, Long Lin, Simiao Niu, Daniel Chenet et al. "Piezoelectricity of single-atomic-layer MoS2 for energy conversion and piezotronics." Nature 514, no. 7523 (2014): 470-474.

[437] Mak, Kin Fai, and Jie Shan. "Photonics and optoelectronics of 2D semiconductor transition metal dichalcogenides." Nature Photonics 10, no. 4 (2016): 216-226.

[438] Wang, Qing Hua, Kourosh Kalantar-Zadeh, Andras Kis, Jonathan N. Coleman, and Michael S. Strano. "Electronics and optoelectronics of two-dimensional transition metal dichalcogenides." Nature nanotechnology 7, no. 11 (2012): 699-712.

[439] Mak, Kin Fai, Changgu Lee, James Hone, Jie Shan, and Tony F. Heinz. "Atomically thin MoS 2: a new direct-gap semiconductor." Physical review letters 105, no. 13 (2010): 136805.

[440] Zhao, Weijie, Ricardo Mendes Ribeiro, Minglin Toh, Alexandra Carvalho, Christian Kloc, A. H. Castro Neto, and Goki Eda. "Origin of indirect optical transitions in few-layer MoS2, WS2, and WSe2." Nano letters 13, no. 11 (2013): 5627-5634.

[441] Steinhoff, A., J-H. Kim, F. Jahnke, M. Rosner, D-S. Kim, Ch Lee, Gang Hee Han, Mun Seok Jeong, T. O. Wehling, and C. Gies. "Efficient excitonic photoluminescence in direct and indirect band gap monolayer MoS2." Nano letters 15, no. 10 (2015): 6841-6847.

[442] He, Keliang, Nardeep Kumar, Liang Zhao, Zefang Wang, Kin Fai Mak, Hui Zhao, and Jie Shan. "Tightly bound excitons in monolayer WSe2." Physical review letters 113, no. 2 (2014): 026803.

[443] Mak, Kin Fai, Keliang He, Changgu Lee, Gwan Hyoung Lee, James Hone, Tony F. Heinz, and Jie Shan. "Tightly bound trions in monolayer MoS2." Nature materials 12, no. 3 (2013): 207-211.

[444] Moody, Galan, Chandriker Kavir Dass, Kai Hao, Chang-Hsiao Chen, Lain-Jong Li, Akshay Singh, Kha Tran et al. "Intrinsic homogeneous linewidth and broadening mechanisms of excitons in monolayer transition metal dichalcogenides." Nature communications 6, no. 1 (2015): 8315.

[445] Wang, Xiaomu, and Fengnian Xia. "Stacked 2D materials shed light." Nature materials 14, no. 3 (2015): 264-265.

[446] Pei, Jiajie, Jiong Yang, Tanju Yildirim, Han Zhang, and Yuerui Lu. "Many-body complexes in 2D semiconductors." Advanced Materials 31, no. 2 (2019): 1706945.

[447] Li, Yilei, Yi Rao, Kin Fai Mak, Yumeng You, Shuyuan Wang, Cory R. Dean, and Tony F. Heinz. "Probing symmetry properties of few-layer MoS2 and h-BN by optical second-harmonic generation." Nano letters 13, no. 7 (2013): 3329-3333.

[448] Zhou, Xu, Jingxin Cheng, Yubing Zhou, Ting Cao, Hao Hong, Zhimin Liao, Shiwei Wu, Hailin Peng, Kaihui Liu, and Dapeng Yu. "Strong second-harmonic generation in atomic layered GaSe." Journal of the American Chemical Society 137, no. 25 (2015): 7994-7997.

[449] Kang, Jun, Sefaattin Tongay, Jian Zhou, Jingbo Li, and Junqiao Wu. "Band offsets and heterostructures of two-dimensional semiconductors." Applied Physics Letters 102, no. 1 (2013).

450 Moody, Galan, John Schaibley, and Xiaodong Xu. "Exciton dynamics in monolayer transition metal dichalcogenides." JOSA B 33, no. 7 (2016): C39-C49.

451 Jakubczyk, Tomasz, Valentin Delmonte, Maciej Koperski, Karol Nogajewski, Clément Faugeras, Wolfgang Langbein, Marek Potemski, and Jacek Kasprzak. "Radiatively limited dephasing and exciton dynamics in MoSe2 monolayers revealed with four-wave mixing microscopy." Nano letters 16, no. 9 (2016): 5333-5339.

452 Mouri, Shinichiro, Yuhei Miyauchi, and Kazunari Matsuda. "Tunable photoluminescence of monolayer MoS2 via chemical doping." Nano letters 13, no. 12 (2013): 5944-5948.

453 Ross, Jason S., Sanfeng Wu, Hongyi Yu, Nirmal J. Ghimire, Aaron M. Jones, Grant Aivazian, Jiaqiang Yan et al. "Electrical control of neutral and charged excitons in a monolayer semiconductor." Nature communications 4, no. 1 (2013): 1474.

454 Yang, Jiong, Tieyu Lu, Ye Win Myint, Jiajie Pei, Daniel Macdonald, Jin-Cheng Zheng, and Yuerui Lu. "Robust excitons and trions in monolayer MoTe2." ACS nano 9, no. 6 (2015): 6603-6609.

455 Amani, Matin, Der-Hsien Lien, Daisuke Kiriya, Jun Xiao, Angelica Azcatl, Jiyoung Noh, Surabhi R. Madhvapathy et al. "Near-unity photoluminescence quantum yield in MoS2." Science 350, no. 6264 (2015): 1065-1068.

456 Miller, R. C., D. A. Kleinman, W. T. Tsang, and A. C. Gossard. "Observation of the excited level of excitons in GaAs quantum wells." Physical Review B 24, no. 2 (1981): 1134.

457 Miller, R. C., and D. A. Kleinman. "Excitons in GaAs quantum wells." Journal of Luminescence 30, no. 1-4 (1985): 520-540.

458 Cheiwchanchamnangij, Tawinan, and Walter RL Lambrecht. "Quasiparticle band structure calculation of monolayer, bilayer, and bulk MoS 2." Physical Review B 85, no. 20 (2012): 205302.

459 Ramasubramaniam, Ashwin. "Large excitonic effects in monolayers of molybdenum and tungsten dichalcogenides." Physical Review B 86, no. 11 (2012): 115409.

460 Park, Soohyung, Niklas Mutz, Thorsten Schultz, Sylke Blumstengel, Ali Han, Areej Aljarb, Lain-Jong Li, Emil JW List-Kratochvil, Patrick Amsalem, and Norbert Koch. "Direct determination of monolayer MoS 2 and WSe 2 exciton binding energies on insulating and metallic substrates." 2D Materials 5, no. 2 (2018): 025003.

461 Zhu, Bairen, Xi Chen, and Xiaodong Cui. "Exciton binding energy of monolayer WS2." Scientific reports 5, no. 1 (2015): 9218.

462 Ugeda, Miguel M., Aaron J. Bradley, Su-Fei Shi, Felipe H. Da Jornada, Yi Zhang, Diana Y. Qiu, Wei Ruan et al. "Giant bandgap renormalization and excitonic effects in a monolayer transition metal dichalcogenide semiconductor." Nature materials 13, no. 12 (2014): 1091-1095.

463 Kübler, J. K. "The exciton binding energy of III–V semiconductor compounds." physica status solidi (b) 35, no. 1 (1969): 189-195.

464 Hao, Kai, Judith F. Specht, Philipp Nagler, Lixiang Xu, Kha Tran, Akshay Singh, Chandriker Kavir Dass et al. "Neutral and charged inter-valley biexcitons in monolayer MoSe2." Nature communications 8, no. 1 (2017): 15552.

465 Ross, Jason S., Philip Klement, Aaron M. Jones, Nirmal J. Ghimire, Jiaqiang Yan, D. G. Mandrus, Takashi Taniguchi et al. "Electrically tunable excitonic light-emitting diodes based on monolayer WSe2 p–n junctions." Nature nanotechnology 9, no. 4 (2014): 268-272.

466 Zhang, Y. J., T. Oka, Ryo Suzuki, J. T. Ye, and Y. Iwasa. "Electrically switchable chiral light-emitting transistor." Science 344, no. 6185 (2014): 725-728.

467 Cheng, Rui, Dehui Li, Hailong Zhou, Chen Wang, Anxiang Yin, Shan Jiang, Yuan Liu, Yu Chen, Yu Huang, and Xiangfeng Duan. "Electroluminescence and photocurrent generation from atomically sharp WSe2/MoS2 heterojunction p–n diodes." Nano letters 14, no. 10 (2014): 5590-5597.

468 Zhu, Yi, Xueqian Sun, Yilin Tang, Lan Fu, and Yuerui Lu. "Two-dimensional materials for light emitting applications: Achievement, challenge and future perspectives." Nano Research 14 (2021): 1912-1936.

469 Jo, Sanghyun, Nicolas Ubrig, Helmuth Berger, Alexey B. Kuzmenko, and Alberto F. Morpurgo. "Mono-and bilayer WS2 light-emitting transistors." Nano letters 14, no. 4 (2014): 2019-2025.

470 Ma, Shuang, Molang Cai, Tai Cheng, Xihong Ding, Xiaoqiang Shi, Ahmed Alsaedi, Tasawar Hayat, Yong Ding, Zhan'ao Tan, and Songyuan Dai. "Two-dimensional organic-inorganic hybrid perovskite: from material properties to device applications." Sci. China Mater 61, no. 10 (2018): 1257-1277.

471 Loryuenyong, Vorrada, Pajaree Thongpon, Sasathorn Saudmalai, and Achanai Buasri. "The Synthesis of 2D CH 3 NH 3 PbI 3 Perovskite Films with Tunable Bandgaps by Solution Deposition Route." International Journal of Photoenergy 2019 (2019).

472 Xue, Yunzhou, Jian Yuan, Jingying Liu, and Shaojuan Li. "Controllable synthesis of 2D perovskite on different substrates and its application as photodetector." Nanomaterials 8, no. 8 (2018): 591.

473 Yang, Xiaolei, Xingwang Zhang, Jinxiang Deng, Zema Chu, Qi Jiang, Junhua Meng, Pengyang Wang, Liuqi Zhang, Zhigang Yin, and Jingbi You. "Efficient green light-emitting diodes based on quasi-two-dimensional composition and phase engineered perovskite with surface passivation." Nature communications 9, no. 1 (2018): 570.

474 Xing, Jun, Fei Yan, Yawen Zhao, Shi Chen, Huakang Yu, Qing Zhang, Rongguang Zeng et al. "High-efficiency light-emitting diodes of organometal halide perovskite amorphous nanoparticles." ACS nano 10, no. 7 (2016): 6623-6630.

475 Mitzi, David B. "Solution-processed inorganic semiconductors." Journal of Materials Chemistry 14, no. 15 (2004): 2355-2365.

476 Kim, Young-Hoon, Himchan Cho, and Tae-Woo Lee. "Metal halide perovskite light emitters." Proceedings of the National Academy of Sciences 113, no. 42 (2016): 11694-11702.

477 Zou, Wei, Renzhi Li, Shuting Zhang, Yunlong Liu, Nana Wang, Yu Cao, Yanfeng Miao et al. "Minimising efficiency roll-off in high-brightness perovskite light-emitting diodes." Nature communications 9, no. 1 (2018): 608.

478 Lin, Kebin, Jun Xing, Li Na Quan, F. Pelayo García de Arquer, Xiwen Gong, Jianxun Lu, Liqiang Xie et al. "Perovskite light-emitting diodes with external quantum efficiency exceeding 20 per cent." Nature 562, no. 7726 (2018): 245-248.

479 Nicolosi, Valeria, Manish Chhowalla, Mercouri G. Kanatzidis, Michael S. Strano, and Jonathan N. Coleman. "Liquid exfoliation of layered materials." Science 340, no. 6139 (2013): 1226419.

480 Hu, Jun, Zhenkun Guo, Peter E. Mcwilliams, John E. Darges, Daniel L. Druffel, Andrew M. Moran, and Scott C. Warren. "Band gap engineering in a 2D material for solar-to-chemical energy conversion." Nano letters 16, no. 1 (2016): 74-79.

481 Zhang, Xiaoyan, Lili Hou, Artur Ciesielski, and Paolo Samorì. "2D materials beyond graphene for high-performance energy storage applications." Advanced Energy Materials 6, no. 23 (2016): 1600671.

482 Anasori, Babak, Maria R. Lukatskaya, and Yury Gogotsi. "2D metal carbides and nitrides (MXenes) for energy storage." Nature Reviews Materials 2, no. 2 (2017): 1-17.

483 Chee, Wei Kit, Hong Ngee Lim, Zulkarnain Zainal, Nay Ming Huang, Ian Harrison, and Y. Andou. "Flexible graphene-based supercapacitors: a review." The Journal of Physical Chemistry C 120, no. 8 (2016): 4153-4172.

484 Palaniselvam, Thangavelu, and Jong-Beom Baek. "Graphene based 2D-materials for supercapacitors." 2D Materials 2, no. 3 (2015): 032002.

485 Wang, Hua, Hongbin Feng, and Jinghong Li. "Graphene and graphene-like layered transition metal dichalcogenides in energy conversion and storage." Small 10, no. 11 (2014): 2165-2181.

486 Zhou, Ding, Yi Cui, and BaoHang Han. "Graphene-based hybrid materials and their applications in energy storage and conversion." Chinese science bulletin 57 (2012): 2983-2994.

487 Yan, Jun, Qian Wang, Tong Wei, and Zhuangjun Fan. "Recent advances in design and fabrication of electrochemical supercapacitors with high energy densities." Advanced Energy Materials 4, no. 4 (2014): 1300816.

488 Yang, Wen, Mei Ni, Xin Ren, Yafen Tian, Ning Li, Yuefeng Su, and Xiaoling Zhang. "Graphene in supercapacitor applications." Current Opinion in Colloid & Interface Science 20, no. 5-6 (2015): 416-428.

489 Chen, Tao, and Liming Dai. "Carbon nanomaterials for high-performance supercapacitors." Materials Today 16, no. 7-8 (2013): 272-280.

490 Zhang, Li Li, and X. S. Zhao. "Carbon-based materials as supercapacitor electrodes." Chemical society reviews 38, no. 9 (2009): 2520-2531.

491 Wang, Guoping, Lei Zhang, and Jiujun Zhang. "A review of electrode materials for electrochemical supercapacitors." Chemical Society Reviews 41, no. 2 (2012): 797-828.

492 Conway, Brian E., V. Birss, and John Wojtowicz. "The role and utilization of pseudocapacitance for energy storage by supercapacitors." Journal of power sources 66, no. 1-2 (1997): 1-14.

493 Xia, Jilin, Fang Chen, Jinghong Li, and Nongjian Tao. "Measurement of the quantum capacitance of graphene." Nature nanotechnology 4, no. 8 (2009): 505-509.

494 El-Kady, Maher F., Veronica Strong, Sergey Dubin, and Richard B. Kaner. "Laser scribing of high-performance and flexible graphene-based electrochemical capacitors." Science 335, no. 6074 (2012): 1326-1330.

495 Wang, Yan, Zhiqiang Shi, Yi Huang, Yanfeng Ma, Chengyang Wang, Mingming Chen, and Yongsheng Chen. "Supercapacitor devices based on graphene materials." The Journal of Physical Chemistry C 113, no. 30 (2009): 13103-13107.

496 Han, Jongwoo, Li Li Zhang, Seungjun Lee, Junghoon Oh, Kyoung-Seok Lee, Jeffrey R. Potts, Junyi Ji, Xin Zhao, Rodney S. Ruoff, and Sungjin Park. "Generation of B-doped graphene nanoplatelets using a solution process and their supercapacitor applications." ACS nano 7, no. 1 (2013): 19-26.

497 El-Kady, Maher F., and Richard B. Kaner. "Scalable fabrication of high-power graphene micro-supercapacitors for flexible and on-chip energy storage." Nature communications 4, no. 1 (2013): 1475.

498 Snook, Graeme A., Pon Kao, and Adam S. Best. "Conducting-polymer-based supercapacitor devices and electrodes." Journal of power sources 196, no. 1 (2011): 1-12.

499 Memon, Mushtaque A., Wei Bai, Jinhua Sun, Muhammad Imran, Shah Nawaz Phulpoto, Shouke Yan, Yong Huang, and Jianxin Geng. "Conjunction of conducting polymer nanostructures with macroporous structured graphene thin films for high-performance flexible supercapacitors." ACS applied materials & interfaces 8, no. 18 (2016): 11711-11719.

500 Wang, Xianfen, Satoshi Kajiyama, Hiroki Iinuma, Eiji Hosono, Shinji Oro, Isamu Moriguchi, Masashi Okubo, and Atsuo Yamada. "Pseudocapacitance of MXene nanosheets for high-power sodium-ion hybrid capacitors." Nature communications 6, no. 1 (2015): 6544.

501 Zhang, Chuanfang, Lorcan McKeon, Matthias P. Kremer, Sang-Hoon Park, Oskar Ronan, Andrés Seral-Ascaso, Sebastian Barwich et al. "Additive-free MXene inks and direct printing of micro-supercapacitors." Nature communications 10, no. 1 (2019): 1795.

502 Zhang, Chuanfang, Sang-Hoon Park, Andrés Seral-Ascaso, Sebastian Barwich, Niall McEvoy, Conor S. Boland, Jonathan N. Coleman, Yury Gogotsi, and Valeria Nicolosi. "High capacity silicon anodes enabled by MXene viscous aqueous ink." Nature communications 10, no. 1 (2019): 849.

503 Lukatskaya, Maria R., Olha Mashtalir, Chang E. Ren, Yohan Dall'Agnese, Patrick Rozier, Pierre Louis Taberna, Michael Naguib, Patrice Simon, Michel W. Barsoum, and Yury Gogotsi. "Cation intercalation and high volumetric capacitance of two-dimensional titanium carbide." Science 341, no. 6153 (2013): 1502-1505.

[504] Levi, Mikhael D., Maria R. Lukatskaya, Sergey Sigalov, Majid Beidaghi, Netanel Shpigel, Leonid Daikhin, Doron Aurbach, Michel W. Barsoum, and Yury Gogotsi. "Solving the capacitive paradox of 2D MXene using electrochemical quartz-crystal admittance and in situ electronic conductance measurements." Advanced Energy Materials 5, no. 1 (2015): 1400815.

[505] Zhang, Chuanfang, Babak Anasori, Andrés Seral-Ascaso, Sang-Hoon Park, Niall McEvoy, Aleksey Shmeliov, Georg S. Duesberg, Jonathan N. Coleman, Yury Gogotsi, and Valeria Nicolosi. "Transparent, flexible, and conductive 2D titanium carbide (MXene) films with high volumetric capacitance." Advanced materials 29, no. 36 (2017): 1702678.

[506] Ghidiu, Michael, Maria R. Lukatskaya, Meng-Qiang Zhao, Yury Gogotsi, and Michel W. Barsoum. "Conductive two-dimensional titanium carbide 'clay'with high volumetric capacitance." Nature 516, no. 7529 (2014): 78-81.

[507] Zhan, Cheng, Michael Naguib, Maria Lukatskaya, Paul RC Kent, Yury Gogotsi, and De-en Jiang. "Understanding the MXene pseudocapacitance." The journal of physical chemistry letters 9, no. 6 (2018): 1223-1228.

[508] Krishnan, Syam G., Arulraj Arunachalam, Priyanka Jagadish, and Mohammad Khalid. "2D Materials for supercapacitor and supercapattery applications." In Adapting 2D Nanomaterials for Advanced Applications, pp. 33-47. American Chemical Society, 2020.

[509] Lin, Zifeng, Daffos Barbara, Pierre-Louis Taberna, Katherine L. Van Aken, Babak Anasori, Yury Gogotsi, and Patrice Simon. "Capacitance of Ti3C2Tx MXene in ionic liquid electrolyte." Journal of Power Sources 326 (2016): 575-579.

[510] Conway, Brian E., V. Birss, and John Wojtowicz. "The role and utilization of pseudocapacitance for energy storage by supercapacitors." Journal of power sources 66, no. 1-2 (1997): 1-14.

[511] Acerce, Muharrem, Damien Voiry, and Manish Chhowalla. "Metallic 1T phase MoS2 nanosheets as supercapacitor electrode materials." Nature nanotechnology 10, no. 4 (2015): 313-318.

[512] Feng, Jun, Xu Sun, Changzheng Wu, Lele Peng, Chenwen Lin, Shuanglin Hu, Jinlong Yang, and Yi Xie. "Metallic few-layered VS2 ultrathin nanosheets: high two-dimensional conductivity for in-plane supercapacitors." Journal of the American Chemical Society 133, no. 44 (2011): 17832-17838.

[513] Zhang, Xiaoyan, Lili Hou, Artur Ciesielski, and Paolo Samorì. "2D materials beyond graphene for high-performance energy storage applications." Advanced Energy Materials 6, no. 23 (2016): 1600671.

[514] Yang, Jiaqin, Xiaochuan Duan, Qing Qin, and Wenjun Zheng. "Solvothermal synthesis of hierarchical flower-like β-NiS with excellent electrochemical performance for supercapacitors." Journal of Materials Chemistry A 1, no. 27 (2013): 7880-7884.

[515] Li, Wei, Shaolan Wang, Lipeng Xin, Ming Wu, and Xiaojie Lou. "Single-crystal β-NiS nanorod arrays with a hollow-structured $Ni_3S_2$ framework for supercapacitor applications." Journal of materials chemistry A 4, no. 20 (2016): 7700-7709.

516 Yang, Jiaqin, Xiaochuan Duan, Wei Guo, Di Li, Huili Zhang, and Wenjun Zheng. "Electrochemical performances investigation of NiS/rGO composite as electrode material for supercapacitors." Nano Energy 5 (2014): 74-81.
517 Sun, Chencheng, Mingze Ma, Jun Yang, Yufei Zhang, Peng Chen, Wei Huang, and Xiaochen Dong. "Phase-controlled synthesis of α-NiS nanoparticles confined in carbon nanorods for high performance supercapacitors." Scientific reports 4, no. 1 (2014): 7054.
518 Gou, Jianxia, Shengli Xie, Zhichao Yang, Yaqing Liu, Yinjuan Chen, Yanru Liu, and Chenguang Liu. "A high-performance supercapacitor electrode material based on NiS/Ni3S4 composite." Electrochimica Acta 229 (2017): 299-305.
519 Wu, Jianghong, Canbin Ouyang, Shuo Dou, and Shuangyin Wang. "Hybrid NiS/CoO mesoporous nanosheet arrays on Ni foam for high-rate supercapacitors." Nanotechnology 26, no. 32 (2015): 325401.
520 Wang, Yaping, Anqiang Pan, Yifang Zhang, Junrong Shi, Jiande Lin, Shuquan Liang, and Guozhong Cao. "Heterogeneous NiS/NiO multi-shelled hollow microspheres with enhanced electrochemical performances for hybrid-type asymmetric supercapacitors." Journal of Materials Chemistry A 6, no. 19 (2018): 9153-9160.
521 Yang, Jun, Mingze Ma, Chencheng Sun, Yufei Zhang, Wei Huang, and Xiaochen Dong. "Hybrid $NiCo_2S_4$@$MnO_2$ heterostructures for high-performance supercapacitor electrodes." Journal of Materials Chemistry A 3, no. 3 (2015): 1258-1264.
522 Xiao, Junwu, Lian Wan, Shihe Yang, Fei Xiao, and Shuai Wang. "Design hierarchical electrodes with highly conductive NiCo2S4 nanotube arrays grown on carbon fiber paper for high-performance pseudocapacitors." Nano letters 14, no. 2 (2014): 831-838.
523 Jinlong, Lv, Liang Tongxiang, Yang Meng, Suzuki Ken, and Miura Hideo. "Performance comparison of NiCo2O4 and NiCo2S4 formed on Ni foam for supercapacitor." Composites Part B: Engineering 123 (2017): 28-33.
524 Wu, Zhibin, Xuli Pu, Xiaobo Ji, Yirong Zhu, Mingjun Jing, Qiyuan Chen, and Feipeng Jiao. "High energy density asymmetric supercapacitors from mesoporous NiCo2S4 nanosheets." Electrochimica Acta 174 (2015): 238-245.
525 Hou, Linrui, Ruiqi Bao, Zhiyi Chen, Muhammad Rehan, Liuniu Tong, Gang Pang, and Changzhou Yuan. "Comparative investigation of hollow mesoporous NiCo2S4 ellipsoids with enhanced pseudo-capacitances towards high-performance asymmetric supercapacitors." Electrochimica Acta 214 (2016): 76-84.
526 Kong, Lingping, Chuangfang Zhang, Jitong Wang, Donghui Long, Wenming Qiao, and Licheng Ling. "Ultrahigh intercalation pseudocapacitance of mesoporous orthorhombic niobium pentoxide from a novel cellulose nanocrystal template." Materials Chemistry and Physics 149 (2015): 495-504.
527 Huang, Jichun, Ting Lei, Xiaopei Wei, Xinwei Liu, Tong Liu, Dianxue Cao, Jinling Yin, and Guiling Wang. "Effect of Al-doped β-Ni (OH) 2 nanosheets on

electrochemical behaviors for high performance supercapacitor application." Journal of power sources 232 (2013): 370-375.

528 Dong, Xinyi, Lei Wang, Dong Wang, Cheng Li, and Jian Jin. "Layer-by-layer engineered Co–Al hydroxide nanosheets/graphene multilayer films as flexible electrode for supercapacitor." Langmuir 28, no. 1 (2012): 293-298.

529 Chen, Hao, Linfeng Hu, Yan Yan, Renchao Che, Min Chen, and Limin Wu. "One-step fabrication of ultrathin porous nickel hydroxide-manganese dioxide hybrid nanosheets for supercapacitor electrodes with excellent capacitive performance." Advanced Energy Materials 3, no. 12 (2013): 1636-1646.

530 Wu, Changzheng, Xiuli Lu, Lele Peng, Kun Xu, Xu Peng, Jianliu Huang, Guihua Yu, and Yi Xie. "Two-dimensional vanadyl phosphate ultrathin nanosheets for high energy density and flexible pseudocapacitors." Nature communications 4, no. 1 (2013): 2431.

531 Kung, Chung-Wei, Hsin-Wei Chen, Chia-Yu Lin, R. Vittal, and Kuo-Chuan Ho. "Synthesis of Co3O4 nanosheets via electrodeposition followed by ozone treatment and their application to high-performance supercapacitors." Journal of Power Sources 214 (2012): 91-99.

532 Chen, Yuanzhen, Yongning Liu, and Wei Yan. "Preparation of porous (Ni, Co) 3 (BO 3) 2/Ni (OH) 2 nanosheet networks as pseudocapacitor materials with superior performance." Journal of Materials Chemistry A 2, no. 16 (2014): 5903-5909.

533 Chen, Xiang, Hong-Jie Peng, Rui Zhang, Ting-Zheng Hou, Jia-Qi Huang, Bo Li, and Qiang Zhang. "An analogous periodic law for strong anchoring of polysulfides on polar hosts in lithium sulfur batteries: S-or Li-binding on first-row transition-metal sulfides?." ACS Energy Letters 2, no. 4 (2017): 795-801.

534 Chen, Jun Song, Cao Guan, Yang Gui, and Daniel John Blackwood. "Rational design of self-supported Ni3S2 nanosheets array for advanced asymmetric supercapacitor with a superior energy density." ACS Applied Materials & Interfaces 9, no. 1 (2017): 496-504.

535 Li, Jian, Xiaotao Yuan, Cong Lin, Yanquan Yang, Le Xu, Xin Du, Jinglin Xie, Jianhua Lin, and Junliang Sun. "Achieving high pseudocapacitance of 2D titanium carbide (MXene) by cation intercalation and surface modification." Advanced Energy Materials 7, no. 15 (2017): 1602725.

536 Mei, Jun, Ting Liao, Liangzhi Kou, and Ziqi Sun. "Two-dimensional metal oxide nanomaterials for next-generation rechargeable batteries." Advanced Materials 29, no. 48 (2017): 1700176.

537 Geng, Jianxin, Shang Chen, and Xin Chen. "State-of-the-art applications of 2D nanomaterials in energy storage." In Adapting 2D Nanomaterials for Advanced Applications, pp. 253-293. American Chemical Society, 2020.

538 Sahoo, Ramkrishna, Anjali Pal, and Tarasankar Pal. "2D materials for renewable energy storage devices: outlook and challenges." Chemical communications 52, no. 93 (2016): 13528-13542.

539 Cong, Lina, Haiming Xie, and Jinghong Li. "Hierarchical structures based on two-dimensional nanomaterials for rechargeable lithium batteries." Advanced Energy Materials 7, no. 12 (2017): 1601906.

540 Zhang, Jintao, Zhenhai Xia, and Liming Dai. "Carbon-based electrocatalysts for advanced energy conversion and storage." Science advances 1, no. 7 (2015): e1500564.

541 Yu, Xin Yao, and Xiong Wen Lou. "Mixed metal sulfides for electrochemical energy storage and conversion." Advanced Energy Materials 8, no. 3 (2018): 1701592.

542 Sun, Zhenyu, Edyta Madej, Christian Wiktor, Ilya Sinev, Roland A. Fischer, Gustaaf Van Tendeloo, Martin Muhler, Wolfgang Schuhmann, and Edgar Ventosa. "One-pot synthesis of carbon-coated nanostructured iron oxide on few-layer graphene for lithium-ion batteries." Chemistry–A European Journal 21, no. 45 (2015): 16154-16161.

543 Du, Guodong, Zaiping Guo, Shiquan Wang, Rong Zeng, Zhixin Chen, and Huakun Liu. "Superior stability and high capacity of restacked molybdenum disulfide as anode material for lithium ion batteries." Chemical communications 46, no. 7 (2010): 1106-1108.

544 Wang, Da, Li-Min Liu, Shi-Jin Zhao, Zi-Yu Hu, and Hao Liu. "Potential application of metal dichalcogenides double-layered heterostructures as anode materials for Li-ion batteries." The Journal of Physical Chemistry C 120, no. 9 (2016): 4779-4788.

545 Wang, Dashuai, Yanhui Liu, Xing Meng, Yingjin Wei, Yingying Zhao, Qiang Pang, and Gang Chen. "Two-dimensional $VS_2$ monolayers as potential anode materials for lithium-ion batteries and beyond: first-principles calculations." Journal of materials chemistry A 5, no. 40 (2017): 21370-21377.

546 Karmakar, Sharmistha, Chandra Chowdhury, and Ayan Datta. "Two-dimensional group IV monochalcogenides: anode materials for Li-ion batteries." The Journal of Physical Chemistry C 120, no. 27 (2016): 14522-14530.

547 Li, Zhangpeng, Xin-Yao Yu, and Ungyu Paik. "Facile preparation of porous Co3O4 nanosheets for high-performance lithium ion batteries and oxygen evolution reaction." Journal of power sources 310 (2016): 41-46.

548 Sun, Wenping, Xianhong Rui, Jixin Zhu, Linghui Yu, Yu Zhang, Zhichuan Xu, Srinivasan Madhavi, and Qingyu Yan. "Ultrathin nickel oxide nanosheets for enhanced sodium and lithium storage." Journal of power sources 274 (2015): 755-761.

549 Sun, Zhenyu, Kunpeng Xie, Zi An Li, Ilja Sinev, Petra Ebbinghaus, Andreas Erbe, Michael Farle, Wolfgang Schuhmann, Martin Muhler, and Edgar Ventosa. "Hollow and Yolk-Shell Iron Oxide Nanostructures on Few-Layer Graphene in Li-Ion Batteries." Chemistry–A European Journal 20, no. 7 (2014): 2022-2030.

550 Kundu, Manab, Cheuk Chi Albert Ng, Dmitri Y. Petrovykh, and Lifeng Liu. "Nickel foam supported mesoporous MnO 2 nanosheet arrays with superior lithium storage performance." Chemical communications 49, no. 76 (2013): 8459-8461.

551 Tian, Xiaocong, Xu Xu, Liang He, Qiulong Wei, Mengyu Yan, Lin Xu, Yunlong Zhao, Chuchu Yang, and Liqiang Mai. "Ultrathin pre-lithiated V6O13 nanosheet cathodes with enhanced electrical transport and cyclability." Journal of Power Sources 255 (2014): 235-241.

552 Sun, Zhenyu, Xing Huang, Martin Muhler, Wolfgang Schuhmann, and Edgar Ventosa. "A carbon-coated TiO 2 (B) nanosheet composite for lithium ion batteries." Chemical communications 50, no. 41 (2014): 5506-5509.

553 Wang, Cen, Gaohui Du, Kenny Ståhl, Haixiao Huang, Yijun Zhong, and J. Z. Jiang. "Ultrathin SnO2 nanosheets: oriented attachment mechanism, nonstoichiometric defects, and enhanced lithium-ion battery performances." The Journal of Physical Chemistry C 116, no. 6 (2012): 4000-4011.

554 Leong, Chon Chio, Hui Pan, and Sut Kam Ho. "Two-dimensional transition-metal oxide monolayers as cathode materials for Li and Na ion batteries." Physical Chemistry Chemical Physics 18, no. 10 (2016): 7527-7534.

555 Zhao, Guixia, Tao Wen, Juan Zhang, Jiaxing Li, Huanli Dong, Xiangke Wang, Yuguo Guo, and Wenping Hu. "Two-dimensional $Cr_2O_3$ and interconnected graphene–$Cr_2O_3$ nanosheets: synthesis and their application in lithium storage." Journal of Materials Chemistry A 2, no. 4 (2014): 944-948.

556 Liu, Yu, Wei Wang, Lin Gu, Yewu Wang, Yulong Ying, Yiyin Mao, Luwei Sun, and Xinsheng Peng. "Flexible CuO nanosheets/reduced-graphene oxide composite paper: binder-free anode for high-performance lithium-ion batteries." ACS applied materials & interfaces 5, no. 19 (2013): 9850-9855.

557 Boland, John B., Andrew Harvey, Ruiyuan Tian, Damien Hanlon, Victor Vega-Mayoral, Beata Szydlowska, Aideen Griffin et al. "Liquid phase exfoliation of MoO 2 nanosheets for lithium ion battery applications." Nanoscale Advances 1, no. 4 (2019): 1560-1570.

558 Wang, Xin, Lanyan Huang, Yan Zhao, Yongguang Zhang, and Guofu Zhou. "Synthesis of mesoporous ZnO nanosheets via facile solvothermal method as the anode materials for lithium-ion batteries." Nanoscale research letters 11 (2016): 1-6.

559 Peng, Lele, Pan Xiong, Lu Ma, Yifei Yuan, Yue Zhu, Dahong Chen, Xiangyi Luo, Jun Lu, Khalil Amine, and Guihua Yu. "Holey two-dimensional transition metal oxide nanosheets for efficient energy storage." Nature communications 8, no. 1 (2017): 15139.

560 Xie, Yu, Michael Naguib, Vadym N. Mochalin, Michel W. Barsoum, Yury Gogotsi, Xiqian Yu, Kyung-Wan Nam, Xiao-Qing Yang, Alexander I. Kolesnikov, and Paul RC Kent. "Role of surface structure on Li-ion energy storage capacity of two-dimensional transition-metal carbides." Journal of the American Chemical Society 136, no. 17 (2014): 6385-6394.

561 Tang, Qing, Zhen Zhou, and Panwen Shen. "Are MXenes promising anode materials for Li ion batteries? Computational studies on electronic properties and Li storage capability of Ti3C2 and Ti3C2X2 (X= F, OH) monolayer." Journal of the American Chemical Society 134, no. 40 (2012): 16909-16916.

562 Luo, Jianmin, Xinyong Tao, Jun Zhang, Yang Xia, Hui Huang, Liyuan Zhang, Yongping Gan, Chu Liang, and Wenkui Zhang. "Sn4+ ion decorated highly conductive Ti3C2 MXene: promising lithium-ion anodes with enhanced volumetric capacity and cyclic performance." ACS nano 10, no. 2 (2016): 2491-2499.

563 Ahmed, Bilal, Dalaver H. Anjum, Yury Gogotsi, and Husam N. Alshareef. "Atomic layer deposition of SnO2 on MXene for Li-ion battery anodes." Nano Energy 34 (2017): 249-256.

564 Ahmed, Bilal, Dalaver H. Anjum, Mohamed N. Hedhili, Yury Gogotsi, and Husam N. Alshareef. "$H_2O_2$ assisted room temperature oxidation of $Ti_2C$ MXene for Li-ion battery anodes." Nanoscale 8, no. 14 (2016): 7580-7587.

565 Zhou, Jie, Xianhu Zha, Xiaobing Zhou, Fanyan Chen, Guoliang Gao, Shuwei Wang, Cai Shen et al. "Synthesis and electrochemical properties of two-dimensional hafnium carbide." ACS nano 11, no. 4 (2017): 3841-3850.

566 Guo, Zhonglu, Jian Zhou, and Zhimei Sun. "New two-dimensional transition metal borides for Li ion batteries and electrocatalysis." Journal of materials chemistry A 5, no. 45 (2017): 23530-23535.

567 Sun, Changlong, Mingzhi Yang, Tailin Wang, Yongliang Shao, Yongzhong Wu, and Xiaopeng Hao. "Graphene-oxide-assisted synthesis of GaN nanosheets as a new anode material for lithium-ion battery." ACS applied materials & interfaces 9, no. 32 (2017): 26631-26636.

568 Huang, Changshui, Yongjun Li, Ning Wang, Yurui Xue, Zicheng Zuo, Huibiao Liu, and Yuliang Li. "Progress in research into 2D graphdiyne-based materials." Chemical reviews 118, no. 16 (2018): 7744-7803.

569 Sun, Chenghua, and Debra J. Searles. "Lithium storage on graphdiyne predicted by DFT calculations." The Journal of Physical Chemistry C 116, no. 50 (2012): 26222-26226.

570 Shang, Hong, Zicheng Zuo, Liang Li, Fan Wang, Huibiao Liu, Yongjun Li, and Yuliang Li. "Ultrathin graphdiyne nanosheets grown in situ on copper nanowires and their performance as lithium-ion battery anodes." Angewandte Chemie International Edition 57, no. 3 (2018): 774-778.

571 Li, Weifeng, Yanmei Yang, Gang Zhang, and Yong-Wei Zhang. "Ultrafast and directional diffusion of lithium in phosphorene for high-performance lithium-ion battery." Nano letters 15, no. 3 (2015): 1691-1697.

572 Zhang, Yuqin, Ningning Dong, Hengcong Tao, Chao Yan, Jiawei Huang, Tengfei Liu, Alex W. Robertson, John Texter, Jun Wang, and Zhenyu Sun. "Exfoliation of stable 2D black phosphorus for device fabrication." Chemistry of Materials 29, no. 15 (2017): 6445-6456.

573 Li, Qing-Fang, Chun-Gang Duan, X. G. Wan, and Jer-Lai Kuo. "Theoretical prediction of anode materials in Li-ion batteries on layered black and blue phosphorus." The Journal of Physical Chemistry C 119, no. 16 (2015): 8662-8670.

574 Zhao, Shijun, Wei Kang, and Jianming Xue. "The potential application of phosphorene as an anode material in Li-ion batteries." Journal of Materials Chemistry A 2, no. 44 (2014): 19046-19052.

575 Guo, Gen-Cai, Da Wang, Xiao-Lin Wei, Qi Zhang, Hao Liu, Woon-Ming Lau, and Li-Min Liu. "First-principles study of phosphorene and graphene heterostructure as anode materials for rechargeable Li batteries." The journal of physical chemistry letters 6, no. 24 (2015): 5002-5008.

576 Zhang, Yu, Huanwen Wang, Zhongzhen Luo, Hui Teng Tan, Bing Li, Shengnan Sun, Zhong Li et al. "An air-stable densely packed phosphorene–graphene composite toward advanced lithium storage properties." Advanced Energy Materials 6, no. 12 (2016): 1600453.

577 Chowdhury, Chandra, Sharmistha Karmakar, and Ayan Datta. "Capping black phosphorene by h-BN enhances performances in anodes for Li and Na ion batteries." ACS Energy Letters 1, no. 1 (2016): 253-259.

578 Dudney, Nancy J., and Juchuan Li. "Using all energy in a battery." Science 347, no. 6218 (2015): 131-132.

579 Yoon, Dohyeon, Kyung Yoon Chung, Wonyoung Chang, Seung Min Kim, Mi Jin Lee, Zonghoon Lee, and Jaehoon Kim. "Hydrogen-enriched reduced graphene oxide with enhanced electrochemical performance in lithium ion batteries." Chemistry of Materials 27, no. 1 (2015): 266-275.

580 Zhao, Dongdong, Lei Wang, Peng Yu, Lu Zhao, Chungui Tian, Wei Zhou, Lei Zhang, and Honggang Fu. "From graphite to porous graphene-like nanosheets for high rate lithium-ion batteries." Nano Research 8 (2015): 2998-3010.

581 Wu, Zhong-Shuai, Wencai Ren, Li Xu, Feng Li, and Hui-Ming Cheng. "Doped graphene sheets as anode materials with superhigh rate and large capacity for lithium ion batteries." ACS nano 5, no. 7 (2011): 5463-5471.

582 Hardikar, Rahul P., Deya Das, Sang Soo Han, Kwang-Ryeol Lee, and Abhishek K. Singh. "Boron doped defective graphene as a potential anode material for Li-ion batteries." Physical Chemistry Chemical Physics 16, no. 31 (2014): 16502-16508.

583 Das, Deya, Rahul P. Hardikar, Sang Soo Han, Kwang-Ryeol Lee, and Abhishek K. Singh. "Monolayer BC 2: an ultrahigh capacity anode material for Li ion batteries." Physical Chemistry Chemical Physics 19, no. 35 (2017): 24230-24239.

584 Lalitha, Murugan, Shivaraja Selva Mahadevan, and Senthilkumar Lakshmipathi. "Improved lithium adsorption in boron-and nitrogen-substituted graphene derivatives." Journal of Materials Science 52 (2017): 815-831.

585 Ma, Xinlong, Guoqing Ning, Chuanlei Qi, Chenggen Xu, and Jinsen Gao. "Phosphorus and nitrogen dual-doped few-layered porous graphene: a high-performance anode material for lithium-ion batteries." ACS Applied Materials & Interfaces 6, no. 16 (2014): 14415-14422.

586 Ai, Wei, Zhimin Luo, Jian Jiang, Jianhui Zhu, Zhuzhu Du, Zhanxi Fan, Linghai Xie, Hua Zhang, Wei Huang, and Ting Yu. "Nitrogen and sulfur codoped graphene: multifunctional electrode materials for high-performance Li-ion batteries and oxygen reduction reaction." Adv. Mater 26, no. 35 (2014): 6186-6192.

587 Huang, Shizheng, Yu Li, Yiyu Feng, Haoran An, Peng Long, Chengqun Qin, and Wei Feng. "Nitrogen and fluorine co-doped graphene as a high-performance anode material for lithium-ion batteries." Journal of Materials Chemistry A 3, no. 46 (2015): 23095-23105.

588 Mo, Runwei, David Rooney, Kening Sun, and Hui Ying Yang. "3D nitrogen-doped graphene foam with encapsulated germanium/nitrogen-doped graphene

yolk-shell nanoarchitecture for high-performance flexible Li-ion battery." Nature communications 8, no. 1 (2017): 13949.

589 Wang, Xiaolei, Ge Li, Min Ho Seo, Fathy M. Hassan, Md Ariful Hoque, and Zhongwei Chen. "Sulfur Atoms Bridging Few-Layered MoS2 with S-Doped Graphene Enable Highly Robust Anode for Lithium-Ion Batteries." Advanced Energy Materials 5, no. 23 (2015): 1501106.

590 Subbaraman, Ram, Dusan Tripkovic, Dusan Strmcnik, Kee-Chul Chang, Masanobu Uchimura, Arvydas P. Paulikas, Vojislav Stamenkovic, and Nenad M. Markovic. "Enhancing hydrogen evolution activity in water splitting by tailoring Li+-Ni (OH) 2-Pt interfaces." Science 334, no. 6060 (2011): 1256-1260.

591 Conway, B. E., and B. V. Tilak. "Interfacial processes involving electrocatalytic evolution and oxidation of H2, and the role of chemisorbed H." Electrochimica acta 47, no. 22-23 (2002): 3571-3594.

592 Zheng, Yao, Yan Jiao, Anthony Vasileff, and Shi-Zhang Qiao. "The hydrogen evolution reaction in alkaline solution: from theory, single crystal models, to practical electrocatalysts." Angewandte Chemie International Edition 57, no. 26 (2018): 7568-7579.

593 Kuang, Yun, Guang Feng, Pengsong Li, Yongmin Bi, Yaping Li, and Xiaoming Sun. "Single-crystalline ultrathin nickel nanosheets array from in situ topotactic reduction for active and stable electrocatalysis." Angewandte Chemie 128, no. 2 (2016): 703-707.

594 Gholamvand, Zahra, David McAteer, Claudia Backes, Niall McEvoy, Andrew Harvey, Nina C. Berner, Damien Hanlon et al. "Comparison of liquid exfoliated transition metal dichalcogenides reveals MoSe2 to be the most effective hydrogen evolution catalyst." Nanoscale 8, no. 10 (2016): 5737-5749.

595 Ma, Lianbo, Yi Hu, Renpeng Chen, Guoyin Zhu, Tao Chen, Hongling Lv, Yanrong Wang et al. "Self-assembled ultrathin NiCo2S4 nanoflakes grown on Ni foam as high-performance flexible electrodes for hydrogen evolution reaction in alkaline solution." Nano Energy 24 (2016): 139-147.

596 Wang, Fengmei, Yuanchang Li, Tofik Ahmed Shifa, Kaili Liu, Feng Wang, Zhenxing Wang, Peng Xu, Qisheng Wang, and Jun He. "Selenium-enriched nickel selenide nanosheets as a robust electrocatalyst for hydrogen generation." Angewandte Chemie International Edition 55, no. 24 (2016): 6919-6924.

597 Gao, Jian, Lu Li, Jiawei Tan, Hao Sun, Baichang Li, Juan Carlos Idrobo, Chandra Veer Singh, Toh-Ming Lu, and Nikhil Koratkar. "Vertically oriented arrays of ReS2 nanosheets for electrochemical energy storage and electrocatalysis." Nano letters 16, no. 6 (2016): 3780-3787.

598 Yuan, Jiangtan, Jingjie Wu, Will J. Hardy, Philip Loya, Minhan Lou, Yingchao Yang, Sina Najmaei et al. "Facile synthesis of single crystal vanadium disulfide nanosheets by chemical vapor deposition for efficient hydrogen evolution reaction." Advanced Materials 27, no. 37 (2015): 5605-5609.

599 Zhao, Wenwen, Bohua Dong, Zenglong Guo, Ge Su, Rongjie Gao, Wei Wang, and Lixin Cao. "Colloidal synthesis of VSe 2 single-layer nanosheets as novel

electrocatalysts for the hydrogen evolution reaction." Chemical communications 52, no. 59 (2016): 9228-9231.

600 Yang, Yang, Huilong Fei, Gedeng Ruan, Yilun Li, and James M. Tour. "Vertically aligned WS2 nanosheets for water splitting." Advanced Functional Materials 25, no. 39 (2015): 6199-6204.

601 Datta, R. S., Farjana Haque, Md Mohiuddin, B. J. Carey, Nitu Syed, A. Zavabeti, Baoyue Zhang et al. "Highly active two dimensional α-$MoO_{3-x}$ for the electrocatalytic hydrogen evolution reaction." Journal of materials chemistry A 5, no. 46 (2017): 24223-24231.

602 Zheng, Tingting, Wei Sang, Zhihai He, Qiushi Wei, Bowen Chen, Hongliang Li, Cong Cao et al. "Conductive tungsten oxide nanosheets for highly efficient hydrogen evolution." Nano letters 17, no. 12 (2017): 7968-7973.

603 Han, Xiaotong, Chang Yu, Juan Yang, Changtai Zhao, Huawei Huang, Zhibin Liu, Pulickel M. Ajayan, and Jieshan Qiu. "Mass and charge transfer coenhanced oxygen evolution behaviors in CoFe-layered double hydroxide assembled on graphene." Advanced Materials Interfaces 3, no. 7 (2016): 1500782.

604 Feng, Jin-Xian, Liang-Xin Ding, Sheng-Hua Ye, Xu-Jun He, Han Xu, Ye-Xiang Tong, and Gao-Ren Li. "Co (OH) 2@ PANI hybrid nanosheets with 3D networks as high-performance electrocatalysts for hydrogen evolution reaction." Advanced Materials 27, no. 44 (2015): 7051-7057.

605 Wang, Lei, Chong Lin, Dekang Huang, Jianmei Chen, Lin Jiang, Mingkui Wang, Lifeng Chi, Lin Shi, and Jian Jin. "Optimizing the volmer step by single-layer nickel hydroxide nanosheets in hydrogen evolution reaction of platinum." ACS Catalysis 5, no. 6 (2015): 3801-3806.

606 Li, Shuang, Ping Tuo, Junfeng Xie, Xiaodong Zhang, Jianguang Xu, Jian Bao, Bicai Pan, and Yi Xie. "Ultrathin MXene nanosheets with rich fluorine termination groups realizing efficient electrocatalytic hydrogen evolution." Nano energy 47 (2018): 512-518.

607 Ang, Huixiang, Hui Teng Tan, Zhi Min Luo, Yu Zhang, Yuan Yuan Guo, Guilue Guo, Hua Zhang, and Qingyu Yan. "Hydrophilic nitrogen and sulfur Co-doped molybdenum carbide nanosheets for electrochemical hydrogen evolution." Small 11, no. 47 (2015): 6278-6284.

608 Zhao, Zhenhuan, Fan Qin, Sashank Kasiraju, Lixin Xie, Md Kamrul Alam, Shuo Chen, Dezhi Wang et al. "Vertically aligned MoS2/Mo2C hybrid nanosheets grown on carbon paper for efficient electrocatalytic hydrogen evolution." ACS Catalysis 7, no. 10 (2017): 7312-7318.

609 Meyer, Simon, Aleksey V. Nikiforov, Irina M. Petrushina, Klaus Köhler, Erik Christensen, Jens Oluf Jensen, and Niels J. Bjerrum. "Transition metal carbides (WC, Mo2C, TaC, NbC) as potential electrocatalysts for the hydrogen evolution reaction (HER) at medium temperatures." international journal of hydrogen energy 40, no. 7 (2015): 2905-2911.

610 Cao, Bingfei, Gabriel M. Veith, Joerg C. Neuefeind, Radoslav R. Adzic, and Peter G. Khalifah. "Mixed close-packed cobalt molybdenum nitrides as non-noble metal electrocatalysts for the hydrogen evolution reaction." Journal of the American Chemical Society 135, no. 51 (2013): 19186-19192.

611 Wang, Zhaoyang, Lin Xu, Fuzhi Huang, Longbing Qu, Jiantao Li, Kwadwo Asare Owusu, Ziang Liu et al. "Copper–nickel nitride nanosheets as efficient bifunctional catalysts for hydrazine-assisted electrolytic hydrogen production." Advanced Energy Materials 9, no. 21 (2019): 1900390.

612 Jin, Huanyu, Xin Liu, Anthony Vasileff, Yan Jiao, Yongqiang Zhao, Yao Zheng, and Shi-Zhang Qiao. "Single-crystal nitrogen-rich two-dimensional Mo5N6 nanosheets for efficient and stable seawater splitting." Acs Nano 12, no. 12 (2018): 12761-12769.

613 Jiang, Nan, Bo You, Meili Sheng, and Yujie Sun. "Electrodeposited cobalt-phosphorous-derived films as competent bifunctional catalysts for overall water splitting." Angewandte Chemie 127, no. 21 (2015): 6349-6352.

614 Xu, You, Rui Wu, Jingfang Zhang, Yanmei Shi, and Bin Zhang. "Anion-exchange synthesis of nanoporous FeP nanosheets as electrocatalysts for hydrogen evolution reaction." Chemical communications 49, no. 59 (2013): 6656-6658.

615 Jiang, Ping, Qian Liu, and Xuping Sun. "NiP2 nanosheet arrays supported on carbon cloth: an efficient 3D hydrogen evolution cathode in both acidic and alkaline solutions." Nanoscale 6, no. 22 (2014): 13440-13445.

616 Song, Bo, Kai Li, Ying Yin, Tao Wu, Lianna Dang, Miguel Cabán-Acevedo, Jiecai Han et al. "Tuning mixed nickel iron phosphosulfide nanosheet electrocatalysts for enhanced hydrogen and oxygen evolution." Acs Catalysis 7, no. 12 (2017): 8549-8557.

617 Gusmão, Rui, Zdeněk Sofer, and Martin Pumera. "Exfoliated layered manganese trichalcogenide phosphite (MnPX3, X= S, Se) as electrocatalytic van der Waals materials for hydrogen evolution." Advanced Functional Materials 29, no. 2 (2019): 1805975.

618 Zheng, Tingting, Wei Sang, Zhihai He, Qiushi Wei, Bowen Chen, Hongliang Li, Cong Cao et al. "Conductive tungsten oxide nanosheets for highly efficient hydrogen evolution." Nano letters 17, no. 12 (2017): 7968-7973.

619 Gusmão, Rui, Zdeněk Sofer, Daniel Bouša, and Martin Pumera. "Pnictogen (As, Sb, Bi) nanosheets for electrochemical applications are produced by shear exfoliation using kitchen blenders." Angewandte Chemie 129, no. 46 (2017): 14609-14614.

620 Park, Hyounmyung, Yuemei Zhang, Jan P. Scheifers, Palani R. Jothi, Andrew Encinas, and Boniface PT Fokwa. "Graphene-and phosphorene-like boron layers with contrasting activities in highly active Mo2B4 for hydrogen evolution." Journal of the American Chemical Society 139, no. 37 (2017): 12915-12918.

621 Zheng, Yao, Yan Jiao, Yihan Zhu, Lu Hua Li, Yu Han, Ying Chen, Aijun Du, Mietek Jaroniec, and Shi Zhang Qiao. "Hydrogen evolution by a metal-free electrocatalyst." Nature communications 5, no. 1 (2014): 3783.

622 Kibsgaard, Jakob, Zhebo Chen, Benjamin N. Reinecke, and Thomas F. Jaramillo. "Engineering the surface structure of MoS2 to preferentially expose active edge sites for electrocatalysis." Nature materials 11, no. 11 (2012): 963-969.

[623] Voiry, Damien, Hisato Yamaguchi, Junwen Li, Rafael Silva, Diego CB Alves, Takeshi Fujita, Mingwei Chen et al. "Enhanced catalytic activity in strained chemically exfoliated WS2 nanosheets for hydrogen evolution." Nature materials 12, no. 9 (2013): 850-855.

[624] Lukowski, Mark A., Andrew S. Daniel, Fei Meng, Audrey Forticaux, Linsen Li, and Song Jin. "Enhanced hydrogen evolution catalysis from chemically exfoliated metallic MoS2 nanosheets." Journal of the American Chemical Society 135, no. 28 (2013): 10274-10277.

[625] Tao, Hengcong, Qun Fan, Tao Ma, Shizhen Liu, Henry Gysling, John Texter, Fen Guo, and Zhenyu Sun. "Two-dimensional materials for energy conversion and storage." Progress in Materials Science 111 (2020): 100637.

[626] Jaramillo, Thomas F., Kristina P. Jørgensen, Jacob Bonde, Jane H. Nielsen, Sebastian Horch, and Ib Chorkendorff. "Identification of active edge sites for electrochemical H2 evolution from MoS2 nanocatalysts." science 317, no. 5834 (2007): 100-102.

[627] Huang, Yufeng, Robert J. Nielsen, William A. Goddard III, and Manuel P. Soriaga. "The reaction mechanism with free energy barriers for electrochemical dihydrogen evolution on MoS2." Journal of the American Chemical Society 137, no. 20 (2015): 6692-6698.

[628] Tsai, Charlie, Karen Chan, Frank Abild-Pedersen, and Jens K. Nørskov. "Active edge sites in MoSe 2 and WSe 2 catalysts for the hydrogen evolution reaction: a density functional study." Physical Chemistry Chemical Physics 16, no. 26 (2014): 13156-13164.

[629] Xie, Junfeng, Hao Zhang, Shuang Li, Ruoxing Wang, Xu Sun, Min Zhou, Jingfang Zhou, Xiong Wen Lou, and Yi Xie. "Defect-rich MoS2 ultrathin nanosheets with additional active edge sites for enhanced electrocatalytic hydrogen evolution." Advanced materials 40 (2013): 5807-5813.

[630] Zhang, Jian, Shaohua Liu, Haiwei Liang, Renhao Dong, and Xinliang Feng. "Hierarchical Transition-Metal Dichalcogenide Nanosheets for Enhanced Electrocatalytic Hydrogen Evolution." Advanced Materials (Deerfield Beach, Fla.) 27, no. 45 (2015): 7426-7431.

[631] Geng, Xiumei, Weiwei Sun, Wei Wu, Benjamin Chen, Alaa Al-Hilo, Mourad Benamara, Hongli Zhu, Fumiya Watanabe, Jingbiao Cui, and Tar-pin Chen. "Pure and stable metallic phase molybdenum disulfide nanosheets for hydrogen evolution reaction." Nature communications 7, no. 1 (2016): 10672.

[632] Yin, Ying, Jiecai Han, Yumin Zhang, Xinghong Zhang, Ping Xu, Quan Yuan, Leith Samad et al. "Contributions of phase, sulfur vacancies, and edges to the hydrogen evolution reaction catalytic activity of porous molybdenum disulfide nanosheets." Journal of the American Chemical Society 138, no. 25 (2016): 7965-7972.

[633] Benson, Eric E., Hanyu Zhang, Samuel A. Schuman, Sanjini U. Nanayakkara, Noah D. Bronstein, Suzanne Ferrere, Jeffrey L. Blackburn, and Elisa M. Miller. "Balancing the hydrogen evolution reaction, surface energetics, and stability of metallic MoS2 nanosheets via covalent functionalization." Journal of the American Chemical Society 140, no. 1 (2018): 441-450.

634 Voiry, Damien, Jieun Yang, and Manish Chhowalla. "Recent strategies for improving the catalytic activity of 2D TMD nanosheets toward the hydrogen evolution reaction." Advanced materials 28, no. 29 (2016): 6197-6206.

635 Ye, Gonglan, Yongji Gong, Junhao Lin, Bo Li, Yongmin He, Sokrates T. Pantelides, Wu Zhou, Robert Vajtai, and Pulickel M. Ajayan. "Defects engineered monolayer MoS2 for improved hydrogen evolution reaction." Nano letters 16, no. 2 (2016): 1097-1103.

636 Tsai, Charlie, Hong Li, Sangwook Park, Joonsuk Park, Hyun Soo Han, Jens K. Nørskov, Xiaolin Zheng, and Frank Abild-Pedersen. "Electrochemical generation of sulfur vacancies in the basal plane of MoS2 for hydrogen evolution." Nature communications 8, no. 1 (2017): 1-8.

637 Ma, Lianbo, Yi Hu, Guoyin Zhu, Renpeng Chen, Tao Chen, Hongling Lu, Yanrong Wang et al. "In situ thermal synthesis of inlaid ultrathin MoS2/graphene nanosheets as electrocatalysts for the hydrogen evolution reaction." Chemistry of Materials 28, no. 16 (2016): 5733-5742.

638 Konkena, Bharathi, Justus Masa, Wei Xia, Martin Muhler, and Wolfgang Schuhmann. "MoSSe@ reduced graphene oxide nanocomposite heterostructures as efficient and stable electrocatalysts for the hydrogen evolution reaction." Nano Energy 29 (2016): 46-53.

639 Tang, Yu-Jia, Yu Wang, Xiao-Li Wang, Shun-Li Li, Wei Huang, Long-Zhang Dong, Chun-Hui Liu, Ya-Fei Li, and Ya-Qian Lan. "Molybdenum disulfide/nitrogen-doped reduced graphene oxide nanocomposite with enlarged interlayer spacing for electrocatalytic hydrogen evolution." Advanced Energy Materials 6, no. 12 (2016): 1600116.

640 Voiry, Damien, Raymond Fullon, Jieun Yang, Cecilia de Carvalho Castro e Silva, Rajesh Kappera, Ibrahim Bozkurt, Daniel Kaplan et al. "The role of electronic coupling between substrate and 2D MoS2 nanosheets in electrocatalytic production of hydrogen." Nature materials 15, no. 9 (2016): 1003-1009.

641 Gao, Min-Rui, Jin-Xia Liang, Ya-Rong Zheng, Yun-Fei Xu, Jun Jiang, Qiang Gao, Jun Li, and Shu-Hong Yu. "An efficient molybdenum disulfide/cobalt diselenide hybrid catalyst for electrochemical hydrogen generation." Nature communications 6, no. 1 (2015): 5982.

642 Xiong, Pan, Xiuyun Zhang, Hao Wan, Shijian Wang, Yufei Zhao, Jinqiang Zhang, Dong Zhou et al. "Interface modulation of two-dimensional superlattices for efficient overall water splitting." Nano letters 19, no. 7 (2019): 4518-4526.

643 Ding, Qi, Bo Song, Ping Xu, and Song Jin. "Efficient electrocatalytic and photoelectrochemical hydrogen generation using MoS2 and related compounds." Chem 1, no. 5 (2016): 699-726.

644 Li, Guoqing, Du Zhang, Yifei Yu, Shengyang Huang, Weitao Yang, and Linyou Cao. "Activating MoS2 for pH-universal hydrogen evolution catalysis." Journal of the American Chemical Society 139, no. 45 (2017): 16194-16200.

645 Liang, Hanfeng, Linsen Li, Fei Meng, Lianna Dang, Junqiao Zhuo, Audrey Forticaux, Zhoucheng Wang, and Song Jin. "Porous two-dimensional nanosheets converted from layered double hydroxides and their applications in

electrocatalytic water splitting." Chemistry of Materials 27, no. 16 (2015): 5702-5711.

646 Wu, Yuanyuan, Yipu Liu, Guo-Dong Li, Xu Zou, Xinran Lian, Dejun Wang, Lei Sun, Tewodros Asefa, and Xiaoxin Zou. "Efficient electrocatalysis of overall water splitting by ultrasmall NixCo3−xS4 coupled Ni3S2 nanosheet arrays." Nano Energy 35 (2017): 161-170.

647 Zhang, Xiao, Zhuangchai Lai, Chaoliang Tan, and Hua Zhang. "Solution-processed two-dimensional MoS2 nanosheets: preparation, hybridization, and applications." Angewandte Chemie International Edition 55, no. 31 (2016): 8816-8838.

648 Zheng, Yao, Yan Jiao, Lu Hua Li, Tan Xing, Ying Chen, Mietek Jaroniec, and Shi Zhang Qiao. "Toward design of synergistically active carbon-based catalysts for electrocatalytic hydrogen evolution." ACS nano 8, no. 5 (2014): 5290-5296.

649 Ito, Yoshikazu, Weitao Cong, Takeshi Fujita, Zheng Tang, and Mingwei Chen. "High catalytic activity of nitrogen and sulfur co-doped nanoporous graphene in the hydrogen evolution reaction." Angewandte Chemie International Edition 54, no. 7 (2015): 2131-2136.

650 Jiao, Yan, Yao Zheng, Kenneth Davey, and Shi-Zhang Qiao. "Activity origin and catalyst design principles for electrocatalytic hydrogen evolution on heteroatom-doped graphene." Nature Energy 1, no. 10 (2016): 1-9.

651 Sathe, Bhaskar R., Xiaoxin Zou, and Tewodros Asefa. "Metal-free B-doped graphene with efficient electrocatalytic activity for hydrogen evolution reaction." Catalysis science & technology 4, no. 7 (2014): 2023-2030.

652 Long, Gui-fa, Kai Wan, Ming-yao Liu, Zhen-xing Liang, Jin-hua Piao, and Panagiotis Tsiakaras. "Active sites and mechanism on nitrogen-doped carbon catalyst for hydrogen evolution reaction." Journal of catalysis 348 (2017): 151-159.

653 Jiang, Hongliang, Yihua Zhu, Yunhe Su, Yifan Yao, Yanyan Liu, Xiaoling Yang, and Chunzhong Li. "Highly dual-doped multilayer nanoporous graphene: efficient metal-free electrocatalysts for the hydrogen evolution reaction." Journal of materials chemistry A 3, no. 24 (2015): 12642-12645.

654 Tian, Ye, Rui Mei, Da-zhong Xue, Xiao Zhang, and Wei Peng. "Enhanced electrocatalytic hydrogen evolution in graphene via defect engineering and heteroatoms co-doping." Electrochimica Acta 219 (2016): 781-789.

655 Ito, Yoshikazu, Yuhao Shen, Daisuke Hojo, Yoji Itagaki, Takeshi Fujita, Linghan Chen, Tsutomu Aida, Zheng Tang, Tadafumi Adschiri, and Mingwei Chen. "Correlation between Chemical Dopants and Topological Defects in Catalytically Active Nanoporous Graphene." Advanced Materials (Deerfield Beach, Fla.) 28, no. 48 (2016): 10644-10651.

656 Kumatani, Akichika, Chiho Miura, Hirotaka Kuramochi, Tatsuhiko Ohto, Mitsuru Wakisaka, Yuki Nagata, Hiroki Ida et al. "Chemical dopants on edge of holey graphene accelerate electrochemical hydrogen evolution reaction." Advanced Science 6, no. 10 (2019): 1900119.

657 Zhang, Jintao, and Liming Dai. "Nitrogen, phosphorus, and fluorine tri-doped graphene as a multifunctional catalyst for self-powered electrochemical water splitting." Angewandte Chemie 128, no. 42 (2016): 13490-13494.

658 Jia, Yi, Longzhou Zhang, Aijun Du, Guoping Gao, Jun Chen, Xuecheng Yan, Christopher L. Brown, and Xiangdong Yao. "Defect graphene as a trifunctional catalyst for electrochemical reactions." (2016): 9532.

659 Shinde, Sambhaji S., Abdul Sami, and Jung-Ho Lee. "Nitrogen-and phosphorus-doped nanoporous graphene/graphitic carbon nitride hybrids as efficient electrocatalysts for hydrogen evolution." ChemCatChem 7, no. 23 (2015): 3873-3880.

660 Tan, Xin, Hassan A. Tahini, and Sean C. Smith. "p-Doped graphene/graphitic carbon nitride hybrid electrocatalysts: unraveling charge transfer mechanisms for enhanced hydrogen evolution reaction performance." ACS Catalysis 6, no. 10 (2016): 7071-7077.

661 Duan, Jingjing, Sheng Chen, Mietek Jaroniec, and Shi Zhang Qiao. "Porous C3N4 nanolayers@ N-graphene films as catalyst electrodes for highly efficient hydrogen evolution." ACS nano 9, no. 1 (2015): 931-940.

662 Wei, Ying, and Zhenyu Sun. "Liquid-phase exfoliation of graphite for mass production of pristine few-layer graphene." Current opinion in colloid & interface science 20, no. 5-6 (2015): 311-321.

663 Liu, Danni, Jiahong Wang, Jiang Lu, Chao Ma, Hao Huang, Zhongzheng Wang, Lie Wu et al. "Direct synthesis of metal-doped phosphorene with enhanced electrocatalytic hydrogen evolution." Small Methods 3, no. 7 (2019): 1900083.

664 Najafi, L., S. Bellani, R. Oropesa-Nuñez, B. Martín-García, M. Prato, and F. Bonaccorso. "Single-/few-layer graphene as long-lasting electrocatalyst for hydrogen evolution reaction." ACS Applied Energy Materials 2, no. 8 (2019): 5373-5379.

665 Kong, Xiangkai, Kun Xu, Changlin Zhang, Jun Dai, Shirin Norooz Oliaee, Lingyan Li, Xiaocheng Zeng, Changzheng Wu, and Zhenmeng Peng. "Free-standing two-dimensional Ru nanosheets with high activity toward water splitting." Acs Catalysis 6, no. 3 (2016): 1487-1492.

666 Zhang, Fang-Shuai, Jia-Wei Wang, Jun Luo, Rui-Rui Liu, Zhi-Ming Zhang, Chun-Ting He, and Tong-Bu Lu. "Extraction of nickel from NiFe-LDH into Ni 2 P@ NiFe hydroxide as a bifunctional electrocatalyst for efficient overall water splitting." Chemical science 9, no. 5 (2018): 1375-1384.

667 Saadi, Fadl H., Azhar I. Carim, Walter S. Drisdell, Sheraz Gul, Jack H. Baricuatro, Junko Yano, Manuel P. Soriaga, and Nathan S. Lewis. "Operando spectroscopic analysis of CoP films electrocatalyzing the hydrogen-evolution reaction." Journal of the American Chemical Society 139, no. 37 (2017): 12927-12930.

668 Scherer, Günther G. "Interfacial aspects in the development of polymer electrolyte fuel cells." Solid State Ionics 94, no. 1-4 (1997): 249-257.

669 Kunze, Julia, and Ulrich Stimming. "Electrochemical Versus Heat-Engine Energy Technology: A Tribute to Wilhelm Ostwald's Visionary Statements." Angewandte Chemie International Edition 48, no. 49 (2009): 9230-9237.

[670] Jin, Huanyu, Chunxian Guo, Xin Liu, Jinlong Liu, Anthony Vasileff, Yan Jiao, Yao Zheng, and Shi-Zhang Qiao. "Emerging two-dimensional nanomaterials for electrocatalysis." Chemical reviews 118, no. 13 (2018): 6337-6408.

[671] Gasteiger, Hubert A., and Nenad M. Marković. "Just a dream—or future reality?." science 324, no. 5923 (2009): 48-49.

[672] Kamarudin, Siti Kartom, F. Achmad, and Wan Ramli Wan Daud. "Overview on the application of direct methanol fuel cell (DMFC) for portable electronic devices." International Journal of hydrogen energy 34, no. 16 (2009): 6902-6916.

[673] Grove, W. R. "London, Edinburgh, and Dublin Philos. Magaz." J. Sci 21 (1842): 417-420.

[674] Mitlitsky, Fred, Blake Myers, Andrew H. Weisberg, Trent M. Molter, and William F. Smith. "Reversible (unitised) PEM fuel cell devices." Fuel Cells Bulletin 2, no. 11 (1999): 6-11.

[675] Ferrari, Andrea C., Francesco Bonaccorso, Vladimir Fal'Ko, Konstantin S. Novoselov, Stephan Roche, Peter Bøggild, Stefano Borini et al. "Science and technology roadmap for graphene, related two-dimensional crystals, and hybrid systems." Nanoscale 7, no. 11 (2015): 4598-4810.

[676] Jia, Yi, Longzhou Zhang, Aijun Du, Guoping Gao, Jun Chen, Xuecheng Yan, Christopher L. Brown, and Xiangdong Yao. "Defect graphene as a trifunctional catalyst for electrochemical reactions." (2016): 9532.

[677] Liu, Yuan, Yuting Shen, Litao Sun, Jincheng Li, Chang Liu, Wencai Ren, Feng Li et al. "Elemental superdoping of graphene and carbon nanotubes." Nature communications 7, no. 1 (2016): 10921.

[678] Shi, Zhiming, Alex Kutana, and Boris I. Yakobson. "How much N-doping can graphene sustain?." The Journal of Physical Chemistry Letters 6, no. 1 (2015): 106-112.

[679] Jiao, Yan, Yao Zheng, Mietek Jaroniec, and Shi Zhang Qiao. "Origin of the electrocatalytic oxygen reduction activity of graphene-based catalysts: a roadmap to achieve the best performance." Journal of the American Chemical Society 136, no. 11 (2014): 4394-4403.

[680] Ma, Ruguang, Xiaodong Ren, Bao Yu Xia, Yao Zhou, Chi Sun, Qian Liu, Jianjun Liu, and Jiacheng Wang. "Novel synthesis of N-doped graphene as an efficient electrocatalyst towards oxygen reduction." Nano Research 9 (2016): 808-819.

[681] Yu, Huijun, Lu Shang, Tong Bian, Run Shi, G. I. Waterhouse, Yufei Zhao, Chao Zhou, Li-Zhu Wu, Chen-Ho Tung, and Tierui Zhang. "Nitrogen-Doped Porous Carbon Nanosheets Templated from g-C3 N4 as Metal-Free Electrocatalysts for Efficient Oxygen Reduction Reaction." Advanced Materials (Deerfield Beach, Fla.) 28, no. 25 (2016): 5080-5086.

[682] Zhang, Chenzhen, Nasir Mahmood, Han Yin, Fei Liu, and Yanglong Hou. "Synthesis of phosphorus-doped graphene and its multifunctional applications for oxygen reduction reaction and lithium ion batteries." Advanced materials (Deerfield Beach, Fla.) 25, no. 35 (2013): 4932-4937.

683 Wu, Jiao, Zhenrong Yang, Qijun Sun, Xiaowei Li, Peter Strasser, and Ruizhi Yang. "Synthesis and electrocatalytic activity of phosphorus-doped carbon xerogel for oxygen reduction." Electrochimica Acta 127 (2014): 53-60.

684 Zhang, Xilin, Zhansheng Lu, Zhaoming Fu, Yanan Tang, Dongwei Ma, and Zongxian Yang. "The mechanisms of oxygen reduction reaction on phosphorus doped graphene: A first-principles study." Journal of Power Sources 276 (2015): 222-229.

685 Lu, Guolong, Youlong Zhu, Kongliang Xu, Yinghua Jin, Zhiyong Jason Ren, Zhenning Liu, and Wei Zhang. "Metallated porphyrin based porous organic polymers as efficient electrocatalysts." Nanoscale 7, no. 43 (2015): 18271-18277.

686 Masa, Justus, Anqi Zhao, Wei Xia, Zhenyu Sun, Bastian Mei, Martin Muhler, and Wolfgang Schuhmann. "Trace metal residues promote the activity of supposedly metal-free nitrogen-modified carbon catalysts for the oxygen reduction reaction." Electrochemistry Communications 34 (2013): 113-116.

687 Wang, Lu, Adriano Ambrosi, and Martin Pumera. "“Metal-free” catalytic oxygen reduction reaction on heteroatom-doped graphene is caused by trace metal impurities." Angewandte Chemie 125, no. 51 (2013): 14063-14066.

688 Sun, Xiaofu, Qinggong Zhu, Xinchen Kang, Huizhen Liu, Qingli Qian, Zhaofu Zhang, and Buxing Han. "Molybdenum–bismuth bimetallic chalcogenide nanosheets for highly efficient electrocatalytic reduction of carbon dioxide to methanol." Angewandte Chemie 128, no. 23 (2016): 6883-6887.

689 Zheng, Yao, Yan Jiao, Lei Ge, Mietek Jaroniec, and Shi Zhang Qiao. "Two-step boron and nitrogen doping in graphene for enhanced synergistic catalysis." Angewandte Chemie International Edition 52, no. 11 (2013).

690 Choi, Chang Hyuck, Min Wook Chung, Han Chang Kwon, Sung Hyeon Park, and Seong Ihl Woo. "B, N-and P, N-doped graphene as highly active catalysts for oxygen reduction reactions in acidic media." Journal of Materials Chemistry A 1, no. 11 (2013): 3694-3699.

691 Liang, Ji, Yan Jiao, Mietek Jaroniec, and Shi Zhang Qiao. "Sulfur and nitrogen dual-doped mesoporous graphene electrocatalysts for oxygen reduction reaction with synergistically enhanced performance." Angewandte Chemie International Edition 51 (2012): 1-6.

692 Chen, Chi, Xiao-Dong Yang, Zhi-You Zhou, Yu-Jiao Lai, Muhammad Rauf, Ying Wang, Jing Pan et al. "Aminothiazole-derived N, S, Fe-doped graphene nanosheets as high performance electrocatalysts for oxygen reduction." Chemical communications 51, no. 96 (2015): 17092-17095.

693 Khan, Karim, Ayesha Khan Tareen, Jia Li, Usman Khan, Adeela Nairan, Yao Yuan, Xuyang Zhang, Minghui Yang, and Zhengbiao Ouyang. "Facile synthesis of tin-doped mayenite electride composite as a non-noble metal durable electrocatalyst for oxygen reduction reaction (ORR)." Dalton Transactions 47, no. 38 (2018): 13498-13506.

694 Peng, Jiahe, Xingzhu Chen, Wee-Jun Ong, Xiujian Zhao, and Neng Li. "Surface and heterointerface engineering of 2D MXenes and their nanocomposites: insights into electro-and photocatalysis." Chem 5, no. 1 (2019): 18-50.

[695] Xie, Xiaohong, Siguo Chen, Wei Ding, Yao Nie, and Zidong Wei. "An extraordinarily stable catalyst: Pt NPs supported on two-dimensional Ti 3 C 2 X 2 (X= OH, F) nanosheets for oxygen reduction reaction." Chemical Communications 49, no. 86 (2013): 10112-10114.

[696] Zhang, Zhiwei, Hanning Li, Guodong Zou, Carlos Fernandez, Baozhong Liu, Qingrui Zhang, Jie Hu, and Qiuming Peng. "Self-reduction synthesis of new MXene/Ag composites with unexpected electrocatalytic activity." ACS Sustainable Chemistry & Engineering 4, no. 12 (2016): 6763-6771.

[697] Srivastava, Ajit, Meinrad Sidler, Adrien V. Allain, Dominik S. Lembke, Andras Kis, and Atac Imamoğlu. "Optically active quantum dots in monolayer WSe2." Nature nanotechnology 10, no. 6 (2015): 491-496.

[698] Grosso, Gabriele, Hyowon Moon, Benjamin Lienhard, Sajid Ali, Dmitri K. Efetov, Marco M. Furchi, Pablo Jarillo-Herrero, Michael J. Ford, Igor Aharonovich, and Dirk Englund. "Tunable and high-purity room temperature single-photon emission from atomic defects in hexagonal boron nitride." Nature communications 8, no. 1 (2017): 705.

[699] Linhart, Lukas, Matthias Paur, Valerie Smejkal, Joachim Burgdörfer, Thomas Mueller, and Florian Libisch. "Localized intervalley defect excitons as single-photon emitters in WSe 2." Physical review letters 123, no. 14 (2019): 146401.

[700] Palacios-Berraquero, Carmen, Dhiren M. Kara, Alejandro R-P. Montblanch, Matteo Barbone, Pawel Latawiec, Duhee Yoon, Anna K. Ott, Marko Loncar, Andrea C. Ferrari, and Mete Atatüre. "Large-scale quantum-emitter arrays in atomically thin semiconductors." Nature communications 8, no. 1 (2017): 15093.

[701] Kianinia, Mehran, Blake Regan, Sherif Abdulkader Tawfik, Toan Trong Tran, Michael J. Ford, Igor Aharonovich, and Milos Toth. "Robust solid-state quantum system operating at 800 K." Acs Photonics 4, no. 4 (2017): 768-773.

[702] Xu, Zai-Quan, Christopher Elbadawi, Toan Trong Tran, Mehran Kianinia, Xiuling Li, Daobin Liu, Timothy B. Hoffman et al. "Single photon emission from plasma treated 2D hexagonal boron nitride." Nanoscale 10, no. 17 (2018): 7957-7965.

[703] He, Yu-Ming, Genevieve Clark, John R. Schaibley, Yu He, Ming-Cheng Chen, Yu-Jia Wei, Xing Ding et al. "Single quantum emitters in monolayer semiconductors." Nature nanotechnology 10, no. 6 (2015): 497-502.

[704] Jungwirth, Nicholas R., and Gregory D. Fuchs. "Optical absorption and emission mechanisms of single defects in hexagonal boron nitride." Physical review letters 119, no. 5 (2017): 057401.

[705] Martínez, L. J., Thomas Pelini, V. Waselowski, J. R. Maze, Bernard Gil, Guillaume Cassabois, and Vincent Jacques. "Efficient single photon emission from a high-purity hexagonal boron nitride crystal." Physical review B 94, no. 12 (2016): 121405.

[706] Li, Xiangzhi, Gabriella D. Shepard, Andrew Cupo, Nicolas Camporeale, Kamran Shayan, Yue Luo, Vincent Meunier, and Stefan Strauf. "Nonmagnetic quantum emitters in boron nitride with ultranarrow and sideband-free emission spectra." ACS nano 11, no. 7 (2017): 6652-6660.

[707] Bourrellier, Romain, Sophie Meuret, Anna Tararan, Odile Stéphan, Mathieu Kociak, Luiz HG Tizei, and Alberto Zobelli. "Bright UV single photon emission at point defects in h-BN." Nano letters 16, no. 7 (2016): 4317-4321.

[708] Camphausen, Robin, Loris Marini, Sherif Abdulkader Tawfik, Toan Trong Tran, Michael J. Ford, and Stefano Palomba. "Observation of near-infrared sub-Poissonian photon emission in hexagonal boron nitride at room temperature." APL Photonics 5, no. 7 (2020).

[709] Aharonovich, Igor, Dirk Englund, and Milos Toth. "Solid-state single-photon emitters." Nature photonics 10, no. 10 (2016): 631-641.

[710] Zhao, Huan, Michael T. Pettes, Yu Zheng, and Han Htoon. "Site-controlled telecom-wavelength single-photon emitters in atomically-thin MoTe2." Nature communications 12, no. 1 (2021): 6753.

[711] He, Yu-Ming, Genevieve Clark, John R. Schaibley, Yu He, Ming-Cheng Chen, Yu-Jia Wei, Xing Ding et al. "Single quantum emitters in monolayer semiconductors." Nature nanotechnology 10, no. 6 (2015): 497-502.

[712] Dastidar, Madhura Ghosh, Immanuel Thekkooden, Pramoda K. Nayak, and Vidya Praveen Bhallamudi. "Quantum emitters and detectors based on 2D van der Waals materials." Nanoscale 14, no. 14 (2022): 5289-5313.

[713] Srivastava, Ajit, Meinrad Sidler, Adrien V. Allain, Dominik S. Lembke, Andras Kis, and Atac Imamoğlu. "Optically active quantum dots in monolayer WSe2." Nature nanotechnology 10, no. 6 (2015): 491-496.

[714] Chakraborty, Chitraleema, Laura Kinnischtzke, Kenneth M. Goodfellow, Ryan Beams, and A. Nick Vamivakas. "Voltage-controlled quantum light from an atomically thin semiconductor." Nature nanotechnology 10, no. 6 (2015): 507-511.

[715] Lu, Xin, Xiaotong Chen, Sudipta Dubey, Qiang Yao, Weijie Li, Xingzhi Wang, Qihua Xiong, and Ajit Srivastava. "Optical initialization of a single spin-valley in charged WSe2 quantum dots." Nature nanotechnology 14, no. 5 (2019): 426-431.

[716] Mukherjee, Arunabh, Kamran Shayan, Lizhong Li, Jie Shan, Kin Fai Mak, and A. Nick Vamivakas. "Observation of site-controlled localized charged excitons in CrI3/WSe2 heterostructures." Nature Communications 11, no. 1 (2020): 5502.

[717] Yu, Leo, Minda Deng, Jingyuan Linda Zhang, Sven Borghardt, Beata Kardynal, Jelena Vuckovic, and Tony F. Heinz. "Site-controlled quantum emitters in monolayer MoSe2." Nano Letters 21, no. 6 (2021): 2376-2381.

[718] Horodecki, Ryszard, Paweł Horodecki, Michał Horodecki, and Karol Horodecki. "Quantum entanglement." Reviews of modern physics 81, no. 2 (2009): 865.

[719] Jozsa, Richard, and Noah Linden. "On the role of entanglement in quantum-computational speed-up." Proceedings of the Royal Society of London. Series A: Mathematical, Physical and Engineering Sciences 459, no. 2036 (2003): 2011-2032.

[720] Bennett, Charles H., Gilles Brassard, Claude Crépeau, Richard Jozsa, Asher Peres, and William K. Wootters. "Teleporting an unknown quantum state via

dual classical and Einstein-Podolsky-Rosen channels." Physical review letters 70, no. 13 (1993): 1895.

721 Lupo, Cosmo, and Stefano Pirandola. "Ultimate precision bound of quantum and subwavelength imaging." Physical review letters 117, no. 19 (2016): 190802.

722 Degen, Christian L., Friedemann Reinhard, and Paola Cappellaro. "Quantum sensing." Reviews of modern physics 89, no. 3 (2017): 035002.

723 Alfieri, Adam, Surendra B. Anantharaman, Huiqin Zhang, and Deep Jariwala. "Nanomaterials for quantum information science and engineering." Advanced Materials 35, no. 27 (2023): 2109621.

724 Bennett, Charles H., and David P. DiVincenzo. "Quantum information and computation." nature 404, no. 6775 (2000): 247-255.

725 Wang, Yuchen, Zixuan Hu, Barry C. Sanders, and Sabre Kais. "Qudits and high-dimensional quantum computing." Frontiers in Physics 8 (2020): 589504.

726 Kretinin, Andrey V., Yuan Cao, Jhih-Sian Tu, G. L. Yu, Rashid Jalil, Konstantin S. Novoselov, Sarah J. Haigh et al. "Electronic properties of graphene encapsulated with different two-dimensional atomic crystals." Nano letters 14, no. 6 (2014): 3270-3276.

727 Recher, Patrik, and Björn Trauzettel. "Quantum dots and spin qubits in graphene." Nanotechnology 21, no. 30 (2010): 302001.

728 Oostinga, Jeroen B., Hubert B. Heersche, Xinglan Liu, Alberto F. Morpurgo, and Lieven MK Vandersypen. "Gate-induced insulating state in bilayer graphene devices." Nature materials 7, no. 2 (2008): 151-157.

729 Tong, Chuyao, Rebekka Garreis, Angelika Knothe, Marius Eich, Agnese Sacchi, Kenji Watanabe, Takashi Taniguchi et al. "Tunable valley splitting and bipolar operation in graphene quantum dots." Nano Letters 21, no. 2 (2021): 1068-1073.

730 Hollmann, Arne, Tom Struck, Veit Langrock, Andreas Schmidbauer, Floyd Schauer, Tim Leonhardt, Kentarou Sawano et al. "Large, tunable valley splitting and single-spin relaxation mechanisms in a Si/Si x Ge 1− x quantum dot." Physical review applied 13, no. 3 (2020): 034068.

731 Banszerus, Luca, Samuel Möller, Corinne Steiner, Eike Icking, Stefan Trellenkamp, Florian Lentz, Kenji Watanabe, Takashi Taniguchi, Christian Volk, and Christoph Stampfer. "Spin-valley coupling in single-electron bilayer graphene quantum dots." Nature communications 12, no. 1 (2021): 5250.

732 Avsar, Ahmet, Jun You Tan, T. Taychatanapat, J. Balakrishnan, G. K. W. Koon, Y. Yeo, J. Lahiri et al. "Spin–orbit proximity effect in graphene." Nature communications 5, no. 1 (2014): 4875.

733 Banszerus, Luca, Samuel Möller, Eike Icking, Kenji Watanabe, Takashi Taniguchi, Christian Volk, and Christoph Stampfer. "Single-electron double quantum dots in bilayer graphene." Nano letters 20, no. 3 (2020): 2005-2011.

734 Kurzmann, Annika, Marius Eich, Hiske Overweg, M. Mangold, František Herman, Peter Rickhaus, Riccardo Pisoni et al. "Excited states in bilayer graphene quantum dots." Physical review letters 123, no. 2 (2019): 026803.

735 Tong, Chuyao, Annika Kurzmann, Rebekka Garreis, Wei Wister Huang, Samuel Jele, Marius Eich, Lev Ginzburg et al. "Pauli blockade of tunable two-electron

spin and valley states in graphene quantum dots." Physical Review Letters 128, no. 6 (2022): 067702.

736 Banszerus, Luca, Samuel Möller, Eike Icking, Corinne Steiner, Daniel Neumaier, Martin Otto, Kenji Watanabe, Takashi Taniguchi, Christian Volk, and Christoph Stampfer. "Dispersive sensing of charge states in a bilayer graphene quantum dot." Applied Physics Letters 118, no. 9 (2021).

737 Mirzakhani, M., F. M. Peeters, and M. Zarenia. "Circular quantum dots in twisted bilayer graphene." Physical Review B 101, no. 7 (2020): 075413.

738 Sun, Luzhao, Zihao Wang, Yuechen Wang, Liang Zhao, Yanglizhi Li, Buhang Chen, Shenghong Huang et al. "Hetero-site nucleation for growing twisted bilayer graphene with a wide range of twist angles." Nature communications 12, no. 1 (2021): 2391.

739 Kim, Gwangwoo, Hyunseob Lim, Kyung Yeol Ma, A-Rang Jang, Gyeong Hee Ryu, Minbok Jung, Hyung-Joon Shin, Zonghoon Lee, and Hyeon Suk Shin. "Catalytic conversion of hexagonal boron nitride to graphene for in-plane heterostructures." Nano Letters 15, no. 7 (2015): 4769-4775.

740 Kim, Gwangwoo, Sung-Soo Kim, Jonghyuk Jeon, Seong In Yoon, Seokmo Hong, Young Jin Cho, Abhishek Misra et al. "Planar and van der Waals heterostructures for vertical tunnelling single electron transistors." Nature communications 10, no. 1 (2019): 230.

741 Volk, Christian, Christoph Neumann, Sebastian Kazarski, Stefan Fringes, Stephan Engels, Federica Haupt, André Müller, and Christoph Stampfer. "Probing relaxation times in graphene quantum dots." Nature Communications 4, no. 1 (2013): 1753.

742 Banszerus, Luca, Katrin Hecker, Eike Icking, Stefan Trellenkamp, Florian Lentz, Daniel Neumaier, Kenji Watanabe, Takashi Taniguchi, Christian Volk, and Christoph Stampfer. "Pulsed-gate spectroscopy of single-electron spin states in bilayer graphene quantum dots." Physical Review B 103, no. 8 (2021): L081404.

743 Kormányos, Andor, Viktor Zólyomi, Neil D. Drummond, and Guido Burkard. "Spin-orbit coupling, quantum dots, and qubits in monolayer transition metal dichalcogenides." Physical Review X 4, no. 1 (2014): 011034.

744 Brooks, Matthew, and Guido Burkard. "Electric dipole spin resonance of two-dimensional semiconductor spin qubits." Physical Review B 101, no. 3 (2020): 035204.

745 Altıntaş, Abdulmenaf, Maciej Bieniek, Amintor Dusko, Marek Korkusiński, Jarosław Pawłowski, and Paweł Hawrylak. "Spin-valley qubits in gated quantum dots in a single layer of transition metal dichalcogenides." Physical Review B 104, no. 19 (2021): 195412.

746 Pawłowski, J., M. Bieniek, and T. Woźniak. "Valley Two-Qubit System in a Mo S 2-Monolayer Gated Double Quantum dot." Physical Review Applied 15, no. 5 (2021): 054025.

747 Yao, Wang, Di Xiao, and Qian Niu. "Valley-dependent optoelectronics from inversion symmetry breaking." Physical Review B 77, no. 23 (2008): 235406.

748 Cao, Ting, Gang Wang, Wenpeng Han, Huiqi Ye, Chuanrui Zhu, Junren Shi, Qian Niu et al. "Valley-selective circular dichroism of monolayer molybdenum disulphide." Nature communications 3, no. 1 (2012): 887.

749 Xiao, Di, Gui-Bin Liu, Wanxiang Feng, Xiaodong Xu, and Wang Yao. "Coupled spin and valley physics in monolayers of MoS 2 and other group-VI dichalcogenides." Physical review letters 108, no. 19 (2012): 196802.

750 Song, Xiang-Xiang, Zhuo-Zhi Zhang, Jie You, Di Liu, Hai-Ou Li, Gang Cao, Ming Xiao, and Guo-Ping Guo. "Temperature dependence of Coulomb oscillations in a few-layer two-dimensional WS2 quantum dot." Scientific Reports 5, no. 1 (2015): 16113.

751 Song, Xiang-Xiang, Di Liu, Vahid Mosallanejad, Jie You, Tian-Yi Han, Dian-Teng Chen, Hai-Ou Li et al. "A gate defined quantum dot on the two-dimensional transition metal dichalcogenide semiconductor WSe2." Nanoscale 7, no. 40 (2015): 16867-16873.

752 Davari, S., J. Stacy, A. M. Mercado, J. D. Tull, R. Basnet, K. Pandey, K. Watanabe, T. Taniguchi, J. Hu, and H. O. H. Churchill. "Gate-defined accumulation-mode quantum dots in monolayer and bilayer w se 2." Physical Review Applied 13, no. 5 (2020): 054058.

753 Zhang, Zhuo-Zhi, Xiang-Xiang Song, Gang Luo, Guang-Wei Deng, Vahid Mosallanejad, Takashi Taniguchi, Kenji Watanabe et al. "Electrotunable artificial molecules based on van der Waals heterostructures." Science advances 3, no. 10 (2017): e1701699.

754 Wang, Ke, Kristiaan De Greve, Luis A. Jauregui, Andrey Sushko, Alexander High, You Zhou, Giovanni Scuri et al. "Electrical control of charged carriers and excitons in atomically thin materials." Nature nanotechnology 13, no. 2 (2018): 128-132.

755 Lee, Kyunghoon, Girish Kulkarni, and Zhaohui Zhong. "Coulomb blockade in monolayer MoS 2 single electron transistor." Nanoscale 8, no. 14 (2016): 7755-7760.

756 Reinhardt, Simon, Luka Pirker, Christian Bäuml, Maja Remškar, and Andreas K. Hüttel. "Coulomb blockade spectroscopy of a MoS2 nanotube." physica status solidi (RRL)–Rapid Research Letters 13, no. 11 (2019): 1900251.

757 Kumar, Pawan, James P. Horwath, Alexandre C. Foucher, Christopher C. Price, Natalia Acero, Vivek B. Shenoy, Eric A. Stach, and Deep Jariwala. "Direct visualization of out-of-equilibrium structural transformations in atomically thin chalcogenides." npj 2D Materials and Applications 4, no. 1 (2020): 16.

758 Wang, Ziqian, Ruichun Luo, Isaac Johnson, Hamzeh Kashani, and Mingwei Chen. "Inlaid ReS2 quantum dots in monolayer MoS2." ACS nano 14, no. 1 (2019): 899-906.

759 Hu, Tao, Guodong Zhao, Heng Gao, Yabei Wu, Jisang Hong, Alessandro Stroppa, and Wei Ren. "Manipulation of valley pseudospin in WSe 2/CrI 3 heterostructures by the magnetic proximity effect." Physical Review B 101, no. 12 (2020): 125401.

760 Khan, Imran, Brahim Marfoua, and Jisang Hong. "Electric field induced giant valley polarization in two dimensional ferromagnetic WSe2/CrSnSe3 heterostructure." npj 2D Materials and Applications 5, no. 1 (2021): 10.

761 Zhao, Pei, Yandong Ma, Chengan Lei, Hao Wang, Baibiao Huang, and Ying Dai. "Single-layer LaBr2: Two-dimensional valleytronic semiconductor with spontaneous spin and valley polarizations." Applied Physics Letters 115, no. 26 (2019).

762 Peng, Rui, Yandong Ma, Xilong Xu, Zhonglin He, Baibiao Huang, and Ying Dai. "Intrinsic anomalous valley Hall effect in single-layer N b 3 I 8." Physical Review B 102, no. 3 (2020): 035412.

763 Lei, Chengan, Xilong Xu, Ting Zhang, Baibiao Huang, Ying Dai, and Yandong Ma. "Nonvolatile controlling valleytronics by ferroelectricity in 2H-VSe2/Sc2CO2 van der Waals heterostructure." The Journal of Physical Chemistry C 125, no. 4 (2021): 2802-2809.

764 Momma, Koichi, and Fujio Izumi. "VESTA 3 for three-dimensional visualization of crystal, volumetric and morphology data." Journal of applied crystallography 44, no. 6 (2011): 1272-1276.

765 Liu, Xiaolong, and Mark C. Hersam. "2D materials for quantum information science." Nature Reviews Materials 4, no. 10 (2019): 669-684.

766 Feynman, Richard P. "Simulating Physics with Computers." International Journal of Theoretical Physics 21, no. 6-7 (1982): 467-488.

767 Abrams, Daniel S., and Seth Lloyd. "Quantum algorithm providing exponential speed increase for finding eigenvalues and eigenvectors." Physical Review Letters 83, no. 24 (1999): 5162.

768 Shor, Peter W. "Algorithms for quantum computation: discrete logarithms and factoring." In Proceedings 35th annual symposium on foundations of computer science, pp. 124-134. Ieee, 1994.

769 Shor, Peter W. "Scheme for reducing decoherence in quantum computer memory." Physical review A 52, no. 4 (1995): R2493.

770 Shor, Peter W. "Polynomial-time algorithms for prime factorization and discrete logarithms on a quantum computer." SIAM review 41, no. 2 (1999): 303-332.

771 Grover, Lov K. "Quantum mechanics helps in searching for a needle in a haystack." Physical review letters 79, no. 2 (1997): 325.

772 Ladd, Thaddeus D., Fedor Jelezko, Raymond Laflamme, Yasunobu Nakamura, Christopher Monroe, and Jeremy Lloyd O'Brien. "Quantum computers." nature 464, no. 7285 (2010): 45-53.

773 DiVincenzo, David P. "The physical implementation of quantum computation." Fortschritte der Physik: Progress of Physics 48, no. 9-11 (2000): 771-783.

774 Arute, Frank, Kunal Arya, Ryan Babbush, Dave Bacon, Joseph C. Bardin, Rami Barends, Rupak Biswas et al. "Quantum supremacy using a programmable superconducting processor." Nature 574, no. 7779 (2019): 505-510.

775 Preskill, John. "Quantum computing in the NISQ era and beyond." Quantum 2 (2018): 79.

[776] Barthelemy, Pierre, and Lieven MK Vandersypen. "Quantum dot systems: a versatile platform for quantum simulations." Annalen der Physik 525, no. 10-11 (2013): 808-826.

[777] Hensgens, Toivo, Takafumi Fujita, Laurens Janssen, Xiao Li, C. J. Van Diepen, Christian Reichl, Werner Wegscheider, Sankar Das Sarma, and Lieven MK Vandersypen. "Quantum simulation of a Fermi–Hubbard model using a semiconductor quantum dot array." Nature 548, no. 7665 (2017): 70-73.

[778] Dehollain, Juan P., Uditendu Mukhopadhyay, Vincent P. Michal, Yao Wang, Bernhard Wunsch, Christian Reichl, Werner Wegscheider, Mark S. Rudner, Eugene Demler, and Lieven MK Vandersypen. "Nagaoka ferromagnetism observed in a quantum dot plaquette." Nature 579, no. 7800 (2020): 528-533.

[779] Prinz, Gary A. "Science 282 1660 Prinz GA 1995." Phys. Today 48 (1998): 58.

[780] Fiederling, R., M. Keim, Get al Reuscher, W. Ossau, G. Schmidt, A. Waag, and L. W. Molenkamp. "Injection and detection of a spin-polarized current in a light-emitting diode." Nature 402, no. 6763 (1999): 787-790.

[781] Kikkawa, J. M., and D. D. Awschalom. "Resonant spin amplification in n-type GaAs." Physical Review Letters 80, no. 19 (1998): 4313.

[782] Ohno, Y., D. K. Young, B. al Beschoten, Fumihiro Matsukura, Hideo Ohno, and D. D. Awschalom. "Electrical spin injection in a ferromagnetic semiconductor heterostructure." Nature 402, no. 6763 (1999): 790-792.

[783] Loss, Daniel, and David P. DiVincenzo. "Quantum computation with quantum dots." Physical Review A 57, no. 1 (1998): 120.

[784] Steane, Andrew. "Quantum computing." Reports on Progress in Physics 61, no. 2 (1998): 117.

[785] Bennett, Charles H., and David P. DiVincenzo. "Quantum information and computation." nature 404, no. 6775 (2000): 247-255.

[786] DiVincenzo, David P., and Daniel Loss. "Quantum computers and quantum coherence." Journal of Magnetism and Magnetic Materials 200, no. 1-3 (1999): 202-218.

[787] Allain, Adrien, Jiahao Kang, Kaustav Banerjee, and Andras Kis. "Electrical contacts to two-dimensional semiconductors." Nature materials 14, no. 12 (2015): 1195-1205.

[788] Dery, H., P. Dalal, Ł. Cywiński, and Lu J. Sham. "Spin-based logic in semiconductors for reconfigurable large-scale circuits." Nature 447, no. 7144 (2007): 573-576.

[789] Behin-Aein, Behtash, Deepanjan Datta, Sayeef Salahuddin, and Supriyo Datta. "Proposal for an all-spin logic device with built-in memory." Nature nanotechnology 5, no. 4 (2010): 266-270.

[790] Bernstein, Kerry, Ralph K. Cavin, Wolfgang Porod, Alan Seabaugh, and Jeff Welser. "Device and architecture outlook for beyond CMOS switches." Proceedings of the IEEE 98, no. 12 (2010): 2169-2184.

[791] Imre, Alexandra, G. Csaba, L. Ji, A. Orlov, G. H. Bernstein, and W. Porod. "Majority logic gate for magnetic quantum-dot cellular automata." Science 311, no. 5758 (2006): 205-208.

792 Carlton, David B., Nathan C. Emley, Eduard Tuchfeld, and Jeffrey Bokor. "Simulation studies of nanomagnet-based logic architecture." Nano letters 8, no. 12 (2008): 4173-4178.

793 Salahuddin, Sayeef, and Supriyo Datta. "Interacting systems for self-correcting low power switching." Applied physics letters 90, no. 9 (2007).

794 Behin-Aein, Behtash, Sayeef Salahuddin, and Supriyo Datta. "Switching energy of ferromagnetic logic bits." IEEE Transactions on Nanotechnology 8, no. 4 (2009): 505-514.

795 Johnson, Mark, and Robert H. Silsbee. "Interfacial charge-spin coupling: Injection and detection of spin magnetization in metals." Physical review letters 55, no. 17 (1985): 1790.

796 Jedema, Friso Jacobus, A. T. Filip, and B. J. Van Wees. "Electrical spin injection and accumulation at room temperature in an all-metal mesoscopic spin valve." Nature 410, no. 6826 (2001): 345-348.

797 Han, Wei, Roland K. Kawakami, Martin Gmitra, and Jaroslav Fabian. "Graphene spintronics." Nature nanotechnology 9, no. 10 (2014): 794-807.

798 Roche, Stephan, Johan Åkerman, Bernd Beschoten, Jean-Christophe Charlier, Mairbek Chshiev, Saroj Prasad Dash, Bruno Dlubak et al. "Graphene spintronics: the European Flagship perspective." 2D Materials 2, no. 3 (2015): 030202.

799 Gmitra, Martin, Sergej Konschuh, Christian Ertler, Claudia Ambrosch-Draxl, and Jaroslav Fabian. "Band-structure topologies of graphene: Spin-orbit coupling effects from first principles." Physical Review B 80, no. 23 (2009): 235431.

800 Boettger, J. C., and S. B. Trickey. "First-principles calculation of the spin-orbit splitting in graphene." Physical Review B 75, no. 12 (2007): 121402.

801 Bolotin, Kirill I., K. J. Sikes, Zhifang Jiang, M. Klima, G. Fudenberg, James Hone, Phaly Kim, and Horst L. Stormer. "Ultrahigh electron mobility in suspended graphene." Solid state communications 146, no. 9-10 (2008): 351-355.

802 Yan, Wenjing, Oihana Txoperena, Roger Llopis, Hanan Dery, Luis E. Hueso, and Felix Casanova. "A two-dimensional spin field-effect switch." Nature communications 7, no. 1 (2016): 13372.

803 Jiang, Shengwei, Lizhong Li, Zefang Wang, Jie Shan, and Kin Fai Mak. "Spin tunnel field-effect transistors based on two-dimensional van der Waals heterostructures." Nature Electronics 2, no. 4 (2019): 159-163.

804 Zhu, Zhiyong Y., Yingchun C. Cheng, and Udo Schwingenschlögl. "Giant spin-orbit-induced spin splitting in two-dimensional transition-metal dichalcogenide semiconductors." Physical Review B 84, no. 15 (2011): 153402.

805 Gong, Cheng, Lin Li, Zhenglu Li, Huiwen Ji, Alex Stern, Yang Xia, Ting Cao et al. "Discovery of intrinsic ferromagnetism in two-dimensional van der Waals crystals." Nature 546, no. 7657 (2017): 265-269.

806 Bonilla, Manuel, Sadhu Kolekar, Yujing Ma, Horacio Coy Diaz, Vijaysankar Kalappattil, Raja Das, Tatiana Eggers, Humberto R. Gutierrez, Manh-Huong Phan, and Matthias Batzill. "Strong room-temperature ferromagnetism in VSe2

monolayers on van der Waals substrates." Nature nanotechnology 13, no. 4 (2018): 289-293.

807 Deng, Yujun, Yijun Yu, Yichen Song, Jingzhao Zhang, Nai Zhou Wang, Zeyuan Sun, Yangfan Yi et al. "Gate-tunable room-temperature ferromagnetism in two-dimensional Fe3GeTe2." Nature 563, no. 7729 (2018): 94-99.

808 Jiang, Shengwei, Jie Shan, and Kin Fai Mak. "Electric-field switching of two-dimensional van der Waals magnets." Nature materials 17, no. 5 (2018): 406-410.

809 Song, Tiancheng, Xinghan Cai, Matisse Wei-Yuan Tu, Xiaoou Zhang, Bevin Huang, Nathan P. Wilson, Kyle L. Seyler et al. "Giant tunneling magnetoresistance in spin-filter van der Waals heterostructures." Science 360, no. 6394 (2018): 1214-1218.

810 Wang, Zhe, Ignacio Gutiérrez-Lezama, Nicolas Ubrig, Martin Kroner, Marco Gibertini, Takashi Taniguchi, Kenji Watanabe, Ataç Imamoğlu, Enrico Giannini, and Alberto F. Morpurgo. "Very large tunneling magnetoresistance in layered magnetic semiconductor CrI3." Nature communications 9, no. 1 (2018): 2516.

811 Klein, Dahlia R., David MacNeill, Jose L. Lado, David Soriano, Efrén Navarro-Moratalla, Kenji Watanabe, Takashi Taniguchi et al. "Probing magnetism in 2D van der Waals crystalline insulators via electron tunneling." Science 360, no. 6394 (2018): 1218-1222.

812 Xiao, Di, Gui-Bin Liu, Wanxiang Feng, Xiaodong Xu, and Wang Yao. "Coupled spin and valley physics in monolayers of MoS2 and other group-VI dichalcogenides." Physical review letters 108, no. 19 (2012): 196802.

813 Pal, Arnab, Shuo Zhang, Tanmay Chavan, Kunjesh Agashiwala, Chao-Hui Yeh, Wei Cao, and Kaustav Banerjee. "Quantum-Engineered Devices Based on 2D Materials for Next-Generation Information Processing and Storage." Advanced Materials 35, no. 27 (2023): 2109894.

814 Aivazian, Grant, Zhirui Gong, Aaron M. Jones, Rui-Lin Chu, Jiaqiang Yan, David G. Mandrus, Chuanwei Zhang, David Cobden, Wang Yao, and Xiaodong Xu. "Magnetic control of valley pseudospin in monolayer WSe 2." Nature Physics 11, no. 2 (2015): 148-152.

815 Schaibley, John R., Hongyi Yu, Genevieve Clark, Pasqual Rivera, Jason S. Ross, Kyle L. Seyler, Wang Yao, and Xiaodong Xu. "Valleytronics in 2D materials." Nature Reviews Materials 1, no. 11 (2016): 1-15.

816 Yuan, Hongtao, Mohammad Saeed Bahramy, Kazuhiro Morimoto, Sanfeng Wu, Kentaro Nomura, Bohm-Jung Yang, Hidekazu Shimotani et al. "Zeeman-type spin splitting controlled by an electric field." Nature Physics 9, no. 9 (2013): 563-569.

817 Zhou, Baozeng, Zheng Li, Jiaming Wang, Xuechen Niu, and Chongbiao Luan. "Tunable valley splitting and an anomalous valley Hall effect in hole-doped WS 2 by proximity coupling with a ferromagnetic MnO 2 monolayer." Nanoscale 11, no. 28 (2019): 13567-13575.

[818] Sui, Mengqiao, Guorui Chen, Liguo Ma, Wen-Yu Shan, Dai Tian, Kenji Watanabe, Takashi Taniguchi et al. "Gate-tunable topological valley transport in bilayer graphene." Nature Physics 11, no. 12 (2015): 1027-1031.

[819] Shimazaki, Yuya, Michihisa Yamamoto, Ivan V. Borzenets, Kenji Watanabe, Takashi Taniguchi, and Seigo Tarucha. "Generation and detection of pure valley current by electrically induced Berry curvature in bilayer graphene." Nature Physics 11, no. 12 (2015): 1032-1036.

[820] Gorbachev, R. V., J. C. W. Song, G. L. Yu, A. V. Kretinin, F. Withers, Y. Cao, A. Mishchenko et al. "Detecting topological currents in graphene superlattices." Science 346, no. 6208 (2014): 448-451.

[821] Yankowitz, Matthew, Jiamin Xue, Daniel Cormode, Javier D. Sanchez-Yamagishi, K. Watanabe, T. Taniguchi, Pablo Jarillo-Herrero, Philippe Jacquod, and Brian J. LeRoy. "Emergence of superlattice Dirac points in graphene on hexagonal boron nitride." Nature physics 8, no. 5 (2012): 382-386.

[822] Ang, Yee Sin, Shengyuan A. Yang, C. Zhang, Zhongshui Ma, and Lawrence K. Ang. "Valleytronics in merging Dirac cones: All-electric-controlled valley filter, valve, and universal reversible logic gate." Physical Review B 96, no. 24 (2017): 245410.

[823] Carr, Stephen, Daniel Massatt, Shiang Fang, Paul Cazeaux, Mitchell Luskin, and Efthimios Kaxiras. "Twistronics: Manipulating the electronic properties of two-dimensional layered structures through their twist angle." Physical Review B 95, no. 7 (2017): 075420.

[824] Luican, A., Guohong Li, A. Reina, J. Kong, R. R. Nair, Konstantin S. Novoselov, Andre K. Geim, and E. Y. Andrei. "Single-layer behavior and its breakdown in twisted graphene layers." Physical review letters 106, no. 12 (2011): 126802.

[825] Sun, Xueqian, Manuka Suriyage, Ahmed Raza Khan, Mingyuan Gao, Jie Zhao, Boqing Liu, Md Mehedi Hasan et al. "Twisted van der Waals Quantum Materials: Fundamentals, Tunability, and Applications." Chemical Reviews (2024).

[826] Nayak, Pramoda K., Yevhen Horbatenko, Seongjoon Ahn, Gwangwoo Kim, Jae-Ung Lee, Kyung Yeol Ma, A-Rang Jang et al. "Probing evolution of twist-angle-dependent interlayer excitons in MoSe2/WSe2 van der Waals heterostructures." ACS nano 11, no. 4 (2017): 4041-4050.

[827] Van Der Zande, Arend M., Jens Kunstmann, Alexey Chernikov, Daniel A. Chenet, YuMeng You, XiaoXiao Zhang, Pinshane Y. Huang et al. "Tailoring the electronic structure in bilayer molybdenum disulfide via interlayer twist." Nano letters 14, no. 7 (2014): 3869-3875.

[828] Li, Siwei, Ke Wei, Qirui Liu, Yuxiang Tang, and Tian Jiang. "Twistronics and moiré excitonic physics in van der Waals heterostructures." Frontiers of Physics 19, no. 4 (2024): 42501.

[829] Zheng, Shoujun, Linfeng Sun, Xiaohao Zhou, Fucai Liu, Zheng Liu, Zexiang Shen, and Hong Jin Fan. "Coupling and interlayer exciton in twist-stacked WS2 bilayers." Advanced Optical Materials 3, no. 11 (2015): 1600-1605.

[830] Zhang, Jing, Jinhuan Wang, Peng Chen, Yue Sun, Shuang Wu, Zhiyan Jia, Xiaobo Lu et al. "Observation of strong interlayer coupling in MoS2/WS2 heterostructures." Advanced Materials 28, no. 10 (2016): 1950-1956.

[831] Huang, Shengxi, Xi Ling, Liangbo Liang, Jing Kong, Humberto Terrones, Vincent Meunier, and Mildred S. Dresselhaus. "Probing the interlayer coupling of twisted bilayer MoS2 using photoluminescence spectroscopy." Nano letters 14, no. 10 (2014): 5500-5508.

[832] Nayak, Pramoda K., Yevhen Horbatenko, Seongjoon Ahn, Gwangwoo Kim, Jae-Ung Lee, Kyung Yeol Ma, A-Rang Jang et al. "Probing evolution of twist-angle-dependent interlayer excitons in MoSe2/WSe2 van der Waals heterostructures." ACS nano 11, no. 4 (2017): 4041-4050.

[833] Geim, Andre K., and Irina V. Grigorieva. "Van der Waals heterostructures." Nature 499, no. 7459 (2013): 419-425.

[834] Chhowalla, Manish, Hyeon Suk Shin, Goki Eda, Lain-Jong Li, Kian Ping Loh, and Hua Zhang. "The chemistry of two-dimensional layered transition metal dichalcogenide nanosheets." Nature chemistry 5, no. 4 (2013): 263-275.

[835] Xu, Kun, Pengzuo Chen, Xiuling Li, Changzheng Wu, Yuqiao Guo, Jiyin Zhao, Xiaojun Wu, and Yi Xie. "Ultrathin nanosheets of vanadium diselenide: a metallic two-dimensional material with ferromagnetic charge-density-wave behavior." Angewandte Chemie (International ed. in English) 52, no. 40 (2013): 10477-10481.

[836] Castellanos-Gomez, Andres, Efrén Navarro-Moratalla, Guillermo Mokry, Jorge Quereda, Elena Pinilla-Cienfuegos, Nicolás Agraït, Herre SJ Van Der Zant, Eugenio Coronado, Gary A. Steele, and Gabino Rubio-Bollinger. "Fast and reliable identification of atomically thin layers of TaSe 2 crystals." Nano Research 6 (2013): 191-199.

[837] Buscema, Michele, Maria Barkelid, Val Zwiller, Herre SJ van der Zant, Gary A. Steele, and Andres Castellanos-Gomez. "Large and tunable photothermoelectric effect in single-layer MoS2." Nano letters 13, no. 2 (2013): 358-363.

[838] Wickramaratne, Darshana, Ferdows Zahid, and Roger K. Lake. "Electronic and thermoelectric properties of few-layer transition metal dichalcogenides." The Journal of chemical physics 140, no. 12 (2014).

[839] Renteria, J., R. Samnakay, S. L. Rumyantsev, C. Jiang, P. Goli, M. S. Shur, and A. A. Balandin. "Low-frequency 1/f noise in MoS2 transistors: Relative contributions of the channel and contacts." Applied Physics Letters 104, no. 15 (2014).

[840] Taniguchi, Kouji, Akiyo Matsumoto, Hidekazu Shimotani, and Hidenori Takagi. "Electric-field-induced superconductivity at 9.4 K in a layered transition metal disulphide MoS2." Applied Physics Letters 101, no. 4 (2012).

[841] Xi, Xiaoxiang, Liang Zhao, Zefang Wang, Helmuth Berger, László Forró, Jie Shan, and Kin Fai Mak. "Strongly enhanced charge-density-wave order in monolayer NbSe2." Nature nanotechnology 10, no. 9 (2015): 765-769.

[842] Goli, Pradyumna, Javed Khan, Darshana Wickramaratne, Roger K. Lake, and Alexander A. Balandin. "Charge density waves in exfoliated films of van der

Waals materials: evolution of Raman spectrum in TiSe2." Nano letters 12, no. 11 (2012): 5941-5945.

843 Grüner, George. "The dynamics of charge-density waves." Reviews of modern physics 60, no. 4 (1988): 1129.

844 Adelman, T. L., S. V. Zaitsev-Zotov, and R. E. Thorne. "Field-effect modulation of charge-density-wave transport in NbSe3 and TaS3." Physical review letters 74, no. 26 (1995): 5264.

845 Moncton, D. E., J. D. Axe, and F. J. DiSalvo. "Study of superlattice formation in 2H-NbSe2 and 2H-TaSe2 by neutron scattering." Physical Review Letters 34, no. 12 (1975): 734.

846 Monney, Claude, Corsin Battaglia, Hervé Cercellier, Philipp Aebi, and Hans Beck. "Exciton Condensation Driving the Periodic Lattice Distortion of 1 T−TiSe 2." Physical review letters 106, no. 10 (2011): 106404.

847 Bovet, M., D. Popović, F. Clerc, Christian Koitzsch, U. Probst, E. Bucher, H. Berger, D. Naumović, and Philipp Aebi. "Pseudogapped Fermi surfaces of 1 T−TaS2 and 1 T− TaSe2: A charge density wave effect." Physical Review B 69, no. 12 (2004): 125117.

848 Hossain, Mongur, Zhaoyang Zhao, Wen Wen, Xinsheng Wang, Juanxia Wu, and Liming Xie. "Recent advances in two-dimensional materials with charge density waves: synthesis, characterization and applications." Crystals 7, no. 10 (2017): 298.

849 Morosan, Emilia, Henny W. Zandbergen, B. S. Dennis, J. W. G. Bos, Y. Onose, Tomasz Klimczuk, A. P. Ramirez, N. P. Ong, and Robert J. Cava. "Superconductivity in Cu x TiSe2." Nature Physics 2, no. 8 (2006): 544-550.

850 Liu, Y., R. Ang, W. J. Lu, 1WH Song, L. J. Li, and Y. P. Sun. "Superconductivity induced by Se-doping in layered charge-density-wave system 1T-TaS2− xSex." Applied Physics Letters 102, no. 19 (2013).

851 Hong-Tao, Wang, Li Lin-Jun, Ye De-Shu, Cheng Xin-Hong, and Xu Zhu-An. "Effect of Te doping on superconductivity and charge-density wave in dichalcogenides 2H-NbSe2-xTex (x= 0, 0.1, 0.2)." Chinese Physics 16, no. 8 (2007): 2471.

852 Ang, R., Y. Tanaka, E. Ieki, K. Nakayama, T. Sato, L. J. Li, W. J. Lu, Y. P. Sun, and T. Takahashi. "Real-Space Coexistence of the Melted Mott State and Superconductivity in Fe-Substituted 1T−TaS2." Physical review letters 109, no. 17 (2012): 176403.

853 Kusmartseva, Anna F., B. Sipos, H. Berger, Laszlo Forro, and Eduard Tutiš. "Pressure Induced Superconductivity in Pristine 1 T− TiSe 2." Physical review letters 103, no. 23 (2009): 236401.

854 Liu, Guanxiong, Bishwajit Debnath, Timothy R. Pope, Tina T. Salguero, Roger K. Lake, and Alexander A. Balandin. "A charge-density-wave oscillator based on an integrated tantalum disulfide–boron nitride–graphene device operating at room temperature." Nature nanotechnology 11, no. 10 (2016): 845-850.

855 Vaskivskyi, I., I. A. Mihailovic, S. Brazovskii, J. Gospodaric, T. Mertelj, D. Svetin, P. Sutar, and D. Mihailovic. "Fast electronic resistance switching

involving hidden charge density wave states." Nature communications 7, no. 1 (2016): 11442.

856 Stojchevska, L., I. Vaskivskyi, T. Mertelj, P. Kusar, D. Svetin, S. Brazovskii, and D. Mihailovic. "Ultrafast switching to a stable hidden quantum state in an electronic crystal." Science 344, no. 6180 (2014): 177-180.

857 Sugai, S., K. Murase, S. Uchida, and S. Tanaka. "Investigation of the Charge Density Waves in 1 T-VSe sub 2 by Raman Scattering." J. Phys.(Orsay) 42 (1981): 740-742.

858 Hirata, T., and F. S. Ohuchi. "Temperature dependence of the Raman spectra of 1T-TaS2." Solid state communications 117, no. 6 (2001): 361-364.

859 Samnakay, R., D. Wickramaratne, T. R. Pope, R. K. Lake, T. T. Salguero, and A. A. Balandin. "Zone-folded phonons and the commensurate–incommensurate charge-density-wave transition in 1 T-TaSe2 thin films." Nano letters 15, no. 5 (2015): 2965-2973.

860 Hajiyev, Parviz, Chunxiao Cong, Caiyu Qiu, and Ting Yu. "Contrast and Raman spectroscopy study of single-and few-layered charge density wave material: 2H-TaSe2." Scientific reports 3, no. 1 (2013): 1-6.

861 Goli, Pradyumna, Javed Khan, Darshana Wickramaratne, Roger K. Lake, and Alexander A. Balandin. "Charge density waves in exfoliated films of van der Waals materials: evolution of Raman spectrum in TiSe2." Nano letters 12, no. 11 (2012): 5941-5945.

862 Wilson, J. A., F. J. Di Salvo, and S. Mahajan. "Charge-density waves in metallic, layered, transition-metal dichalcogenides." Physical review letters 32, no. 16 (1974): 882.

863 Maclean, D., and M. H. Jericho. "Effect of the charge-density-wave transition on the thermal expansion of 2H-TaSe 2, NbSe 3, and o-TaS 3." Physical Review B 47, no. 24 (1993): 16169.

864 Tada, Haruna, Amy E. Kumpel, Richard E. Lathrop, John B. Slanina, Patricia Nieva, Paul Zavracky, Ioannis N. Miaoulis, and Peter Y. Wong. "Thermal expansion coefficient of polycrystalline silicon and silicon dioxide thin films at high temperatures." Journal of applied physics 87, no. 9 (2000): 4189-4193.

865 Qiu, Dong, Chuanhui Gong, SiShuang Wang, Miao Zhang, Chao Yang, Xianfu Wang, and Jie Xiong. "Recent advances in 2D superconductors." Advanced Materials 33, no. 18 (2021): 2006124.

866 Guo, Yang, Yan-Feng Zhang, Xin-Yu Bao, Tie-Zhu Han, Zhe Tang, Li-Xin Zhang, Wen-Guang Zhu et al. "Superconductivity modulated by quantum size effects." Science 306, no. 5703 (2004): 1915-1917.

867 Nishio, Takahiro, Masanori Ono, Toyoaki Eguchi, Hideaki Sakata, and Yukio Hasegawa. "Superconductivity of nanometer-size Pb islands studied by low-temperature scanning tunneling microscopy." Applied physics letters 88, no. 11 (2006).

868 Qin, Shengyong, Jungdae Kim, Qian Niu, and Chih-Kang Shih. "Superconductivity at the two-dimensional limit." Science 324, no. 5932 (2009): 1314-1317.

869 Zhang, Tong, Peng Cheng, Wen-Juan Li, Yu-Jie Sun, Guang Wang, Xie-Gang Zhu, Ke He et al. "Superconductivity in one-atomic-layer metal films grown on Si (111)." Nature Physics 6, no. 2 (2010): 104-108.

870 Ye, J. T., S. Inoue, K. Kobayashi, Y. Kasahara, H. T. Yuan, H. Shimotani, and Y. Iwasa. "Liquid-gated interface superconductivity on an atomically flat film." Nature materials 9, no. 2 (2010): 125-128.

871 Tan, Shiyong, Yan Zhang, Miao Xia, Zirong Ye, Fei Chen, Xin Xie, Rui Peng et al. "Interface-induced superconductivity and strain-dependent spin density waves in FeSe/SrTiO3 thin films." Nature materials 12, no. 7 (2013): 634-640.

872 Biscaras, J., N. Bergeal, S. Hurand, C. Grossetête, A. Rastogi, R. C. Budhani, D. LeBoeuf, Cyril Proust, and J. Lesueur. "Two-dimensional superconducting phase in LaTiO 3/SrTiO 3 heterostructures induced by high-mobility carrier doping." Physical review letters 108, no. 24 (2012): 247004.

873 Saito, Yu, Yasuharu Nakamura, Mohammad Saeed Bahramy, Yoshimitsu Kohama, Jianting Ye, Yuichi Kasahara, Yuji Nakagawa et al. "Superconductivity protected by spin–valley locking in ion-gated MoS 2." Nature Physics 12, no. 2 (2016): 144-149.

874 Lu, J. M., O. Zheliuk, Inge Leermakers, Noah FQ Yuan, Uli Zeitler, Kam Tuen Law, and JT3445699 Ye. "Evidence for two-dimensional Ising superconductivity in gated MoS2." Science 350, no. 6266 (2015): 1353-1357.

875 Wu, J., O. Pelleg, G. Logvenov, A. T. Bollinger, Y. J. Sun, G. S. Boebinger, M. Vanević, Z. Radović, and I. Božović. "Anomalous independence of interface superconductivity from carrier density." Nature Materials 12, no. 10 (2013): 877-881.

876 Bozovic, Ivan. "Atomic-layer engineering of superconducting oxides: yesterday, today, tomorrow." IEEE transactions on applied superconductivity 11, no. 1 (2001): 2686-2695.

877 Shiogai, J., Y. Ito, T. Mitsuhashi, T. Nojima, and A. Tsukazaki. "Electric-field-induced superconductivity in electrochemically etched ultrathin FeSe films on SrTiO3 and MgO." Nature Physics 12, no. 1 (2016): 42-46.

878 He, Shaolong, Junfeng He, Wenhao Zhang, Lin Zhao, Defa Liu, Xu Liu, Daixiang Mou et al. "Phase diagram and electronic indication of high-temperature superconductivity at 65 K in single-layer FeSe films." Nature materials 12, no. 7 (2013): 605-610.

879 Fiory, A. T., A. F. Hebard, R. H. Eick, P. M. Mankiewich, R. E. Howard, and M. L. O'Malley. "Metallic and superconducting surfaces of YBa 2 Cu 3 O 7 probed by electrostatic charge modulation of epitaxial films." Physical review letters 65, no. 27 (1990): 3441.

880 Xi, X. X., C. Doughty, A. Walkenhorst, C. Kwon, Qi Li, and T. Venkatesan. "Effects of field-induced hole-density modulation on normal-state and superconducting transport in YBa2Cu3O7− x." Physical review letters 68, no. 8 (1992): 1240.

881 Ichinokura, Satoru, Katsuaki Sugawara, Akari Takayama, Takashi Takahashi, and Shuji Hasegawa. "Superconducting calcium-intercalated bilayer graphene." Acs Nano 10, no. 2 (2016): 2761-2765.

882 Xue, Mianqi, Genfu Chen, Huaixin Yang, Yuanhua Zhu, Duming Wang, Junbao He, and Tingbing Cao. "Superconductivity in potassium-doped few-layer graphene." Journal of the American Chemical Society 134, no. 15 (2012): 6536-6539.

883 Cao, Yang, A. Mishchenko, G. L. Yu, Ekaterina Khestanova, A. P. Rooney, Eric Prestat, A. V. Kretinin et al. "Quality heterostructures from two-dimensional crystals unstable in air by their assembly in inert atmosphere." Nano letters 15, no. 8 (2015): 4914-4921.

884 Tedrow, P. M., and R. Meservey. "Critical magnetic field of very thin superconducting aluminum films." Physical Review B 25, no. 1 (1982): 171.

885 Yazdani, Ali, and Aharon Kapitulnik. "Superconducting-insulating transition in two-dimensional a-MoGe thin films." Physical review letters 74, no. 15 (1995): 3037.

886 Gambetta, Jay M., Jerry M. Chow, and Matthias Steffen. "Building logical qubits in a superconducting quantum computing system." npj quantum information 3, no. 1 (2017): 2.

887 Huang, He-Liang, Dachao Wu, Daojin Fan, and Xiaobo Zhu. "Superconducting quantum computing: a review." Science China Information Sciences 63 (2020): 1-32.